CHEMISTRY OF PIGMENTS AND FILLERS

CHEMISTRY OF PIGMENTS AND FILLERS

D. H. SOLOMON

and

D. G. HAWTHORNE

Division of Applied Organic Chemistry
CSIRO
Melbourne, Australia

KRIEGER PUBLISHING COMPANY
MALABAR, FLORIDA
1991

Original Edition 1983
Reprint Edition 1991

Printed and Published by
KRIEGER PUBLISHING COMPANY
KRIEGER DRIVE
MALABAR, FLORIDA 32950

Library of Congress Cataloging-In-Publication Data
Solomon, D.H. (David Henry)
    Chemistry of pigments and fillers / D.H. Solomon and D.G. Hawthorne.
      p.   cm.
    Reprint. Originally published: New York : Wiley, 1983.
    Includes bibliographical references and index.
    ISBN 0-89464-620-6
    1. Pigments. 2. Fillers (in paper, paint, etc.) I. Hawthorne, D.G. (David Geoffrey) II. Title.
TP936.S7613   1991
667'.29--dc20   91-16995
                                CIP

10   9   8   7   6   5   4   3   2

FOREWORD

In 1967 Dr. David H. Solomon authored an outstanding book entitled *The Chemistry of Organic Film Formers.* Now he and his co-author, Dr. D. Geoff Hawthorne, have written this book, which is an exceedingly valuable reference for those working in the field of polymer science. Also, it should be used by mineralogists, industrial chemists, and engineers who have been puzzled by certain interactions between polymers and mineral fillers. The term *filler* is actually a misnomer because in the majority of instances the mineral addition performs a function.

In addition to his excellent research work at CSIRO, Dr. Solomon has applied his knowledge in industry. He worked for fourteen years in resin and polymer research with Balm Paints Pty. Ltd. (now Dulux Australia Ltd.), spent one year with ICI in England in polymer research which resulted in two patents, and stayed one year with Georgia Kaolin Company in the United States working on surface-modified minerals for use in polymeric composites. It was at Georgia Kaolin Company (where I was executive vice president) that I really discovered the extent to which Dr. Solomon could apply his vast knowledge in polymer chemistry to practical industrial chemical problems.

Dr. Hawthorne has been a senior collaborator with Dr. Solomon for many years, and their complementary skills have enabled them to make original contributions in the field of mineral organic interactions.

This book is very well organized and covers the structures and chemical activities of mineral fillers and pigments in the first three chapters. In the next three chapters the reactions between both unmodified and chemically modified mineral surfaces and organic chemicals, with particular emphasis on polymers, are treated.

Manufacturers of polymers and producers of mineral fillers alike can learn many fundamental aspects concerning the interactions between minerals and polymers by reading this outstanding book. It will enable the polymer producer to better select functional fillers and pigments and offers the mineral producer the opportunity to treat and modify the surfaces of the mineral so that they can be specifically tailored for certain functional applications. Another exceptionally valuable aspect of this book is the large number of reference literature citations at the end of each chapter.

HAYDN H. MURRAY

Chairman, Department of Geology
Indiana University
Bloomington, Indiana

PREFACE

The past 15 years have seen significant changes in our understanding of the chemical and physical interactions that can occur between organic entities and the surfaces of mineral or inorganic fillers and pigments. Although the chemistry and sorptive properties of many of these materials in aqueous systems have been extensively studied and are generally recognized, it is often not appreciated that many common minerals have surface activity in nonaqueous media that can affect the chemistry of adsorbed monomers or polymers and, indirectly, the serviceability of filled or pigmented paints, plastics, and elastomers. Such chemical activity is often wrongly believed to be confined to some exotic inorganic materials used as industrial catalysts or to the finely particulate silicas and carbon blacks.

The growing recognition that other minerals, particularly the aluminosilicate clay minerals, pigmentary titanias, and related materials, can react chemically with organic molecules at near-ambient temperatures has resulted in a quickening of interest in mineral–organic chemistry. Many advances of theoretical and practical importance have been made in recent years. These advances range from esoteric hypotheses concerning prebiotic syntheses and the origin of life on this earth to practical applications of the surface activity of aluminosilicates and titanias; the latter include the commercial use of minerals or modified minerals as catalysts for a variety of synthetic organic and polymerization reactions and as reactive additives, for example, in pressure-sensitive copying papers and in many novel polymer composites.

This book provides a critical review of the chemistry of mineral–organic systems, with special reference to the chemistry of clay mineral and titania pigments and fillers. It has been written with the specific aim of providing an understanding of the nature of mineral surfaces, methods for their modification, and the effects that the surface activity or modifying treatments could have on the properties of commercial formulations. The book should be of interest to students, scientists, and technologists in the fields of surface-coatings, inks, plastics and rubber, paper, petroleum, and soils, and to those who use minerals as additives in commercial formulations. It also contains material of interest to those engaged in studies of catalysis, in colloid chemistry and adsorption processes, or in the geochemical transformations of organic compounds.

We first provide a broad description of the structure and general surface chemistry of the important clay minerals and pigmentary titanias and of some other inorganic materials of commercial interest. We show that much of the chemical activity may arise from common surface features of the various minerals and examine how this activity can be affected by variations in the underlying lattice structure. With this background, we then discuss a number of methods for the surface modification of minerals by treatment with organic compounds and the properties and commercial applications of these modified materials. Finally, we review many of the chemical reactions that have been reported to occur on natural or modified minerals and discuss these with the aim of promoting a comprehension of the reactions, rather than simply providing a tabulated record of their occurrence.

We trust that the background provided by this book will help give a rational chemical explanation for the properties of many important mineral–organic systems presently in use. More significantly, it is our aim that the information provided should lead to the development of new and commercially useful systems, and that the discussion will help (1) to bridge the gap between the respective disciplines of the polymer technologist and the inorganic chemist and (2) to prompt further studies of a more fundamental nature into the surface chemistry of minerals in general. We believe the book to be very timely in view of the increasing cost of synthetic polymers and the natural desire to control this factor by the use of suitable mineral additives.

We would like to express our thanks for the assistance provided by our colleagues in the CSIRO Division of Applied Organic Chemistry: R. A. Brett and J. C. Gregg, in the preparation of the illustrations, and Mrs. J. Tydens, in the preparation of the indexes and in checking the manuscript.

D. H. SOLOMON
D. G. HAWTHORNE

Melbourne, Australia

CONTENTS

1

STRUCTURES AND CHEMICAL ACTIVITY OF SILICATE MINERALS, FILLERS, AND PIGMENTS

Silicate minerals comprise one of the largest groups of mineral fillers and pigments, both in terms of variety and in tonnage of production. Sales during 1980 of one clay mineral, kaolin, amounted to over five million tons in the United States alone; this amount does not include the kaolin used in the ceramics and related industries. While most of this kaolin was used in the manufacture of paper, approximately 25% of production went to the plastics, paint, and rubber industries. Another silicate mineral, talc, is used in toiletries and as a dry lubricant, while bentonite and its derivatives are widely used as viscosity control agents in drilling muds, paints, and lubricating greases [1]. Other silicate minerals, attapulgite and the zeolites, are used as sorbents and catalysts and have applications as fillers.

In this chapter we first outline the basic structural features of the clay minerals, with particular emphasis on the hydrated lamellar magnesium and aluminum silicates, which are the most commonly used as fillers. Two important chemical concepts, layer electronic charge and ion exchange of adsorbed cations, are introduced in Sections 1.2.1 and 1.3.5, respectively. This is followed by a discussion on the sources of aspects of their chemical and potential catalytic activity, with emphasis on aspects relevant to their use as fillers and pigments. We deal briefly with the crystallography and mineralogical chemistry of the silicate minerals. More detailed information is available in references 2–5 and in various review articles cited elsewhere in this chapter. Discussion of the colloidal properties of aqueous clay systems is limited to a few aspects that are relevant to the chemistry of fillers and pigments in organic media.

1.1   CLAY MINERALS: AN INTRODUCTION

Some of the silicate minerals used as fillers or pigments occur in massive crystalline forms, for example, asbestos, feldspar, or mica, but most are obtained in finely divided forms as "clays." The term "clay" strictly refers to mineral sediments having particles with dimensions of less than 2 μm. The clay minerals chiefly comprise magnesium and aluminum silicates having layered lattice structures [2,3]. The term "clay" may be loosely used as a mineralogical description of the dominant component of the deposits of a particular region. In the United States, the term "clay" is often used as a synonym for kaolin, whereas in Japan and to a lesser extent in Europe, "clay" is a common name for a different aluminosilicate material, bentonite. The individual components of the clays may be referred to in the literature by their "rock" names or by their "mineral" names. The two most important examples in the present context are the minerals kaolinite and montmorillonite, which are the respective principal components of the kaolin and bentonite clays (rocks).

The silicate fillers and pigments comprise a diverse group of minerals or modified minerals, most of which have layered crystalline structures. These minerals are derived from condensed forms of silicic acid, H_4SiO_4, in which each silicon atom is surrounded by four oxygen atoms which form the apices of a tetrahedral structure. These silicate tetraheda are linked together in regular arrays by the sharing of common oxygen atoms to form chains like those of the pyroxine minerals (Figure 1.1a) or extended two-

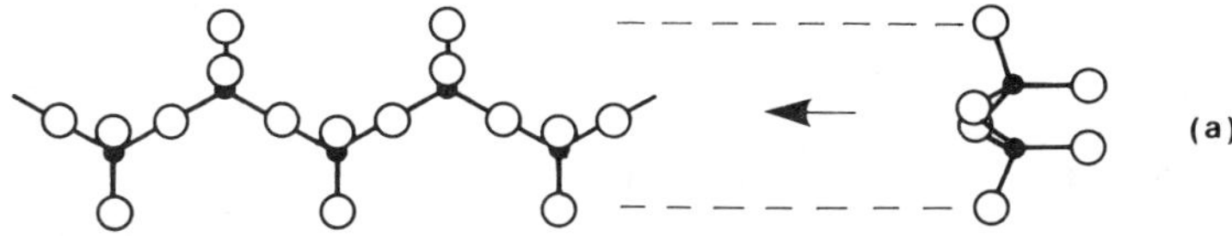

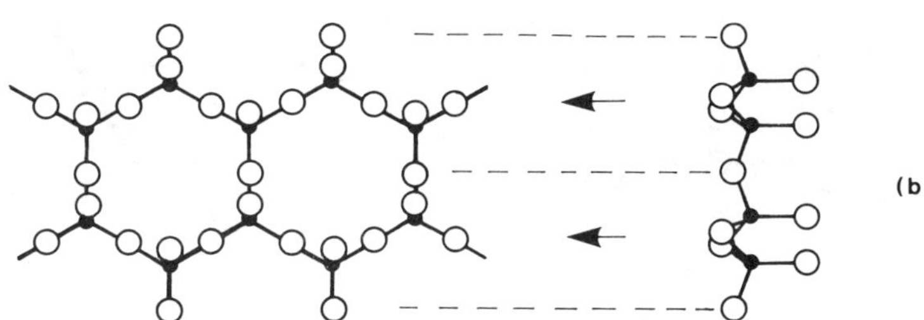

Figure 1.1   Silicate chains of (a) pyroxines and (b) amphiboles.

dimensional sheets or layers consisting of annellated cyclic groups of six silicate tetrahedra like those of the amphibole minerals (Figure 1.1b). The tetrahedra in the silicate layers are usually oriented so that the three oxygens (the basal oxygens) of the tetrahedra lie on a common plane, with the fourth oxygen (the apical oxygen) lying on a second common plane.

Some metal hydroxides, notably those of aluminum and magnesium, may also condense to form two-dimensional layered structures. Each layer of brucite, $Mg(OH)_2$, consists of a sheet of magnesium ions sandwiched between sheets of hydroxide ions. Each magnesium ion is surrounded by an octahedral arrangement of six hydroxide ions, while each hydroxide ion is shared by three magnesiums. Brucite has a trioctahedral structure, that is, a structure in which all octahedral sites are occupied by metal ions. Gibbsite, $Al(OH)_3$, layers have a similar structure but one-third of the octahedral sites are vacant. Gibbsite has a dioctahedral structure, that is, one in which only two-thirds of the octahedral sites are occupied by metal ions.

The dimensions of the silicate tetrahedral and metal hydroxide octahedral layers are sufficiently matched to enable the two layers to condense to form a composite hydrated metal silicate layer in which the apical silicate oxygens replace a proportion of the octahedral hydroxyl groups. The tetrahedral and octahedral layers may be paired to form a 1:1 layer structure like that of kaolin (Figure 1.2), or two tetrahedral layers may flank each octahedral layer to form a 2:1 layer structure like that of the magnesium silicate, talc (Figure 1.3a) or the aluminum silicate, pyrophyllite (Figure 1.3b). Various degrees of mismatch between the tetrahedral and octahedral layers are present in each mineral structure. This mismatch may be accommodated by distortion of the arrangements of tetrahedral and octa-

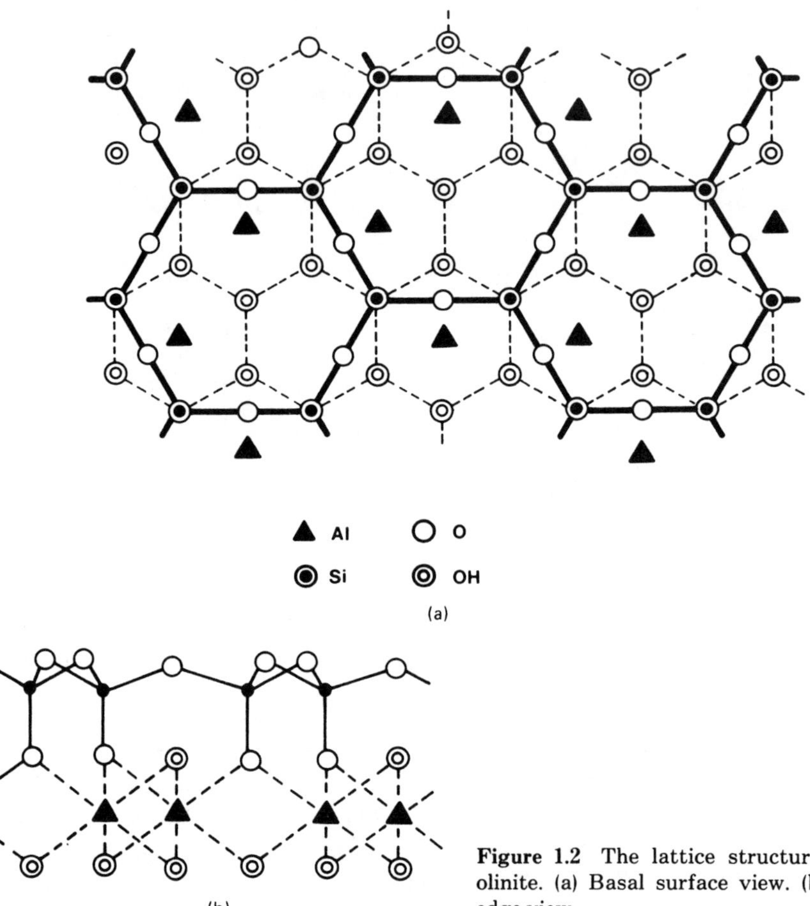

Figure 1.2 The lattice structure of kaolinite. (a) Basal surface view. (b) Layer edge view.

hedral units within the structure and may also cause the morphology and chemistry of the mineral to differ from that of the parent species. The composite layers are described as trioctahedral or dioctahedral, depending on the structure of the octahedral layer component.

The layer silicate crystals are formed by the stacking of a number of these composite layers, often with intermediate layers of hydrated metal cations to compensate for charge discrepancies arising from structural defects in the silicate layers. Most of the silicate minerals used as fillers have crystal lattices formed from layered structures or from related structures containing arrays of abbreviated metal silicate sheets, or ribbons. Attapulgite is an example of a mineral containing a silicate ribbon lattice (see Figure 1.12).

The hydrated metal silicate minerals used as fillers or pigments can be divided into several groups based on their crystalline structure. These groups, which will be discussed in turn, are:

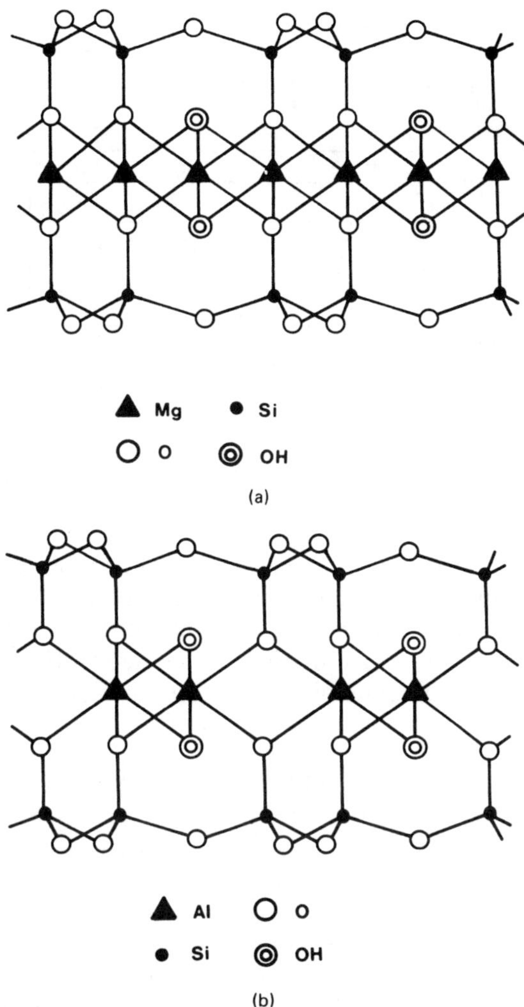

Figure 1.3 The conceptual structures of 2:1 layer silicate minerals. (a) The trioctahedral structure of talc. (b) The dioctahedral structure of pyrophyllite. (Figure 1.3a adapted, with permission, from G. Brown, Ed., *X-Ray Identification and Crystal Structures of Clay Minerals*, copyright © Mineralogical Society, London, 1961, p. 57.)

(i)   *Kaolinite and related 1:1-layer minerals,* which include halloysite and one asbestos mineral, chrysotile.

(ii) *Smectites and related 2:1-layer minerals,* which include montmorillonite, hectorite (Laponite®), vermiculite, talc, the various chlorites, illites, and micas.

(iii) *The amphiboles, pyroxines, and related fibrous minerals having silicate ring, ribbon, or chain structures.* These include wollastonite, attapulgite (palygorskite), sepiolite, and the remaining asbestos minerals.

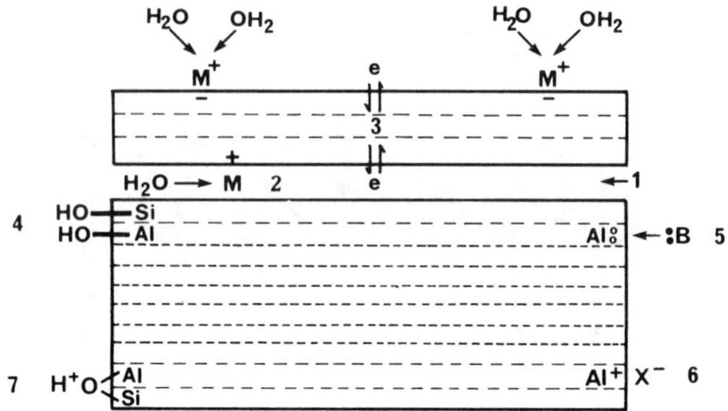

Figure 1.4   The principal surface features of a lamellar silicate mineral.

(iv)   *The zeolites,* which have a variety of three-dimensional silicate cage structures.

Amorphous silicates are discussed as a group in Chapter 3, principally in reference to their use in the surface modification of titanias and other pigments. However, some important aspects of their chemistry are introduced in this chapter because of their relevance to the chemistry of the crystalline silicates. Vitreous silica, glasses, and silica gels are also discussed in Chapter 3.

1.1.1   A Model Lamellar Silicate

Given the diversity of the compositions and structures of the clay minerals, one would expect their chemical properties to be equally diverse. While there are certainly quantitative differences between the properties of the various minerals, much of the surface chemistry is characteristic of the group as a whole or of certain main subdivisions. This is because the sources of the various forms of activity are particular types of surface site or adsorbed species which may be common to many members of the group. The nature of the underlying silicate lattice has an important bearing on the magnitude of these effects, but the qualitative similarities can be of considerable assistance in obtaining a broad understanding of clay-mineral organic chemistry.

The surface chemistry of a particular mineral filler or pigment is of greater interest to the paint or plastics technologist than its detailed lattice structure. It is possible to describe most of this chemistry by the use, as a simplifying and unifying model, of a hypothetical model silicate. Its structure can be represented by a "box," shown in Figure 1.4, the surface of which bears the essential reactive sites or species and other structural features which are of relevance to the mineral organic chemistry of the group.

The model structure contains a series of silicate layers with interlamellar spaces (1) containing intercalated hydrated exchangeable cations (2) These cations and similar exchangeable cations on the exterior surface may have acidic potential; the acidity may be protonic (Bronsted) or coordinative (Lewis). The model structure also contains variable valence species (3) that can act as a source or sink for single-electron transfer with adsorbed or intercalated materials. The exterior surface contains reactive hydroxylic species, such as AlOH and SiOH groups (4), while the layer edges contain Lewis acidic sites (5), anion exchange sites (6), and bridged-hydroxyl groups (7). Not all the features of this hypothetical structure are necessarily present in any one mineral; kaolinites and a number of other silicate fillers, for example, lack accessible interlamellar spaces.

1.1.2   Additional Advantages of Using "Box" Model Structures

The use of "box" model structures in describing the chemistry of the clays or of the oxide pigments has additional advantages as the surface of a mineral is, inevitably, anomalous and often does not follow the structure of the crystal lattice. This is particularly true of modified materials such as the alumina–silica-coated titanias described in Chapter 2; in these, while the coating accounts for only a few percent of the total composition, it may largely govern the surface chemistry of the pigment. The various classes of minerals may bear similar or analogous surface species which can give rise to similarities in some aspects of their chemistry. Surface hydroxy species are found on most practical mineral samples and may be formed as surface "anomalies," either by hydration of exposed metal ions or by hydrolysis of the crystal lattice of the mineral. The presence of these hydroxylic species, as a common feature of most pigments and fillers, is of considerable relevance to the sorptive properties and surface modifications described in Chapter 4.

1.2   KAOLINITE AND RELATED MINERALS

Kaolinite is a dioctahedral hydrated aluminosilicate containing alumina, silica, and water in a molecular ratio of 1:2:2. X-ray crystallography shows that the mineral has a 1:1-layer silicate structure. The unit cell of the kaolinite lattice has the composition $[Si_2Al_2O_5(OH)_4]$. The structure is shown in Figure 1.2. The ideal kaolinite crystal consists of an array of hexagonal basal aluminosilicate layers, stacked like the pages in a book (Figure 1.5), the individual layers having an effective thickness or basal spacing of about 7.2 Å. They are bound together by hydrogen bonds between the hydroxylic gibbsite-like surface of the octahedral layers and the oxygen sheet of the adjacent tetrahedral silicate layers.

The particle sizes of commercial kaolins and of a number of other pig-

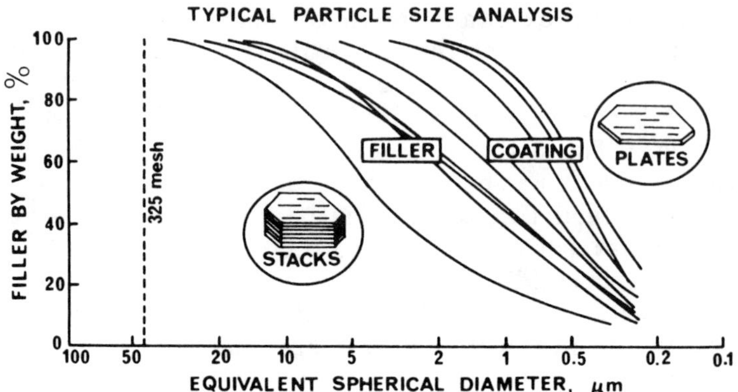

Figure 1.5 Particle size distributions of some commercial kaolins. Copyright © 1976, TAPPI. (Reprinted, with permission, from H. H. Murray, Clays, in *Paper Coating Pigments*, R. W. Hagenmeyer, Ed., TAPPI Press, Atlanta, 1976, p. 88.)

ments are usually expressed in terms of their equivalent spherical diameter, that is, the diameter of a spherical particle having the same density which, according to Stokes law, would sediment at the same rate as the dispersed clay particles. The particle size distributions for some typical commercial kaolins are shown in Figure 1.5. A typical kaolin used as a filler for plastics contains particles having basal cross sections of 1 to 3 μm and thickness of about 0.2 μ. It should be noted that specific surface areas of the minerals obtained from adsorption studies are often considerably greater than the values calculated from their particulate equivalent spherical diameters. This arises from the less streamlined shapes of the mineral particles and porosity associated with the irregular lattice layer edges or with the presence of interlamellar spaces.

Electron micrographs of an extensive range of kaolins show a great diversity of particulate forms [6]. Kaolins contain particles that can differ markedly in their crystallinity and size. For example, the smaller particles of kaolins from Georgia (United States), have greater crystalline perfection than the larger particles [7], although this is not necessarily true of the finer fractions from other deposits. One of the principal crystalline defects that occurs in kaolinites is turbostratic disorder, or irregularity in the alignment of adjacent aluminosilicate layers. Highly disordered kaolinites, for example, those of ball-clays used in the manufacture of earthenware, are termed turbostratic. The shape and dimensions of the particles and the degree of crystallinity can have a marked influence on the colloidal properties of the clays.

Halloysite (10 Å) and Halloysite (7 Å) are hydrated variants of kaolinite that may contain up to four intercalated water molecules per ideal unit cell, accommodated in loosely bound layers between those of the aluminosilicate. The fully hydrated halloysite, also known as endellite, has a basal spacing of 10.1 Å. This interlamellar water is readily eliminated, and an incompletely hydrated halloysite consists of an interstratified mixture of the hydrated (10 Å) form and the second, largely dehydrated form of halloysite; the latter has a basal spacing of approximately 7.2 Å [5]. Halloysite has a lath-like or fibrous morphology in which the aluminosilicate sheets may be curved or coiled to form tubular structures. This curvature arises from internal strain caused by the mismatch between the tetrahedral and octahedral layers [4]. Halloysites swell on treatment with ethylene glycol, with an increase in their basal spacing. This results from the intercalation of the larger glycol molecules that displace the interlamellar water molecules. Halloysite can be readily and irreversibly dehydrated, forming a turbostratic kaolin that retains the morphology of the original halloysite particles.

Kaolins and dehydrated halloysites are not swollen by treatment with ethylene glycol. The kaolinite crystalline lattice can be swollen by the intercalation of small polar molecules that are capable of forming strong hydrogen bonds with the hydroxyl sheets. These intercalating molecules include dimethyl sulfoxide [8], hydrazine, ammonium acetate, and urea [9]. Washing of the expanded kaolinite intercalation complexes with water results in removal of the interlamellar polar molecules and the formation of a turbostratic kaolinite. Treatment of highly crystalline Cornish kaolin with a dimethyl sulfoxide–ammonium fluoride mixture can result in the formation of an intercalation complex which, on washing with water, forms a 10-Å hydrate. This hydrate structure is stabilized because of the partial isomorphous replacement of basal surface hydroxyls by fluoride ions which have less affinity for the opposing silicate surfaces [10].

The serpentine minerals comprise the magnesium analogues of kaolinite. The dioctahedral aluminum hydroxide layers are replaced by trioctahedral magnesium hydroxide layers, forming a lattice which has the ideal unit cell composition $[Si_2Mg_3O_5(OH)_4]$ and which corresponds to that of the asbestos mineral, chrysotile. The magnesium silicate layers of chrysotile are coiled to form tubular fibrils, with the hydroxide layer providing the outermost surface. The fibrous rock consists of loosely demented aggregates of these micron-sized particles.

1.2.1 Isomorphous Substitution and Layer Charge

The crystalline lattice of a layer silicate is largely that of an array of bulky oxide and hydroxide anions with smaller lithium, magnesium, aluminum, or silicon cations fitting in the interstices between the oxygen atoms. Cations at particular locations can be replaced by a variety of other cations having similar ionic radii, without greatly altering the dimensions or arrangement of the layer structure. Such isomorphous substitution with ions of different valence, for example, aluminum for silicon, magnesium for aluminum, or lithium for magnesium, frequently occurs in the structures of the clay mineral lattices. If the replacement cation has a lower valence, the substitution will result in the development of a net anionic layer charge which must be compensated for, either by the inclusion of additional cations or by an opposing matched isomorphous substitution with ions of higher valence elsewhere in the structure. An example of this latter mode of internal compensation is the neutralization of charge arising from Al-for-Si substitution in the vermiculite lattice (see p. 21) by the isomorphous substitution of an equivalent number of octahedral Mg^{2+} by Al^{3+} ions. Isomorphous substitution frequently involves the replacement of ions by others of the same valence, for example, Fe^{III}-for-Al, or Fe^{II}-for-Mg, which does not alter the net layer charge. Isomorphous substitution by larger cations can result in considerable distortion of the mineral lattice. This may be reflected in changes in the chemical activity and other properties of the mineral surface; Fe^{III}-for-Al substitution, for example, can impart oxidant properties to the mineral (Section 1.7.2).

The location of the additional compensating cations depends on the charge density of the silicate layers. They may be adequately accommodated by adsorption on the external crystalline surfaces of minerals that have low degrees of isomorphous substitution and charge discrepancy, or they may have to be accommodated within the crystal lattice of minerals that have high degrees of isomorphous substitution and high layer charge densities. In the higher-charged minerals such as the montmorillonites and micas, these compensating cations plus associated water or other ligand molecules are largely located in the interlamellar spaces.

Most pure kaolinites have a low level of isomorphous substitution and their cation exchange capacities are usually less than 10 μequiv/g, although kaolins containing appreciable impurities may have significantly greater cation exchange capacities. The compensating cations of a pure kaolinite can be accommodated on the external surface of the crystallites and in scattered defects within the lattice. Isomorphous substitution in the serpentine minerals occurs with greater frequency, but is internally compensated for by matching substitution of the octahedral and tetrahedral layers.

Limited isomorphous substitution can occur between fluoride ions and hydroxyl groups in the octahedral layers of clay minerals. This substitution has already been noted in the case of kaolinite and can be induced in the 2:1-layer minerals, although exchange with fluoride ions is largely limited to accessible hydroxyls exposed at the layer edges. Fluorinated smectites can be synthesized by other means.

The individual particles of the layer silicates are inherently defective because of the disruption of the tetrahedral and octahedral layers at the crystal edges. The edges of kaolinite particles have the combined characteristics of silica, alumina, and a silica–alumina mixed oxide. When kaolinite particles are suspended in neutral or acidic media, ionization of water associated with the AlOH groups at the layer edges causes these to develop a positive charge (see Figure 1.10). At the same time, the solvated compensating cations can diffuse from the particle surfaces, causing the basal surfaces to develop a negative charge. The development of charge in aqueous media has important consequences in the colloidal chemistry of the aluminosilicate clay minerals [11–13]. One characteristic of kaolin suspensions, in both aqueous and organic media, is the formation of thixotropic, porous structures of mineral platelets. These structures have been shown to result from ionic interaction and edge–face, face–face, and edge–edge aggregation of the kaolinite particles [14].

The presence of the positive charged sites at the edges of kaolinite particles can be shown by the preferential adsorption of negative charged gold sol on the nonbasal surfaces [15]. The location of the negative charged cation exchange sites on practical kaolinite surfaces is still not clarified. Weiss and co-workers [16,17] have demonstrated the presence of cation exchange sites on the external basal surfaces of kaolinite by electron microscopy of oblique-shadowed kaolinite particles treated with a positive charged silver iodide sol, by autoradiography of a radionickel exchanged kaolinite, and by adsorption of dye-forming aromatic ammonium ions; the latter two experiments were done using a rare specimen of massive crystalline kaolin. Weiss and co-workers found that the cation exchange sites were located only on the external silicate basal surface of their kaolinite samples. They also suggested that the isomorphous substitution was largely confined to this external silicate layer.

The exchange sites on other kaolin clay surfaces do not appear to be localized in this manner. Recent esr studies of a Cu- and Mn-exchanged kaolinite indicate that the exchangeable cations are evenly distributed over the crystal surfaces, with an average separation of 11 to 12 Å, and that they have a high degree of mobility [18]. The kaolin samples used in these esr studies appeared to have a random Mg-for-Al isomorphous substitution, unlike the apparently localized Al-for-Si substitution in the massive crystalline specimens used by Weiss. It should be noted that the location of the cation exchange sites is complicated by the facile hydrolysis of the kaolinite surface which can occur in aqueous media or on exposure to moisture; this hydrolysis is described in Section 1.7.1.

1.2.2 Origins and Commercial Exploitation of Kaolins

Kaolins are the product of the hydrolytic transformation or weathering of feldspar and granite, although they may also result from the transformation of other aluminosilicate clays [19,20]. The deposits may have been transported from other locations, as in the case of the waterborne secondary Georgia (United States) kaolins or, as in the case of the primary Cornwall deposits, they may have formed from upwelling solutions resulting from the decomposition of subterranean granite and deep surface weathering. The natural kaolins may contain significant amounts of other minerals, including mica, quartz, titania, goethite, tourmaline, and 2:1-layer minerals. As previously noted, the particle size distribution and degree of crystallinity of the kaolins can vary significantly, even in neighboring areas of the same deposit. These factors can affect the suitability of the materials for particular applications.

Kaolins intended for use as fillers require some beneficiation, even if only to remove grit. Most grades used in the plastics, paints, and paper industries have been subjected to some form of wet-process beneficiation to improve color and dispersion, to enable some classification by particle size, and to provide uniformity of properties between batches. Beneficiation processes for kaolin have been reviewed elsewhere [e.g., 19,20], but are briefly outlined here as some of the chemical treatments involved can greatly modify the surface chemistry of the beneficiated materials. In a typical process, the crude kaolin is dispersed in aqueous soda or an alkaline phosphate solution, and then subjected to various physical separation and classification processes. The purified kaolin dispersion may be flocculated by treatment with acid or aluminum salt solutions, and the slurry dewatered and then dried at elevated temperatures. The color of the kaolin is often improved by the leaching of ferric oxides which may be present, together with silica and alumina, as a coating on the clay particles. The ferric oxides are not readily removed unless the kaolin is first treated to reduce the iron to the ferrous state, which can be removed by treatment with dilute acids. Magnetic separation techniques are also frequently used to remove iron-containing impurities from the crude clay. For some applications, the dewatered slurry may be redispersed using a small amount of alkaline polyphosphate, and the mixture spray-dried to produce a neutralized, aqueous-dispersible grade of kaolin; dispersed kaolins are also sold as slurries.

The kaolins may be "delaminated" by milling their dispersions with nylon pellets, glass beads, or other abrasive or nonabrasive grinding aids, to produce grades that have the greater brightness and gloss desirable for paper-coating [21]. Because of the strong interaction between the kaolinite layers, these delamination processes can cause considerable reduction in the basal dimensions of the particles. However, the intercalation complexes of kaolinite have weaker interlayer bonding and can be more readily delaminated. Processes for the delamination of kaolin complexes have been patented [22,23], but they do not appear to be in commercial use, although they may have unknowingly been practiced in the past. For example, the ancient Chinese secret for the manufacture of extremely thin-walled, hand-molded porcelain partly lay in the use of urine-steeped kaolins; the urea–kaolin complexes now believed to have been formed during the steeping process would have been readily delaminated during working of the clay prior to molding, thus improving its thixotropy and plasticity [24]. The archetypical kaolin from the Kauling, China deposit was apparently not a kaolinite, but a dehydrated halloysite [25].

The density and light reflectivity of the kaolins can be increased by calcination at about 650°C. This results in the partial dehydration of the kaolin lattice and its conversion into the pseudocrystalline "metakaolinite" [4]. Complete dehydration does not occur until the kaolin is heated above 900°C, when it decomposes to form other mineral phases. Calcined halloysite has also been used as a filler and pigment.

The surface properties of the kaolin fillers may be modified by coating the particulates with inorganic or organic compositions. Kaolins having increased compatibility with nonpolar organic media and having reduced surface acidity can be prepared by the use of a magnesium–silicate gel coating [26], while other products having increased compatibility with polar organic media, for example, polyester resins, can be obtained by coating the particles with alumina or silica–alumina gels [27]. Oleophilic kaolins particularly suited for incorporation in organic media may be obtained by the treatment of kaolins or calcined kaolins with organic compounds; these organic treatments are discussed in Chapter 4.

Chrysotile is the only other 1:1-layer silicate mineral that is widely used as a filler and reinforcement. Chrysotile occurs in the form of fibrous aggregates in a nonfibrous matrix that requires milling, sieving, and air classification to separate the asbestos fibers. The adhesion between the tubular fibrils of chrysotile is relatively weak, and the massive forms can be readily disrupted to colloidal microcrystalline state by milling in the presence of surfactants.

1.3 SMECTITES AND RELATED MINERALS

The smectites are a group of 2:1-layer minerals that includes the hydrated aluminum silicate, montmorillonite. For convenience, we have chosen to include under this heading other clay minerals having 2:1-layer structures such as talc, pyrophyllite, the vermiculites, micas, and chlorites. In these minerals, each octahedral layer is paired with two tetrahedral silicate layers (Figure 1.3).

The smectite group can be broadly divided into dioctahedral (aluminum silicate) and trioctahedral (magnesium silicate) minerals. These groups can be subdivided into those in which the layer charge arises predominantly from isomorphous substitution in the octahedral layer and those with charge arising from tetrahedral layer substitution; further subdivisions can be made on the basis of layer charge density. While the different minerals of the smectite group could form a continuous multidimensional series, they tend to fall into discrete populations differentiated mainly by the source and density of their layer charge [4,5].

The principal octahedral layer ions of most smectites are Al^{3+} or Mg^{2+}. These may be largely replaced by other cations, particularly Fe^{3+} and Fe^{2+}, while still retaining the basic 2:1-layer lattice structure and many of the chemical characteristics of the parent smectites. These derived minerals (and their substituent ions) include nontronite and glauconite (Fe), sauconite (Zn), the less abundant medmontite (Cu), and pimelite (Ni) [4].

The trioctahedral and dioctahedral parents of the group are the minerals talc and pyrophyllite (Figure 1.3), with the respective crystallographic unit cell compositions $[Si_8Mg_6O_{20}(OH)_4]$ and $[Si_8Al_4O_{20}(OH)_4]$. The basal spacing or effective thickness of the talc or pyrophyllite unit layers is about 9.2 Å. These minerals are anomalous because they contain little isomorphous substitution in their crystalline lattices and lack the interlamellar layers of compensating cations that are characteristic of the smectites. Montmorillonite and hectorite are more representative species of the smectite group of minerals.

1.3.1 Montmorillonites

An ideal montmorillonite may be defined as having the unit cell composition shown below [4], in which the subscripts (IV) and (VI) in the formula denote the respective tetrahedral and octahedral layer cations, and (M^+) represents a univalent or equivalent compensating cation.

$$[(Si_8)^{IV} \cdot (Al_{3.33}Mg_{0.67})^{VI} \cdot O_{20}(OH)_4]\, 0.67M^+$$

The isomorphous substitution consists predominantly of Mg-for-Al in the octahedral layer, resulting in a net anionic charge of 0.67 units per unit cell. The Mg-for-Al substitution has been shown to result in incomplete neutralization of the negative charge on the apical oxygens and hydroxide groups coordinated to the magnesiums.

The anionic charge of the aluminosilicate layers is neutralized by the intercalation of compensating cations and their coordinated water molecules. The montmorillonite crystal structure thus consists of superimposed aluminosilicate layers, each of which is interleaved with a "layer" of hydrated, exchangeable compensating cations. These cations can alternatively be described as occupying the interlamellar spaces or the regions between the basal surfaces of opposing silicate layers. Although the compensating cations are normally located in the regions adjacent to the points of anionic charge on the basal surfaces, small anhydrous cations, principally Li^+ or H^+ ions, are capable of migrating through the basal surface oxygen sheet to the neighborhood of the isomorphous substitution sites. The protons appear to associate with the octahedral hydroxyl groups, instead of forming hydroxyl groups by reacting with the incompletely neutralized oxygens [28].

Most native montmorillonites or bentonites have compositions close to that of the ideal mineral, but may contain additional octahedral isomorphous substitution, for example, octahedral Fe^{III}-for-Al, as well as some tetrahedral Al-for-Si substitution and other structural abnormalities, all of which can affect the surface chemistry. Na^+, Ca^{2+}, Mg^{2+}, and Al^{3+} ions are the principal compensating cations in the natural bentonites, the dominant ion depending on the origin of the particular deposit. These cations may be exchanged with other ions, either during weathering of the deposit or during processing of the clay.

Wyoming bentonite, which is believed to have been formed by the weathering of volcanic ash in an ancient sea, with little subsequent leaching, has Na^+ ions as the principal compensating cations. The cation exchange capacities of montmorillonites from this source are about 0.8 to 1.0 mequiv/g. The Texas bentonites have a slightly different genesis, and their constituent montmorillonites have a higher cation exchange capacity and a greater degree of tetrahedral isomorphous substitution than the Wyoming types. As a result of weathering and natural ion exchange, the Texas bentonites contain hydrated Ca^{2+} ions as the compensating cations, not all of which are readily exchangeable. There are other subtle differences in the colloidal and thermochemical properties of the two bentonites, apart from differences in the dominant exchangeable cation, which, for example, can affect their suitability as binders for foundry sands [29].

Minerals from different sources generally have variations in structure or structural order that may result in considerable differences in their colloidal and chemical properties. A survey of over 300 different montmorillonites has shown that only 10% have a regular structure [30]. At one time it was thought that the montmorillonites could be divided into two classes, the Wyoming type and the Cheto type [31]. These classes were originally distinguished by differences in their chemical transformations which were apparent at high temperatures during thermal analysis. The two forms have also been differentiated by use of electron microscopy and by refractive index measurements [32]. The Cheto type was once believed to contain a proportion of an alternative, but now discounted montmorillonite structure, the Edelmann–Favejee structure [4], in which the layer charge was postulated to arise from ionization of basal surface SiOH groups, resulting from inversion of silicate tetrahedra. Authentic structural differences between the two types appear to be slight, the Cheto type having a higher proportion of its charge originating in the octahedral layer [32], while electron diffraction studies indicate that Wyoming montmorillonite particles have better crystallinity [33]. Montmorillonites from Otay (California), Chambers (Arizona), and New Mexico have been classed as Cheto types; these minerals have less than 20% of their charge originating in the tetrahedral layers. Montmorillonites from Amory (Mississippi), which have approximately 40% of the layer charge originating in the tetrahedral layers, could not be classified as either type [32]. Basal surface SiOH groups can occur in montmorillonites that contain tetrahedral cationic vacancies. These may occur in some synthetic montmorillonites [4] and may result from the leaching of Al^{3+} or Fe^{3+}/Fe^{2+} tetrahedral substituent ions from native minerals in acidic environments.

Wyoming bentonite fractions appear to have differences in their cation exchange capacity and selectivity, the smaller particles having a smaller exchange capacity, a more homogeneous charge distribution [34], and a lower capacity for exchange with Ca^{2+} compared to Na^+ ions [35]. Demixing of Ca^{2+} and Na^+ ions appears to occur within zones of a particular particle [36]. This could be related to a nonhomogeneous anionic layer charge distribution. Various studies have shown that most other montmorillonites have some nonuniformity of charge distribution which, in extreme cases can be bimodal, the mineral containing interstratified layers of low-charge and high-charge montmorillonite [37]. The homogeneity of the distribution of charge can also affect the colloidal properties, nonhomogeneous charged minerals remaining peptized in more concentrated salt solutions than the homogeneous charged minerals [38]. Coagulation of a dispersed mixture of high-charge and low-charge montmorillonites with concentrated salt solutions results in the formation of an interstratified montmorillonite if the particles have basal dimensions smaller than 0.1 μm. Larger particles having basal dimensions of 0.1 to 2.0 μm selectively coagulate, reforming a mixture of discrete low-charge and high-charge montmorillonite particles [39].

The commercial bentonites and their derivatives can be divided into four categories: the swelling (Na^+) bentonites, for example, the natural Wyoming mineral or ion-exchanged bentonites from other sources; the nonswelling (predominantly Ca^{2+}) bentonites, which are the most widely distributed natural form; the organophilic bentonites; and the acid bentonites or bleaching clays, used as catalysts and as decolorizing agents in the treatment of vegetable and mineral oils.

The major applications for the bentonites are as additives in the iron ore pelletizing, metal foundry, and oil-drilling industries. For many of these uses, the processing of the bentonite need only consist of drying the raw clay, which may contain 40% water, pulverizing the dry product (15% water), and classification of the powdered clay in air cyclones. The nonswelling bentonites may be converted to the swelling sodium form by treatment with sodium carbonate or other salts that insolubilize the Ca^{2+} present in the raw clay. The bentonites can also be converted to the swelling, dispersable Na^+ form by conventional ion-exchange techniques which may form part of a wet beneficiation process. In a typical wet process, the raw clay is dispersed in a solution containing sodium carbonate and polyphosphate, and the nondispersible mineral fraction allowed to settle; the purified clay is then recovered by centrifuging the dispersion. The removal of Ca^{2+} may be assisted by passing the bentonite suspensions through a column of Na^+ ion-exchange resin, while adsorbed iron oxides can be removed and the tetrahedral Fe^{III} content decreased by treatment with a weakly acidic solution of sodium dithionite. The commercial organophilic bentonites are usually alkylammonium bentonites, prepared from the Na^+ form of the clay by ion exchange; these materials are described in Sections 1.3.5 and 4.4.

The synthetic bleaching earths or acid clays are prepared by heating a mixture of bentonite and moderately concentrated sulfuric or hydrochloric acid with steam, followed by washing to remove the byproduct salts; the resultant degraded bentonite is then dried and activated by heating at moderate to high temperatures. Raw clays with low iron content and high (octahedral layer) Mg^{II} content are preferred for the manufacture of acid clays.

1.3.2 Beidellites, Hectorites, and Saponites

Up to 40% of the charge of some natural montmorillonites may result from tetrahedral Al-for-Si substitution. This can result in differences in the properties of the minerals [40], particularly their swelling in polar liquids, their ion-exchange behavior, and the acidity of their surfaces. Montmorillonites in which more than 50% of charge originates in the tetrahedral layers are strictly classed as beidellites.

Beidellite is a dioctahedral analogue of montmorillonite and contains predominant tetrahedral isomorphous substitution. Contrary to some reports, beidellites do occur in nature, although varieties containing only traces of octahedral substitution are rare [5]; however, they can be synthesized. Most natural beidellites contain appreciable octahedral as well as tetrahedral isomorphous substitution. Nontronite, for example, is a ferric beidellite that has both Al-for-Si, Fe^{III}-for-Si, and Fe^{II}/Fe^{III}-for-Al substitution, with approximately 25% of the charge originating in the octahedral layer. A beidellite has the idealized unit cell composition [4]:

$$[(Si_{7.33}Al_{0.67})^{IV} \cdot (Al_4)^{VI} \cdot O_{20}(OH)_4]\, 0.67M^+$$

Hectorite is the trioctahedral analogue of montmorillonite and contains predominant octahedral Li-for-Mg substitution, while saponite is the trioctahedral analogue of beidellite. Their idealized unit cell compositions [4] are

Hectorite: $[(Si_8)^{IV} \cdot (Mg_{5.33}Li_{0.67})^{VI} \cdot O_{20}(OH)_4]\, 0.67M^+$

Saponite: $[(Si_{7.33}Al_{0.67})^{IV} \cdot (Mg_6)^{VI} \cdot O_{20}(OH)_4]\, 0.67M^+$

The commercial filler, Laponite® is a synthetic low-charge hectorite which, unlike the natural mineral, can be obtained with a negligible iron content. Another grade of Laponite® consists of a fluorohectorite in which the octahedral lattice hydroxyl groups have been replaced with fluoride ions.

Montmorillonite and hectorite have differences in chemical properties that have been ascribed to differences in misfit between the octahedral and tetrahedral layers. One reported example is the lower activation energy for dehydrogenation of adsorbed long-chain quaternary amines on hectorite [41]. Hectorite lacks the potentially acidic, exposed aluminum ions that are partly responsible for the acidic character of the surfaces of the montmorillonites. However, both hectorite and montmorillonite contain other acidic species adsorbed on the basal surfaces and in the interlamellar spaces; these acidic species are activated water molecules, and the origin of this acidity is discussed in Section 1.6.

1.3.3 Swelling of Smectites

The compensating cations, together with their associated water molecules, form separate layers between the opposing silicate sheets, that is, in the interlamellar spaces of the smectites. The compensating cations are generally capable of rapid exchange with other ions, although the minerals may have pronounced affinity for particular cations. The cation exchange capacities of typical smectites have values of 0.4 to 1.0 mequiv/g; this includes about 0.1 to 0.2 mequiv/g of exchange capacity arising from anionic sites on the silicate layer edges [40,42]. The basal spacings of these smectites are larger than the 9.2-Å thickness of the uncharged silicate layers because of the presence of the hydrated interlamellar cations and may have values of 14 Å or greater, depending on the mineral, the compensating cation, and the partial pressure of water vapor in equilibrium with the specimens.

The basal spacing of a montmorillonite containing exchangeable Na^+ ions can vary from 9.5 Å for the dehydrated material up to 15.4 Å for a sample exposed to air at 100% relative humidity, containing 20% adsorbed water. The basal spacings increase in steps with increasing water content, the observed basal spacings of 12.4, 15.4, and 18.6 Å corresponding to the respective presence of one, two, or three layers of hydration water in the interlamellar spaces. Further swelling can occur when the Na-montmorillonite is immersed in aqueous sodium chloride, the interlayer distance increasing with decreasing ionic strength of the media. Interlayer distances in excess of 120 Å occur in 0.1 N NaCl, while in salt-free media the interlayer spacing is effectively infinite, that is, the aluminosilicate layers become dispersed. The basal spacing of Ca-montmorillonite, on the other hand, does not increase beyond 18 Å in aqueous media.

The driving force for the intercalation of water has been likened to osmosis and is only limited by the Coulombic interaction between the silicate layers and the interlamellar cations [43]. The effective ionic strength in the interlamellar medium of aqueous montmorillonite suspensions has been estimated to be about 20 mequiv/l [44]. The swelling of Ca-montmorillonite in water containing a small percentage of acetone (less than 20%) can be considerably greater than 18 Å. This has been ascribed to the formation of mutually repulsive charged double layers at each of the opposing silicate surfaces. At higher concentrations of acetone, collapse of the interlamellar spacing results from increased Coulombic interaction in media having a lower dielectric constant [45].

The smectites, like kaolinite, develop negative charges on their basal surfaces, and positive charges on their layer edges when suspended in neutral or weakly acidic aqueous media. Montmorillonite suspensions in aqueous or ionizing media can form thixotropic porous structures which can be sufficiently extensive to gel the mixture. The structures in media of low salinity (e.g., 10^{-3} N NaCl), and having pH values of 4 to 11, result from extensive edge–face and edge–edge interactions of coagulated single-layer platelets [46]. In more saline media, face–face coagulation can occur, resulting in the breakdown of the thixotropic structure [39].

The swelling of dried beidellites and saponites, on exposure to moisture, is considerably less than that of the analogous montmorillonites and hectorites. This results from the enhanced interaction between the interlamellar water molecules and silicate surface sites associated with tetrahedral substitution. These sites have a greater localization of anionic charge than those associated with octahedral substitution. The swelling of Na-saponite is limited to the equivalent of two to three layers of intercalated water molecules per interlamellar space, the water molecules forming hydrogen-bonded bridges which can link the opposing silicate layers [47]. Restricted swelling of K-, Ca-, and Mg-beidellites has also been associated with the enhanced ionic character of the beidellite surface [48].

The origin of the layer charge affects other properties of the interlamellar water molecules, for example, water adsorbed adjacent to tetrahedral sites being less acidic (Section 1.6) and more strongly bound than that adsorbed adjacent to octahedral sites. This is shown by the incomplete removal of interlamellar water from a Cu-saponite (or vermiculite) on washing with methanol, or from a tetramethylammonium saponite on outgassing, by heating under vacuum, under conditions which would result in complete removal of the interlamellar water from the analogous montmorillonite derivatives [49].

The smectites can intercalate other polar liquids, most notably alcohols such as ethylene glycol or glycerol, ammonia and other amines, ketones, and nitriles; these mineral organic systems are discussed in Section 1.7 and in Chapter 4. The swelling and increase in basal spacings on treatment with polar liquids can be used as a means for the identification of montmorillonite and similar minerals. The swelling in glycerol of a Li-exchanged clay before and after calcination is a classic, although not infallible means for the identification of tetrahedral and octahedral isomorphous substitution in minerals [4,50]. The basal spacing of a Li-montmorillonite collapses irreversibly to about 9.5 Å when the mineral is heated at 200 to 300°C, and the product can no longer swell when treated with glycerol. The collapse of the octahedral-substituted minerals is accompanied by a reduction in their cation exchange capacities [40]. In contrast, Li-exchanged beidellites, nontronites, or saponites are unaffected by such heat treatment and can swell to about 17.7 Å on treatment with glycerol both before and after heating [4]. On dehydration of a Li-montmorillonite, the anhydrous Li^+ ions are believed to diffuse from the basal surfaces to the immediate vicinity of the substituent ions, where they are no longer accessible for ion exchange or for solvation; the consequential reduction of the charge at the silicate surface also reduces the adsorption of polar liquids. In contrast, Li^+ ions associated with tetrahedral sites remain in the vicinity of the basal silicate surface, and are available for ion exchange and for solvation and reswelling of the mineral.

The effects of heat on other smectites depend on the mineral and the nature of the exchangeable cations. In general, the interlamellar water may be readily eliminated on heating at temperatures up to 200°C, with reversible collapse of the smectite structure. Structural water molecules, that is, octahedral hydroxyl groups, are not eliminated until the minerals are heated above 400°C. The elimination of structural water is accompanied by an irreversible collapse of the smectite layer structure; this occurs at 400 to 500°C for nontronites, at 500 to 700°C for montmorillonites and beidellites, at 700 to 900°C for hectorites [4], and at still higher temperatures for pyrophyllite and talc.

1.3.4 Micas and Vermiculites

Other minerals related to montmorillonite may have greater degrees of isomorphous substitution and higher layer charge density which, because of the strong Coulombic interaction between the silicate surfaces and the compensating cations, greatly reduces the ability of the mineral to swell in the presence of polar liquids or even to readily undergo cation exchange. The highly charged montmorillonite analogues are the micas, a typical example, muscovite, having the unit cell composition:

$$[(Si_6Al_2)^{IV} \cdot (Al_4)^{VI} \cdot O_{20}(OH)_4]K_2$$

The trioctahedral analogue of muscovite is the mineral, phlogopite.

Despite the high content of compensating cations, the absence of swelling in aqueous media restricts the interlamellar diffusion of hydrated ions and limits the exchange capacity of the finely divided mineral to about 80 μequiv/g. This value corresponds to the estimated concentration of exchange sites on the external surfaces of the particles; only ions adsorbed at these sites can be exchanged with other inorganic cations at significant rates. Illites are poorly crystalline fine-particle micaceous clays [2] that also have high layer charge densities, low swelling, and very slow rates of ion exchange.

The vermiculite minerals have an intermediate charge density, and by suitable treatment, can undergo ion exchange and swelling. The vermiculites are classed as trioctahedral and largely contain tetrahedral isomorphous substitution. The compensating cations in the native minerals are usually Mg^{2+} ions that are accompanied by a regular array of hydration water molecules; the vermiculites are sometimes known as hydrated micas. Vermiculites have an approximate idealized unit cell composition [4].

$$[(Si_{6.5}Al_{1.5})^{IV} \cdot (Mg_6)^{VI} \cdot O_{20}(OH)_4]\, 0.75Mg^{2+} \cdot xH_2O$$

Many vermiculites contain more extensive Al-for-Si substitution than the idealized composition. This is partly compensated for by octahedral Al-for-Mg or Fe^{III}-for-Mg substitution. Batavite, an iron-free vermiculite, for example, has a unit cell composition [51].

$$[(Si_{5.98}Al_{2.02})^{IV} \cdot (Mg_{5.28}Al_{0.66})^{VI} \cdot O_{20}(OH)_4]\, 0.68Mg^{2+} \cdot 4.7H_2O$$

The cation exchange capacities of the hydrated vermiculites are about 1.1 to 1.2 mequiv/g, or about 1.7 mequiv/g for the dehydrated mineral. The basal spacing of an anhydrous vermiculite is about 9.0 Å, while those of hydrated vermiculites are about 11.6 or 14.8 Å, corresponding to the respective presence of one and two interlamellar layers of hydration water. A fully hydrated vermiculite can contain 20 waters per interlamellar cation and, on rapid heating, the pressure generated by the vaporization of water trapped by the initial collapse of the layer edges can cause the mineral particles to expand remarkably (exfoliate) into worm-like (vermicular) shapes that are used as insulation and as porous sorbents. The vermiculites can undergo ion exchange when treated with concentrated LiCl solutions, and the Li-vermiculite, unlike the other inorganic forms, can be swollen and dispersed in salt-free water. Because of the large basal dimensions of typical dispersed vermiculite layers, the dispersions have a nacreous appearance and, on settling and evaporation, can form mechanically strong, flexible sheets [52].

The chlorites are a group of minerals related to both the montmorillonites and the vermiculites. In these minerals the hydrated Mg^{2+} interlamellar layers of the vermiculite have been replaced by brucite-like layers, an ideal magnesium chlorite consisting of alternating brucite and talc layers. Most magnesium chlorites contain iron and have tetrahedral Al-for-Si isomorphous substitution that is compensated for by an equivalent Al-for-Mg or Fe^{III}-for-Mg substitution in the brucite layers. Chlorites are nonswelling minerals, but may exist as partially swelling, mixed-layer chlorite–montmorillonite or chlorite–vermiculite species [4]. Aluminum chlorites are also known.

1.3.5 Ion Exchange

The accessible compensating cations of the kaolinites, montmorillonites, and related minerals have been shown to be capable of undergoing exchange with other cationic species in a reversible process (Figure 1.6). The compensating cations in the native minerals can be exchanged with other hydrated inorganic cations, with those associated with other ligands, for example, $[Co(NH_3)_6]^{3+}$, and with a wide variety of organic cations. The organic cations most frequently used are those of amine or quaternary ammonium salts, but may also include oxonium, sulfonium, and phospho-

$$[\text{Mont.}]^{-}\,\text{Na}^{+} + (\text{H}^{+})_{aq.} \rightleftharpoons [\text{Mont.}]^{-}\,\text{H}^{+} + (\text{Na}^{+})_{aq.}$$

Figure 1.6

nium ions, and more complex cationic species like Methylene Blue and other cationic dyestuffs. Exchange by the hydronium ion, H_3O^+, can also occur readily. This has important consequences in respect to the hydrolytic stability and weathering processes of the layer silicates.

Not all compensating cations are exchangeable; in some minerals they may be inaccessible because of their location within the silicate layer structure, like Li^+ ions in some lithiated montmorillonites or anhydrous K^+ or Ca^{2+} ions in highly charged micas. The cations are trapped in a compact lattice having interlamellar spaces too small to allow for significant rates of diffusion and exchange with other hydrated ions. Some polyvalent compensating cations, for example, Al^{3+}, may be tenaciously adsorbed as hydrolyzed, polymeric oxycationic species which cannot be readily displaced by other cations.

The kinetics and equilibria of the inorganic cation exchange process in layer minerals are described in many texts and reviews on clay colloidal chemistry [e.g., 13,53]. These aspects of the ion exchange process are of limited relevance to the applications of the minerals as fillers and pigments and will only be briefly summarized here. Some of the qualitative features are:

(i)   The rate of exchange is governed in part by accessibility of the interlamellar cations and may be very slow for nonswelling smectites and micas.

(ii)   Exchange equilibrium favors adsorption of cations of higher valence. Replacement of one multivalent cation by another is not a practical process for preparing homoionic minerals. These are usually prepared from minerals first exchanged with monovalent ions.

(iii)   The exchange equilibrium generally favors the larger cation of a particular valence, although the rate of exchange may favor the smaller cations such as Li^+.

(iv)   Certain cations, most notably K^+, Ba^{2+}, and NH_4^+, can have a particular affinity for the layer silicates as they can be accommodated in the recesses of the silicate basal oxygen sheet and coordinate with the surrounding oxygen atoms.

Ion exchange of minerals with organic ammonium salts is frequently used in the preparation of gellants, fillers, and pigments for organic media. Important practical examples and consequences of this treatment are discussed in Chapter 4, but the basis of the exchange reaction will be discussed here, using n-alkylammonium ions as examples.

Alkylammonium ions can readily exchange with the interlamellar inorganic cations of the smectites, vermiculites, and illites, forming intercalated organomineral species. In the case of the short-chain ammonium derivatives, for example, methyl-, ethyl-, propyl-, allyl-, or butylammonium smectites, the exchanged minerals may remain sufficiently hydrophilic to form stable aqueous dispersions. However, the longer-chain ammonium salts form mineral derivatives that are increasingly hydrophobic. Aqueous mineral dispersions undergo rapid exchange and flocculation when treated with solutions containing these ions. The long-chain ammonium derivatives readily intercalate neutral polar or nonpolar organic molecules, and the layers of treated montmorillonites or vermiculites can be dispersed in organic media.

Long-chain primary (RNH_3^+), secondary ($R_2NH_2^+$), tertiary (R_3NH^+), and quaternary (R_4N^+) ammonium ions are preferentially intercalated. The stability of the alkylammonium-exchanged smectites increases with increase in the alkyl chain length and with increasing substitution on the ammonium ion. This has been ascribed to the favorable effects of van der Waals forces between amine chains in the resultant intercalates [54,55] and to the reduced solvent shielding of the ions in the interlamellar environment [56]. Protonation of organic bases in the interlamellar spaces of aqueous clay suspensions also occurs to a greater extent than that governed by the pH of the suspending medium. This has been ascribed to the effects of stabilization of the protonated amine by interaction with the basal oxygen sheets, the degree of stabilization depending on the magnitude of the layer charge [57].

The basal spacings of the solvent-free alkylammonium derivatives of the higher-charged smectites, vermiculites, and illites increase in a stepwise manner with increasing alkyl chain length, the rate of increase depending on the charge density or exchange site density of the mineral surface (Figure 1.7) [58]. This is a consequence of the ordered arrangement of the longer-chain alkylammonium ions in the interlamellar space, in which opposing alkyl chains having an all-*trans* conformation alternate in an inclined, closely packed array (Figure 1.8c). Short-chain alkylammonium ions may be accommodated in minerals of low-charge density by adsorption with the bulk of their alkyl chains lying parallel to the silicate layers (Figures 1.8b and c). The alternative inclined arrangement of chains occurs when the area per exchange site is insufficient to accommodate the longer alkyl chains in the "flat" arrangement. Alkylammonium chains in both arrangements have their ionic centers similarly keyed into the silicate surfaces, the optimum configuration being obtained by rotation of the alkyl chains about their C1–C2 bonds [59].

The approximate basal surface areas per cation exchange site on the montmorillonite and vermiculite anionic sites are 75 and 60 $\mathring{A}^2$, respectively, and the inclined configuration commences in an ethylammonium vermiculite or a *n*-butylammonium montmorillonite. This is not necessarily true for all specimens of alkylammonium montmorillonites. The basal

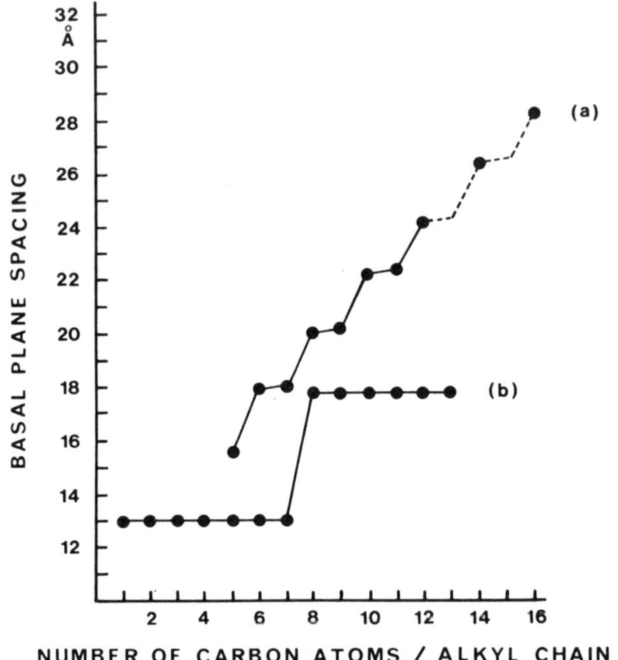

Figure 1.7   Typical basal spacing progressions of some of n-alkylammonium smectites. (a) High-charge micas. (b) Low-charge montmorillonites.

spacings of long-chain alkylammonium-exchanged derivatives of the lower charged Wyoming bentonites, for example, show that the alkyl chains may be preferentially adsorbed in a "flat" arrangement, each interlamellar space containing one, two, or three layers of amine, depending on the degree of exchange and alkyl chain length. The basal spacings of the alkylammonium smectites are apparently independent of the origin of the layer charge. Alkylammonium chains in high-charge minerals like the vermiculites can have metastable arrangements of alkyl chains which give rise to anomalous basal spacings [51]. These are caused by the rotation about one or more C–C bonds per chain, resulting in breaking of the all-*trans* conformation and the formation of blocks of kinked chains.

The alkylammonium montmorillonites can intercalate additional molecules of the alkylammonium salt or of neutral swelling agents, for example, alcohols, in a bimolecular layered arrangement (Figure 1.9). The basal spacings of these complexes may decrease with increasing temperature owing to the reversible formation of blocks of kinked chains in the bimolecular layered complexes [60,61].

α,ω-Alkylene diammonium montmorillonites may also contain inclined arrays of interlamellar alkyl chains which can link the opposing silicate layers, thereby limiting the additional swelling and the degree of absorption which can occur on intercalation of other molecules.

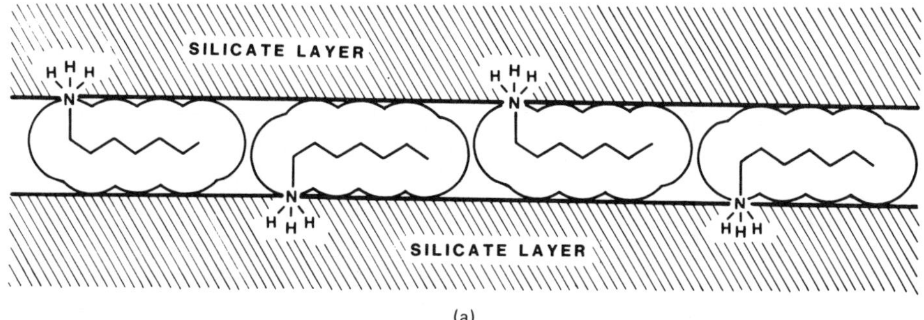

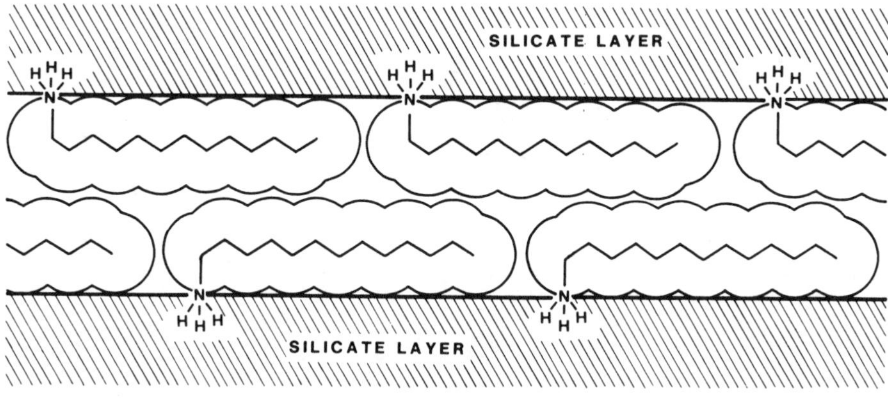

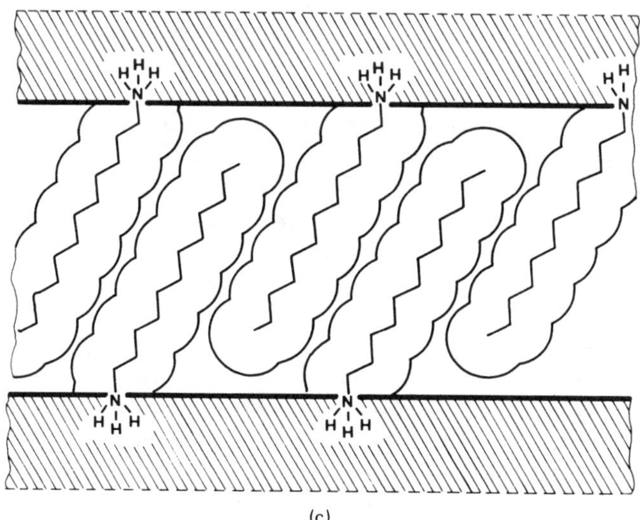

Figure 1.8 Conceptual arrangements of *n*-alkylammonium chains in the smectite interlamellar spaces. (a) Prone configurations of short-chain and (b) long-chain amines in a low-charge smectite. (c) Erect configuration, in a high-charge smectite.

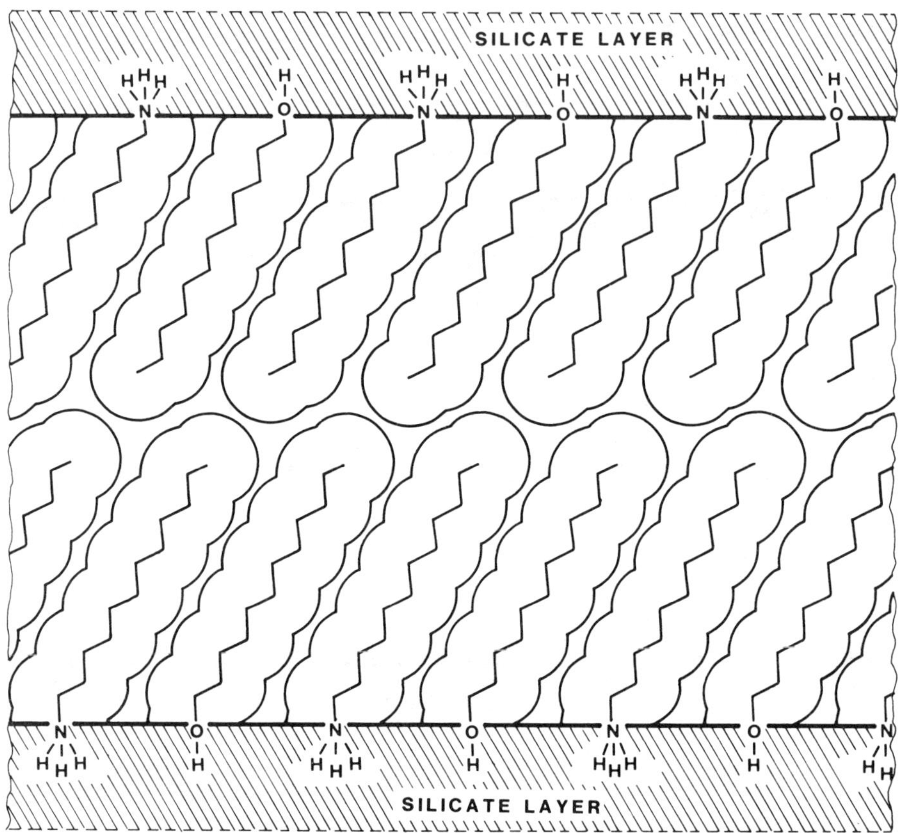

Figure 1.9   A conceptual arrangement of unkinked alkyl chains in an alcohol–alkylammonium smectite complex.

The structures of secondary, tertiary, or quaternary ammonium montmorillonites resemble those of the primary alkylammonium derivatives, provided the alkyl chains are short. If the cation contains more than one long alkyl chain, the structure of the organoclay may be more complex, and the formation of closely packed, regular arrays of intercalated molecules may not be possible [17].

The montmorillonites and related minerals, such as the kaolinites, have potential anion exchange sites at the silicate layer edges. This exchange may take the form of isomorphous F-for-OH substitution [62] or some other form of ligand exchange on the exposed octahedral metal ions, but in the general case it is associated with the ionization of water adsorbed on, or in the vicinity of the octahedral layer edges [63] (Figure 1.10a). In contrast, the adsorption of phosphate ions, which can form chelates with exposed layer–edge aluminums, results in the displacement of the hydroxo- and aquo-ligands associated with the octahedral metal ions (Figure 1.10b). The anion exchange capacity of the clays is pH-dependent, and its measure-

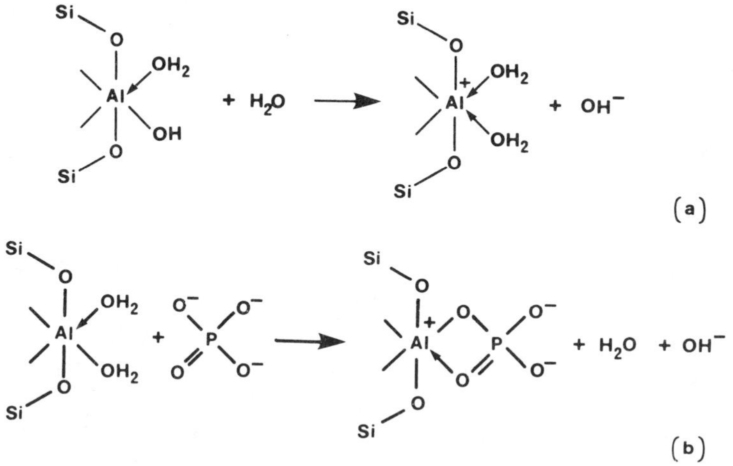

Figure 1.10

ment is complicated by the negative adsorption of anions from the suspension medium in the vicinity of the negative charged surfaces.

1.4 PYROXENES, AMPHIBOLES, AND RELATED MINERALS

The pyroxenes have a crystalline lattice constructed from chains of silicate tetrahedra, the tetrahedra of an individual chain being similarly oriented (Figure 1.1a), and the basal oxygens lying on a common plane. The lattice is formed from opposed pairs of silicate chains linked by metal ions. Wollastonite, $CaSiO_3$, has a related structure in which every third silicate tetrahedron is inverted [64].

The asbestos minerals, apart from serpentine and chrysotile, have an amphibole structure, with an ideal unit cell composition:

$$[(Si_8)^{IV}{\cdot}(Mg_5)^{VI}{\cdot}O_{22}(OH)_2]Mg_2$$

The amphibole lattice is constructed from ribbons of silicate tetrahedra, each ribbon formed by the union of two pyroxene chains giving an annellated cyclic silicate chain (Figure 1.1b). In the magnesium amphiboles, opposing pairs of these ribbons are united by a ribbon of hydrated magnesia to form talc-like strands, adjacent strands being linked together by additional magnesium ions (Figure 1.11) [64]. The various asbestos minerals are derived by isomorphous substitution, the principal octahedral and linking cations of the five common minerals being anthophyllite, $(Mg,Fe^{II})Mg_2$; amosite, $(Fe_5)Mg_2$; crocidolite, (Fe^{II},Fe^{III},Na); tremolite, $(Mg_5)Ca_2$; and actinolite $(Mg,Fe^{II})Ca_2$ [65]. Aluminium amphiboles are also known. In contrast to chrysotile, the amphibole particles develop a negative surface charge when suspended in neutral or weakly acidic media.

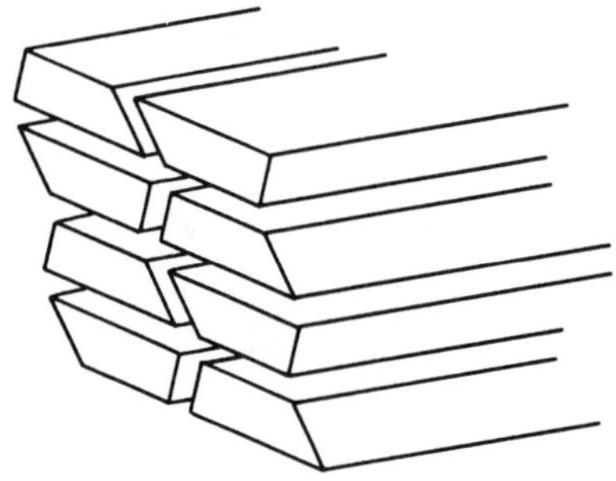

Figure 1.11 The arrangement of the silicate chains in an amphibole asbestos.

The attapulgite (palygorskite) lattice is also constructed from amphibole-like strands, but with adjacent silicate ribbons condensed so that their basal oxygens form a common planar sheet, with the apical oxygens of alternate ribbons lying above and below the basal oxygen sheet

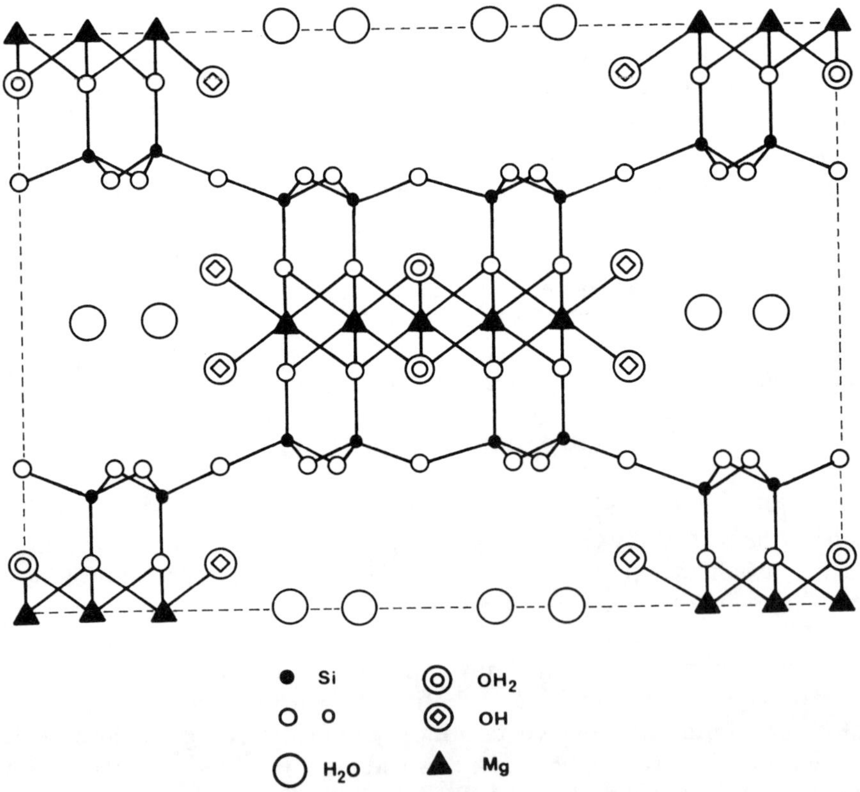

Figure 1.12 The lattice structure of attapulgite; view along axis of silicate ribbons. (Adapted, with permission, from G. Brown, Ed., *X-Ray Identification and Crystal Structure of Clay Minerals*, copyright © Mineralogical Society, London, 1961, p. 345.)

(Figure 1.12) [4]. As a consequence, the mineral lattice has a porous structure, containing channels, with approximate cross sections of 4 by 5 Å that can accommodate absorbed zeolitic water molecules. An ideal magnesium attapulgite would have the half-unit cell composition shown below, in which (OH_2) represents bound water (protonated hydroxyl groups) coordinated with the octahedral layer Mg^{2+} ions at the ribbon edges, and H_2O represents unbound, zeolitic water.

$$[(Si_8)^{IV} \cdot (Mg_5)^{VI} \cdot O_{20}(OH)_2(OH_2)_4] \cdot 4H_2O$$

The octahedral ribbons of typical attapulgites may contain approximately equal numbers of Al^{3+} and Mg^{2+} ions plus Fe^{3+} ions and ionic vacancies; these vacancies are compensated for by replacement of linking oxide ions by hydroxides [4]. The strands carry a net negative charge that is compensated for by the presence of exchangeable cations located in the zeolitic channels. The cation exchange capacities of the typical attapulgites range from 0.1 to 1.2 mequiv/g.

The zeolitic water is expelled on heating the mineral at temperatures up to 200°C, leaving vacant channels which can accommodate absorbed polar and nonpolar organic molecules. The bound waters can be eliminated on heating above 250°C, and the porous structure undergoes irreversible collapse above 400°C.

Sepiolite (meerschaum) also has a silicate ribbon structure similar to

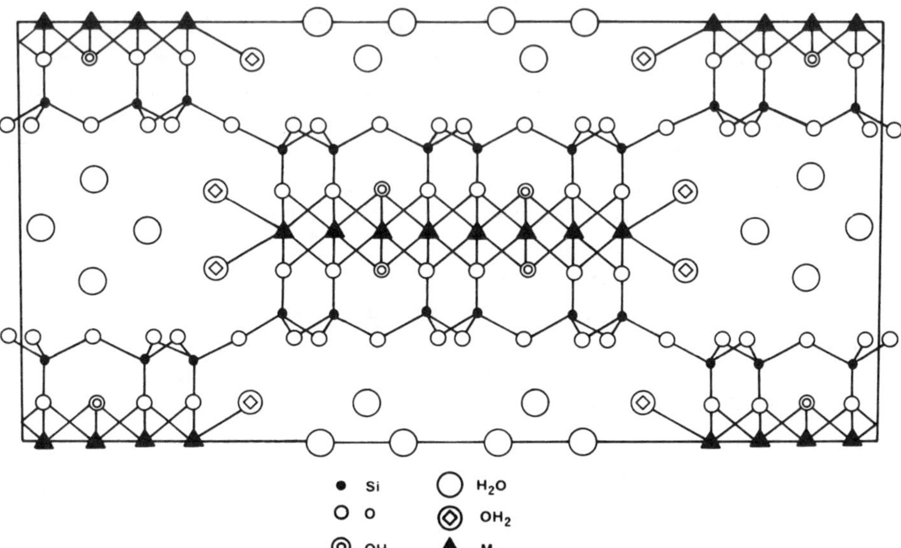

Figure 1.13   The lattice structure of sepiolite; view along axis of silicate ribbons. (Adapted, with permission, from G. Brown, Ed., *X-Ray Identification and Crystal Structure of Clay Minerals*, copyright © Mineralogical Society, London, 1961, p. 329.)

that of attapulgite, with each ribbon formed by the union of two pyroxene chains. However, additional pairs of silicate tetrahedra are added at regular intervals to alternate sides of the ribbon, forming an annellated bicyclic silicate structure (Figure 1.13) [5]. This results in the formation of wider channels in the mineral lattice. Sepiolite is a magnesium silicate, with an ideal half-unit cell composition.

$$[(Si_{12})^{IV} \cdot (Mg_8)^{VI} \cdot O_{30}(OH)_4(OH_2)_4] \cdot 8H_2O$$

Sepiolites usually contain Al-for-Si or Fe-for-Si and Fe-for-Mg substitution and require compensating cations. These may be exchangeable and are accommodated with the zeolitic water in the channels; cation exchange capacities are typically 0.1 to 0.4 mequiv/g. The zeolitic water is eliminated on heating at temperatures up to 200°C, irreversible dehydration occurs above 250°C, and the porous structure collapses at 350°C.

Sepiolite and attapulgite have acicular particles that can intermesh to form thixotropic structures in aqueous media. These minerals are useful as gellants for highly saline media that coagulate bentonite suspensions.

1.5 ZEOLITES

The zeolites comprise a large and diverse group of aluminosilicates that have complex, porous, regular, three-dimensional, cage-like structures. These minerals have limited use as fillers or pigments, although some aspects of their acidic, catalytic activity are relevant to that which may be developed by the lamellar aluminosilicates. Accordingly, we only briefly describe the structure of one form, faujasite, by way of illustration. Details of other zeolite structures can be obtained elsewhere [e.g., 66]; the monograph [67] is especially recommended.

The zeolites are constructed from silicate and aluminate (AlO_4) tetrahedra which are united to form one or more basic three-dimensional cyclic structures. These structures, the primary structures, are hollow polyhedra formed aluminosilicate rings having four, five, six, eight, twelve, or more sides. These primary structures are stacked in regular arrays to form the secondary structures, usually with large internal cavities or porosity. These cavities are connected with each other and the exterior surface by channels between the polyhedra. This porosity of the zeolites can be used for the adsorption of polar or nonpolar molecules that are smaller than the cross section of the channels within the secondary structure. It should be noted that these polyhedral substructures are conceptual and serve as one means of describing the framework topology of the zeolite structures. Stereodiagrams of the framework of the principal zeolites have been prepared by Meier and Olson [68].

Faujasite can be considered to be formed from the linking of truncated

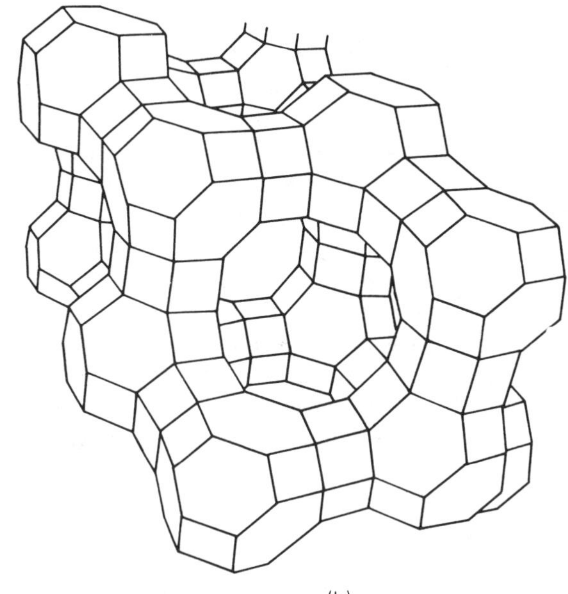

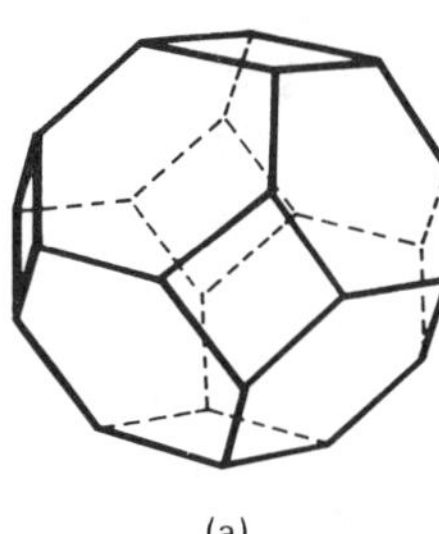

(a)   (b)

Figure 1.14   The topological framework of faujasite. (a) The sodalite structural unit. (b) The arrangement of sodalite units forming the faujasite zeolitic cavities and channels. (Copied, with permission, from L. A. Bursill, J. M. Thomas, and K. J. Rao, *Nature*, **289**, no. 5794, p. 157, copyright © Macmillan Journals Limited, 1981.)

octahedra (sodalite units; Figure 1.14) in a diamond-like pattern through their six-membered rings (Figure 1.14). This structure has large cavities of about 12-Å diameter that can be accessed through channels defined by the distorted twelve-membered rings formed by the linked sodalite units; these channels have an effective diameter of about 8 Å. An alternative cubic stacking arrangement via the four-membered rings occurs in Zeolite A (Linde sieve A), forming smaller cavities with an access diameter of about 4 Å. The nonzeolite mineral, sodalite, has a cubic arrangement formed from shared four-membered rings. Access to the internal porosity of the sodalite primary units can only occur through the six-membered rings which have an access diameter of about 2 Å.

The zeolite structure is anionic and requires the inclusion of compensating cations in equivalent numbers to the aluminate tetrahedra. These cations are usually exchangeable, and are accommodated in the cavities and channels of the zeolite structure. The cations are preferentially located at sites that are determined by their coordination requirements and by their interaction with the electric fields arising from the anionic sites and neighboring cations; they may migrate to different positions during the adsorption of organic species. The proportions of silicon to aluminum in the zeolites can vary widely; the faujasites can have Si:Al ratios that range from about 3:1 to 1:1. Much higher proportions of silica are present in some other zeolite forms such as the large-pored mordenites and the recently developed ZSM zeolites. The catalytic and sorbent properties of the zeolites can be tailored for specific applications by suitable choice of the zeolite structure, the Al:Si ratio in the reaction medium during the zeolite synthesis, the compensating cations, and subsequent treatments which may include ion exchange with catalytic metal ions, dealuminization, or decationization.

Although most types of zeolitic structures can occur in natural minerals, the commercial zeolites are usually synthetic. The natural zeolites appear to be the product of hydrothermal transformations of volcanic debris in strongly alkaline, strongly saline media. The synthetic zeolites are prepared by the crystallization of silica–alumina anionic gels in aqueous alkaline media. The structure of the zeolite or zeolite mixture formed is governed by the type of alkali (M_2O) or potential compensating cations present (these may include bulky quaternary ammonium ions for the synthesis of some wide-pore zeolites [69,70]), the relative proportions of the oxides M_2O, Al_2O_3, SiO_2, and H_2O in the reaction system, the crystallization temperature, and the presence of modifiers such as chelating anions. Crystallization of the zeolite from the amorphous aluminosilicate gel is a slow but autocatalytic process and is often accelerated by addition of seed nuclei of the desired form. Factors governing the crystallization of zeolites are described in greater detail in various chapters of references 66 and 67.

1.6   ACIDITY AND BASICITY OF THE SILICATE MINERALS

One of the most important properties of the many aluminosilicates used as pigments or fillers is the strong surface acidity which can develop on the removal of the bulk of their adsorbed or intercalated water. The surface acidity of dried minerals may exceed that of concentrated sulfuric acid. This can result in the catalysis of many desirable, undesirable, or unexpected reactions when these materials are incorporated in organic media. This acidity is related in origin, and is often similar in strength, to that of the acidic alumina–silica and zeolite cracking catalysts used in the petrochemical industry. Clays containing adsorbed water may also have sufficient surface acidity to catalyze many reactions in soils, including the degradation of pesticides and the transformation of natural organic molecules.

The nature of the acid catalytic species has been a contentious issue, and some historical aspects of a particular controversy, concerning the surface sites involved in the formation of radical-cations from aromatic hydrocarbons adsorbed on aluminosilicate surfaces, are summarized in Section 5.4. The origins and measurement of the surface acidity of solid catalysts, including the aluminosilicate clay minerals, has been frequently reviewed; some useful reviews are contained in references 71 and 72.

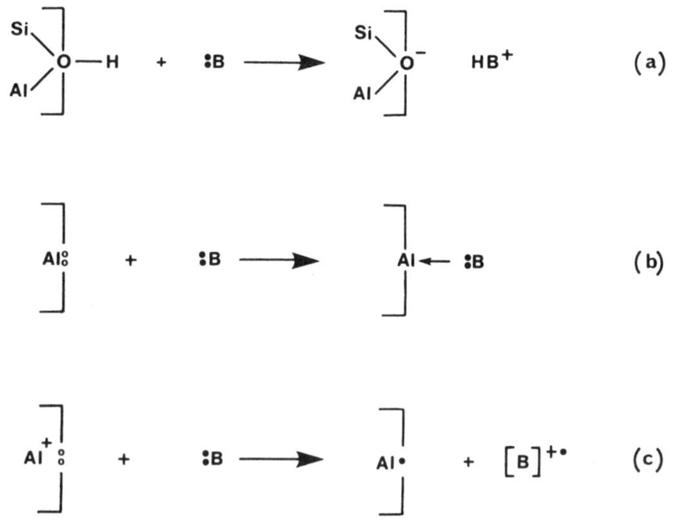

Figure 1.15   Reactions illustrating the interaction of bases with the various classes of surface acid sites. (a) Bronsted site. (b) Lewis site; by coordination. (c) Lewis site; by oxidation.

There are two general classes of acidic species: those that can act as proton donors for basic molecules (Figure 1.15a) and those that can accept electron pairs from basic molecules, forming coordination bonds (Figure 1.15b). The former are known as Bronsted acids and include familiar protonic acids such as sulfuric acid; the latter are known as Lewis acids and include anhydrous polyvalent salts such as aluminum chloride. Lewis acids may form Bronsted acids on hydration, provided the polarization of the H–O bonds of the coordinated water molecules is sufficient to significantly enhance their dissociation (see Figure 1.17).

The surfaces of the silicate minerals may contain both Lewis and Bronsted acidic species, although the Bronsted acidity is probably the more significant in catalysis of reactions by the minerals, either at ordinary temperatures or during the conventional processing and utilization of mineral–polymer composites.

1.6.1   Bronsted Acidic Centers

The Bronsted acidity of the aluminosilicate minerals can arise from several sources. Potential acidic species include the weakly acidic SiOH groups exposed at the layer lattice edges or at basal surface defects, and the strongly acidic bridged-hydroxyl groups (Figure 1.16) which are present in dehydrated amorphous silica alumina (p. 91) and decationized (hydronium) zeolites [73–75], or which may be formed at lattice layer edges, particularly at sites containing tetrahedral Al-for-Si substitution.

The most important sources of Bronsted acidity in the layer aluminosilicates are polarized water molecules adsorbed on the mineral surfaces, par-

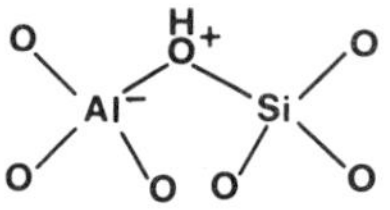

Figure 1.16

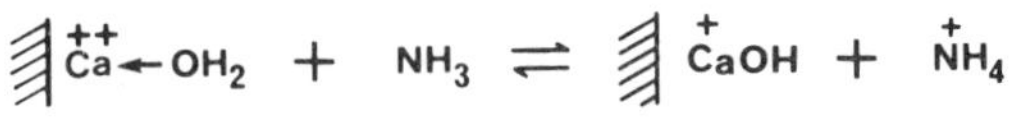

Figure 1.17

ticularly those waters associated with polyvalent compensating cations such as Ca^{2+} or Al^{3+}. The acidity can be correlated with the polarizing power of the cations. In the case of a series of homoionic montmorillonites, this is shown by the protonation of adsorbed ammonia, which increases with change of compensating cation in the order $K^+ < Na^+ < Li^+ < Ca^{2+} < Mg^{2+} < Al^{3+}$ [76]. In the case of Ca-montmorillonite, the yield of NH_4^+ produced by interlamellar protonation can be approximately equal to the number of Ca^{2+} ions, but in other systems the yield may approach the cation exchange capacity of the mineral with decrease in water content [76] (Figure 1.17).

The dissociation of adsorbed water in a Ca-montmorillonite at ambient temperatures has been found to be up to 10^7 times that of the bulk liquid [77]. The dissociation increases with increasing dehydration of the mineral, reflecting the increase in the polarizing effects of the cations with the decrease in the number of coordinated water molecules. The acidity of the residual water on a dried Mg-kaolinite has been reported to be greater than that of 71% sulfuric acid [78], while the surface acidity of a dry Na-kaolinite is equivalent to 48% sulfuric acid [79]. The development of surface acidity on drying of an Al^{3+}-flocculated kaolin is shown in Figure 1.18 [80].

The dissociation of water coordinated to compensating cations in zeolites is similar to that of the interlamellar cations in smectites [81]. However, on dehydration of the zeolites, protonation of the lattice may also occur, with the formation of bridged-hydroxylic species [67].

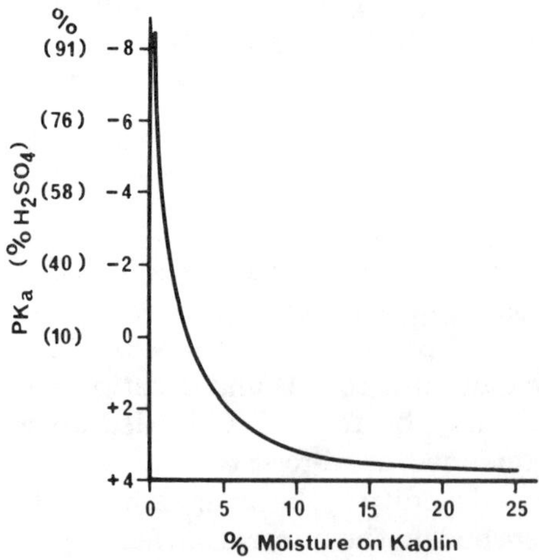

Figure 1.18 The surface acidity–hydration relationship of a typical alum-flocced kaolin. (Reprinted, by courtesy of Marcel Dekker Inc., from D. H. Solomon, J. D. Swift, and A. J. Murphy, *J. Macromol. Sci. Chem.*, 5, p. 589, copyright © Marcel Dekker, New York, 1971.)

Measurement of the acidity of the minerals in aqueous suspension does not give a valid indication of their potential acidity because of the effect of excess water on the nature of the mineral surface and because of the leveling effect of the solvent on the activity of the strongly acidic species. The titrated acidity of a K-kaolinite, for example, was found to vary with the nature of the solvent and to range in value from nil in water to 43 μequiv/g in acetone. This difference reflects that of the activity of the water molecules coordinated with layer–edge aluminum ions [82].

The acidity of moderately acidic minerals can be measured by study of adsorption equilibria, for example, of ammonia in reaction with partially dehydrated montmorillonites [76]. However, the simplest method for determining the acidity of strongly acidic surfaces has been the use of Hammett adsorption indicators [83]. Use of these indicators enables bracketing values of acidity, the so-called Hammett acidity (H_0), to be assigned to the surface species. These bracketing H_0 values correspond to the pK_a values of particular pairs of indicators, only one of which, the more basic indicator, is protonated when adsorbed on the mineral.

The relation between Hammett acidity and activity of the surface protonic species (a_{H^+}) is described by the equation below, in which f_B and f_{BH^+} represent the respective activity coefficients of the unprotonated and pronated forms of the adsorbed indicator:

$$H_0 = -\log(a_{H^+} \times f_B/f_{BH^+})$$

The Hammett acidity is a measure of the ability of the surface to convert an adsorbed neutral base (B) to its conjugate acid (BH^+). The Hammett indicators should not be used for measurement of the acidity of Lewis acids, although these acids may form coordinated species with the Hammett indicators which mimic the colors of the protonated species. The Bronsted basicity of the indicators only parallels, to a limited extent, their ability to form coordination complexes with Lewis acids.

Some frequently used Hammett indicators are listed in Table 1.1 [84] and the Hammett acidities of some dried clay minerals are given in Table 1.2 [80]. The validity of some of the published measurements of surface acidity have been questioned because the commonly used indicators for the more acidic species ($H_0 < -3.7$), benzalacetophenone and anthraquinone, can form physisorbed colored species which could be mistaken for the corresponding protonated species [85].

The quantitative distribution of the acidic species can be determined by titration of a suspension of the mineral in a hydrocarbon solvent with a solution of organic base such as n-butylamine, using a series of these indicators [86,87]. The Hammett acidity distributions for a selection of dried aluminosilicate minerals are shown in Table 1.3 [87]. The validity of this method is dependent on the establishment of an adsorption equilibrium between the indicators, the titrant, and the acidic sites during the titration

Table 1.1   Some Common Hammett Indicators[a]

Indicator	Basic Color	Acidic Color	pK_a	H_2SO_4 Equivalence
Neutral red	Yellow	Red	$+6.8$	$8 \times 10^{-8}\%$
Phenylazonaphthylamine	Yellow	Red	$+4.0$	$5 \times 10^{-5}\%$
Butter yellow	Yellow	Red	$+3.3$	$3 \times 10^{-4}\%$
Benzeneazodiphenylamine	Yellow	Purple	$+1.5$	$2 \times 10^{-2}\%$
Dicinnamalacetone	Yellow	Red	-3.0	48%
Benzalacetophenone	Colorless	Yellow	-5.6	71%
Anthraquinone	Colorless	Yellow	-8.2	90%

[a]Reprinted with permission from H. Benesi, *J. Amer. Chem. Soc.*, **78**, 5490 (1956), copyright © American Chemical Society.

process. Recent studies have shown that this equilibration seldom occurs and that a true acidity distribution is unlikely to be obtained by these titration techniques; an alternative method has been described in reference 88.

Although the surfaces of kaolins and bentonites, particularly their aluminum- or hydronium-exchanged forms, may carry a proportion of strongly acidic catalytic species, the surface densities of these active species usually does not match those on the acidic zeolites or silica–aluminas. However, the aluminosilicate clays can be converted to efficient catalysts by treatment with acid, followed by calcination. These modified clays are frequently described as acid kaolinites or montmorillonites, but are essentially amorphous silica–aluminas.

Alkali-metal ion-exchanged montmorillonites are generally regarded as having low Bronsted acidity. However, the dried, Li-, Cs-, and Rb-exchanged minerals can convert triphenylcarbinol to the triphenylcarbo-

Table 1.2   Acidity of Some Dried Minerals[a]

Mineral	Hammett Acidity[b]	Total Acidity[c]
Kaolinite	-8.2	0.04
Montmorillonite	-5.6 to -8.2	0.2
Attapulgite	-8.2	0.2
Talc	4.0 to 3.3	0.005
Titania (silica–alumina coated)	3.3 to 2.8	0.005
Kaolin (Calgon® treated)	1.5 to -3.0	0.04
Silica–alumina catalyst	-8.2	0.2

[b]H_{0-} range of the strongest sites.
[c]Total acidity to H_0 4.0 (mequiv amine/g mineral).
[a]Reprinted with permission from D. H. Solomon and H. H. Murray, *Clays Clay Miner.*, **20**, 135 (1972), copyright © Pergamon Press.

Table 1.3   Acidity Distribution of Some Dried Clays[a]

Clay (specific surface, m^2/g)	Butylamine titer (moles/g) in H_0 range			
	+ 3.3 to + 1.5	+ 1.5 to − 3.0	− 3.0	Total
Kaolinite (25)	0.004	0.004	0.014	0.022
Attapulgite (125)	0.05	0.08	0.01	0.14
Montmorillonites				
Natural (44)	0.00	0.09	0.00	0.09
Na-form (41)	0.03	0.01	0.00	0.04
H-form (52)	0.10	0.00	0.55	0.65

[a]Adapted with permission from H. Benesi, *J. Phys. Chem.*, **61**, 970 (1957), copyright © American Chemical Society.

nium ion, indicating the presence of moderately strong Bronsted surface acidity. The Na- and K-minerals are inactive. This may be due to the ability of Na^+ and K^+ ions to enter the hexagonal recesses in the basal oxygen sheet during dehydration of the mineral. These embedded ions are stripped of their potentially acidic ligand waters and coordinate instead with the surrounding silicate oxygens [89].

The Bronsted acidity of the smectites is dependent on the location of the layer charge. If the anionic charge of a montmorillonite exchange site is considered to be distributed over the basal oxygens of the silicate tetrahedra whose apical oxygens are most closely linked to an octahedral substitution site, then the charge from a single site will be distributed over at least ten oxygens. In contrast, the charge resulting from tetrahedral substitution will be largely localized on the three basal oxygens directly bound to the substituent ion [90]. This delocalization of surface charge in the montmorillonites and hectorites promotes the ionization of water molecules associated with the interlamellar cations, whereas the localization of charge in the saponites results in preferential hydrogen bond formation between the strongly basic basal oxygens at the exchange sites and water molecules coordinated with the compensating cations [90,91]; these surface-bonded waters cannot readily protonate weaker bases like ammonia or the organic amines.

The reduced acidity of interlamellar water associated with tetrahedral substituent sites has been demonstrated by a comparison of the absorption behavior of ammonia in Ca-saponite and Ca-montmorillonite [92]. In the former case, the ammonia is largely adsorbed in a coordinated form, whereas in the latter case, it is absorbed in a protonated form, that is, as ammonium ions. The degree of protonation and the stability of a pyridine–(Ca-saponite) complex is also less than that of the analogous montmorillonite complex [93]. The properties of hydronium-exchanged smectites are also dependent on the location of the isomorphous substitution; these derivatives are described in Section 1.7.1.

The acidic clays most frequently used to catalyze organic reactions are the aluminosilicates, kaolin, bentonite, attapulgite, or their derivatives. However, magnesium clays, like saponite and hectorite, containing highly electronegative exchangeable ions, such as Al^{3+} or Cr^{3+}, can be used as acidic catalysts for the alkylation of aromatic hydrocarbons [94]. Mg −, Ca −, or Zn-exchanged forms of these minerals have been used as the acidic receptors in the manufacture of carbonless copying papers [95], the chemistry of which is discussed in Chapter 5. The catalytic activity of the aluminosilicates can also be promoted by exchange with Al^{3+}, by treatment with acids or, in some instances [96], by reactions with byproducts from the catalyzed reaction.

1.6.2 Lewis Acidic Centers

The dominant Lewis acidic species on aluminosilicate mineral surfaces are usually coordinately unsaturated Al^{3+} ions, either those exposed at the edges of lattice layers, or those which can be formed by dehydration of adsorbed oligomeric oxyaluminum cations or of amorphous silica–alumina contaminants. In the presence of water, these Lewis acidic species are hydrated and their acidity masked, although if the Al^{3+} ions have electronegative substituents, such as halide or silicate ions, the coordinated waters may be sufficiently polarized to form strong Bronsted acidic species.

Estimation of the strength of Lewis acidity is often difficult and typically relies on infrared or ultraviolet spectroscopic measurements of the equilibrated adsorption of Lewis bases, such as amines, with the mineral. The protonated and coordinated amines usually can be distinguished, although the spectroscopic methods are often insensitive to the presence of small amounts of either species. The spectra of p-dimethylaminoazobenzene, when adsorbed on kaolin previously dried by heating at 20 to 100°C shows an absorption peak at 470 nm due to the amine adsorbed on neutral sites and a peak at 500 to 520 nm due to the protonated amine. An adsorption peak at 560 nm appears in the spectrum of the amine adsorbed on kaolin dried at 200 to 300°C; this absorption has been assigned to amine coordinated to Lewis acidic species [97]. These methods have been used to show that Bronsted acidic species can persist on the surfaces of kaolinites that have been calcined at temperatures up to 900°C.

^{13}C nmr spectroscopy of adsorbed amines has been proposed for the simultaneous quantitative determination of the surface concentrations of Bronsted and Lewis acidic centers on mineral surfaces [98]. An alternative procedure for measuring the distribution of Bronsted and Lewis acidity consists of titrating the mineral with n-butylamine, using a range of Hammett indicators to obtain values for the apparent combined acidity. A second set of titrations with arylmethanol indicators is then used to separately determine the Bronsted acidity distribution, and the Lewis acidity distribution is found by differentiation [99].

$$R—CH=CH_2 \quad \xrightarrow{-e} \quad R\overset{+}{C}H—\overset{\bullet}{C}H_2$$

Figure 1.19

Lewis acidic species can adsorb Hammett indicators, forming colored species which, to the eye, may mimic the protonated species. The apparent acidity cannot be expressed in terms of the Hammett acidity function (which only applies for Bronsted acidity), and the measurements are of limited value when used for comparison between different Lewis acids [71]. Despite the inevitable difference in pK_a values for the distinctly different reactions of a given indicator with Bronsted and Lewis acids, the indicators have been used in the past in the determination of acidity values and acidity distributions of the mineral surfaces without apparent regard for the nature of the acidic species supposedly involved.

Lewis acidic species on aluminosilicate or silica–alumina surfaces can accept single electrons from donor molecules with low ionization potentials, and they can also coordinate organic radicals. Strong Lewis acids may be capable of abstracting electrons from vinylic monomers, thereby initiating polymerization via radical-cationic intermediates [100] (Figure 1.19).

1.6.3 Basic Centers

The basicity of the common silicate mineral fillers and pigments is generally of less consequence than their acidity. The surface basicity can be titrated with organic acids, such as chloroacetic acid in hydrocarbon solvents, using Hammett indicators, and thereby placed on a common scale with the Hammett acidity [101]. The principal basic surface species are the AlOH or MgOH groups exposed at the layer lattice edges which can accept protons from water or other Bronsted acids. Magnesium silicates such as talc or chrysotile, which essentially lack exchangeable cations, form weakly alkaline aqueous suspensions by ionization of edge or surface MgOH groups. The silicate oxygen sheets can act as Lewis basic centers for the adsorption of electron-deficient species, which include protons, organic carbonium ions, and radicals. However, this Lewis basicity is not a readily measureable attribute of the clay silicate surfaces.

1.7 SURFACE REACTIONS OF SILICATE MINERALS

In this section we consider only the reactions of silicate minerals relevant to the location and identification of active species, other than acids or bases, which may be present on the mineral surfaces and which can participate in mineral–organic reactions. The various organic reactions and interactions which can occur at these sites will be discussed in more detail in Chapters 4 to 6.

1.7.1   Hydrolysis and Weathering

So far we have considered only idealized mineral crystallites containing surface species that could be predicted from an examination of their lattice structures. Most silicate minerals, whether natural, commercially beneficiated, or "purified homoionic" laboratory specimens, are likely to have anomalous surfaces that may result from the various spontaneous reactions that can occur during chemical processing or natural weathering.

Superficial degradation of the crystallites through hydrolysis is probably the dominant source of anomalous surface species in otherwise pure materials. The various minerals described in this chapter are, as a group, not readily degraded by mineral acids other than hydrofluoric acid, although the rates of dissolution of the aluminosilicates are increased by the presence of octahedral iron or magnesium. Slow hydrolysis does occur in aqueous media. This hydrolysis is important in the geological transformation of the minerals. The hydrolysis is accelerated by reduction in the ionic strength of the suspending medium, which can occur during ion-exchange resin treatment, dialysis of aqueous clay suspensions, or in washing of salts from salt-exchanged homoionic minerals [102]. This is an indirect result of Donnan hydrolysis of the exchangeable ions, that is, their diffusion from the mineral surface and equilibrated replacement with H_3O^+ ions. Even in saline suspensions that approach neutrality, Donnan hydrolysis causes the surface of the clays to be acidic, the surface proton concentration increasing with decreasing salinity of the media [103].

Studies using kaolinite [104], montmorillonite [105], and attapulgite [106] have shown that leaching of the minerals can result in a rapid loss of the original exchangeable ions, particularly those of the alkali metals, and their replacement by H_3O^+ and Al^{3+} ions. (See Figure 1.20.) The hydronium-clays are unstable, as the solvated protons can readily diffuse over the mineral surface to the layer edges, where they can displace exposed octahedral Al^{3+} ions [72]. This process is accelerated by the presence of water, but occurs less rapidly in presence of alcohols, or in dried, refrigerated samples.

Hydrolysis of kaolins in neutral or weakly acidic media results in the formation of adsorbed, exchangeable Al^{3+} ions [107], the liberation of soluble hydrated silica and alumina species [104], and the subsequent redeposition of an amorphous, alumina-rich alumina–silica gel on the surface of the mineral particles [63]. All natural kaolin particles appear to have such gel coatings [108], although the thickness of the coating may only be of the order of 15 Å on a "clean surfaced" crystallite. Although treatments in aqueous media are commonly used for the removal of surface contaminants from kaolinite, most of the methods also cause hydrolytic degradation of the mineral surface and the probable redeposition of gel, particularly during washing to remove the excess reagents or soluble products [104].

$$\left[\text{Mont. Al}\right]^- \text{Na}^+ \;+\; \text{H}^+_{aq.} \;\rightleftharpoons\; \left[\text{Mont. Al}\right]^- \text{H}^+ \;+\; \text{Na}^+_{aq.}$$

$$\left[\text{Mont. Al}\right]^- \text{H}^+ \;\longrightarrow\; \left[\text{Mont. H}\right]^- \text{Al}^+(\text{OH})_2$$

Figure 1.20

Hydrolysis of montmorillonite can result in dealuminization of the mineral edges and, if the medium is not strongly acidic, the intercalation of oligomeric oxyaluminum ions and the gradual "chloritization" of the mineral. The products are similar to those formed from the adsorption and hydrolysis of aluminum salt solutions in the presence of montmorillonite [109]. The rate of dissolution of the smectites in dilute acid can be correlated with the content of Mg and Fe^{III} in the aluminosilicate lattice [110]. The surface hydrolysis of kaolinites and montmorillonites results in an increase in the acidity of the mineral, because of the presence of adsorbed H_3O^+ ions or potentially acidic hydrated Al^{3+} ions and an increase in the amount of surface hydroxylic species. The natural acidic bentonites are often extensively degraded, almost amorphous minerals.

Hydrolysis of beidellites and vermiculites has not been extensively studied, although their hydronium derivatives have been shown to be unstable. In contrast to the behavior of exchangeable H_3O^+ ions adsorbed on the basal surfaces of a montmorillonite adjacent to the octahedral substitution sites, those adsorbed at the tetrahedral Al-for-Si substitution sites of a vermiculite or beidellite can react with the Al–O–Si bonds to form weakly acidic surface SiOH groups [72]. Further attack, resulting in displacement of Al^{3+} ions, can occur in the presence of a substantial source of hydronium ions, such as a H^+-exchange resin, or in moderately concentrated acidic media. This loss of tetrahedral Al^{3+} substituent ions may not result in any marked reduction in the apparent cation exchange capacity of the minerals, as the charge discrepancy can be transferred to the octahedral layer [111].

Hydronium-exchanged minerals may be produced by heating their ammonium-exchanged forms at temperatures above 300°C. This results in the dissociation of the ammonium ions, with the evolution of ammonia and the formation of a "decationized" mineral. The residual protons of decationized zeolites and silica–aluminas can form highly acidic bridged-hydroxyl groups (see Figure 1.16). In the case of beidellites and montmorillonites, the acidity of the decationized minerals and their activity as catalysts for the dehydration of ethanol was found to be proportional to the amounts of tetrahedral isomorphous substitution originally present [112]. This probably results from the diffusion of the balance of the liberated protons through the lattice to the neighborhood of the octahedral substituent sites during the thermal desorption of the ammonia. These embedded protons are no longer accessible for catalysis, whereas protons

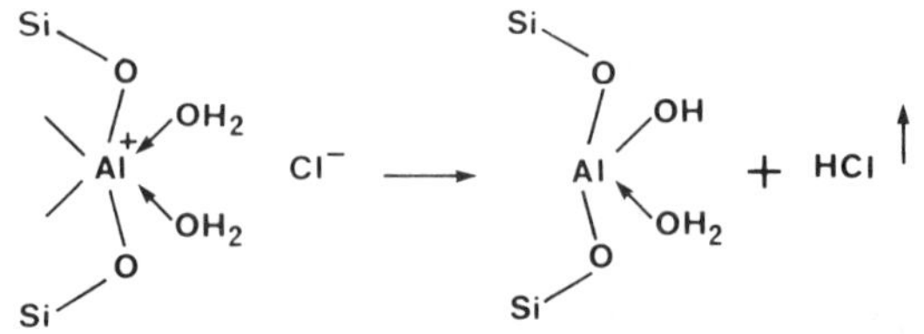

Figure 1.21   The thermal hydrolysis of adsorbed chloride ions from the layer edges of an aluminosilicate.

associated with tetrahedral sites remain on the surface, possibly in the form of bridged-surface hydroxyls like those of the acidic zeolites of silica–aluminas.

The magnesium silicates are more susceptible to acid degradation than the analogous aluminosilicates. Hydrolysis of these minerals suspended in, or contacted with media of low ionic strength occurs readily, although the leached magnesia may not be redeposited on the parent mineral surfaces [113]. Hydrolysis of hectorite, for example, involves the leaching of Mg^{2+} ions from the layer edges, accompanied by sloughing of the residual silicic acid to expose fresh surfaces [114]. The surface composition of talc to a depth of 20 Å has also been found to be similar to that of the bulk material [115].

The aluminum and magnesium silicates, including talc, can also undergo a form of thermal hydrolysis, with the evolution of hydrogen halides, when suspensions of these minerals containing residual alkali halides are dried at 200 to 300 °C. The acid is believed to evolve from halide ions adsorbed at anion exchange sites at the octahedral layer edges [116] (Figure 1.21). The mechanism of its formation appears to be different from that of the acid that is formed during the elimination of structural water when kaolinite or montmorillonite is heated at 350 to 500 °C in the presence of alkali halides [117].

1.7.2   Electron Transfer

The silicate clay minerals can act as oxidants or reductants in a number of organic reactions [100], the most important of which are the dye-forming and related reactions that are discussed in Chapter 5. The principal sources of this activity are Fe^{2+} and Fe^{3+} ions or other variable-valence transition metal ions; these may be present as exchangeable species or as isomorphous substituents. In these reactions, electrons appear to be able to diffuse or tunnel to octahedral sites from the lattice layer edges or basal surfaces [118]. Mossbauer studies of the hydrazine reduction of nontronite have demonstrated the reversible formation of octahedral Fe^{II} in the reduced mineral [119]. Other potential oxidizing species include Fe^{3+} ions adsorbed at the layer edges or occupying tetrahedral layer vacancies;

these are less readily reduced, and the resultant Fe^{2+} ions can be readily leached from the mineral by acid. Oxidation by adsorbed Fe^{3+} ions often may only occur on freshly prepared samples because oligomeric oxyferric species formed by the spontaneous hydrolysis of the hydrated Fe^{3+} ions are ineffective as oxidants (see Section 5.3). The oxidizing power of montmorillonites appears to be largely due to octahedral Fe^{III} [120]. Oxidation of octahedral Fe^{II} in biotite is not accompanied by any marked change in cation exchange capacity. This has been ascribed to the loss of an equivalent number of structural protons, probably those of hydroxyl groups linked to Fe^{II} substituents in the octahedral layers [121].

Lamellar silicate minerals can also catalyze the aerial oxidation of organic materials, such as the benzidine salts, either by the action of mineral oxidant sites which can be regenerated by aerial oxidation or through activation of the organic species by adsorption on the silicate surface. In the case of the benzidine oxidation, formation of the characteristic blue color of the radical-cation is dependent on the presence of the silicate surface to stabilize the radical and prevent its further oxidation. Development of the color requires the presence of octahedral Fe^{III}, adsorbed Fe^{3+} ions being ineffective [122,123].

Some oxidations are promoted by the presence of exchangeable Cu^{2+} or Fe^{3+} ions, as in the case of hydroquinone adsorbed on montmorillonite. In this example, the rate of oxidation was found to increase with the increasing cation-exchange capacity of the mineral and with the decrease in its Li-fixing capacity, that is, with apparent increase in the beidellite character of the montmorillonite [50]. Fe^{2+} ions adsorbed on the surface of montmorillonite undergo aerial oxidation at a rate two orders of magnitude greater than that of Fe^{2+} ions in solution, indicating that the crystal lattice can act as an electron acceptor and oxidation catalyst; the catalytic species are probably octahedral Fe^{II}/Fe^{III} [122].

Electron transfer can also occur with incompletely coordinated aluminum ions that may be exposed at the layer edges of aluminosilicates or present in adsorbed oxyaluminum species or amorphous aluminosilicate contaminants. Organic molecules that have low ionization potentials, such as the aromatic hydrocarbon, perylene, can undergo charge transfer to form adsorbed radical-cations, but the identity of the electron acceptors is not known with certainty. Silicate minerals can tenaciously adsorb oxygen, particularly if they are calcined to remove adsorbed water or if the dried minerals are ground in contact with air. This oxygen is chemisorbed and cannot be readily removed by outgassing in vacuum. The dehydration of hydrated silica or aluminosilicates has also been shown to involve the elimination of hydrogen and the formation of oxygen radical-anionic species as a side-reaction [124]. Chemisorbed oxygen, oxygen radical-anions, and their protonated derivatives are strong oxidants. These, rather than coordinately unsaturated aluminums, have been proposed as the active species responsible for the oxidation of polycyclic hydrocarbons on calcined aluminosilicates [125].

1.7.3   Esterification and Alkylation

Kaolinites, hydrated silicas, silica–aluminas, and other minerals with hydroxylic surfaces can form organic derivatives with esterifying or alkylating reagents; reaction with acetyl chloride or diazomethane, for example, can cause surface hydroxyl groups to be substituted, respectively, by covalently bound acetate or methoxide groups. The hydroxyl groups can react with organometallic reagents, such as the lithium alkyls, with the abstraction of active hydrogen, the liberation of the corresponding alkane, and formation of a lithiated surface. The hydroxyl groups can also react with thionyl chloride to produce surface chlorides; these can undergo metathetical reactions with organometallic reagents to form surface alkyl or aryl species. (See Figure 1.22.) All of these reactions have been well authenticated [126].

Similar reactions have been reported to occur on montmorillonite surfaces, to a greater extent than would be expected from the reaction of edge hydroxyls alone. Much of the early work was due to Deuel and his co-workers [127–129], who believed their H^+-exchanged clays to have basal surface SiOH groups, in accordance with the then-current Edelmann–Faverjee structure of montmorillonite (see p. 15). Subsequent explanations were based on the hypothesis that the organic reagents could disrupt the silicate sheets in the vicinity of the exchange sites, to enable the formation of covalently bound surface adducts [130].

The formation of these covalently bound "organosmectite" derivatives has been disputed [131,132]. The products, for the most part, have never been adequately characterized, and the published data serve largely to show that there was interlamellar retention of halogen- or of carbon-containing species, without indicating the presence or nature of any mineral–organic bonds. It would appear that the organic component of these adducts consisted largely of strongly adsorbed, possibly polymeric species, or derivatives of edge hydroxyls or adsorbed oligomeric oxy-aluminum cations. The reaction of diazomethane with montmorillonite, for example, has been shown to produce a strongly hydrophobic product. The hydrophobicity, however, results from the formation of polymethylene in the interlamellar spaces, by reactions catalyzed by water associated with the exchangeable cations, rather than from the formation of a methylated mineral surface [133]. Reaction between alkylating or halogenating reagents and the basal oxygen sheet may be possible in minerals that contain tetrahedral layer vacancies formed, for example, by leaching of isomorphous substituent ions.

Acid leaching of clay minerals results in the extraction of the Al^{3+} or Mg^{2+} ions and the eventual formation of an amorphous hydrated silica. However, reaction with acid in the presence of a silylating reagent, for example, trimethylsilyl chloride or hexamethyldisiloxane, can result in the formation of trimethylsilylated polysilicic acids; these may be cyclic oligo-

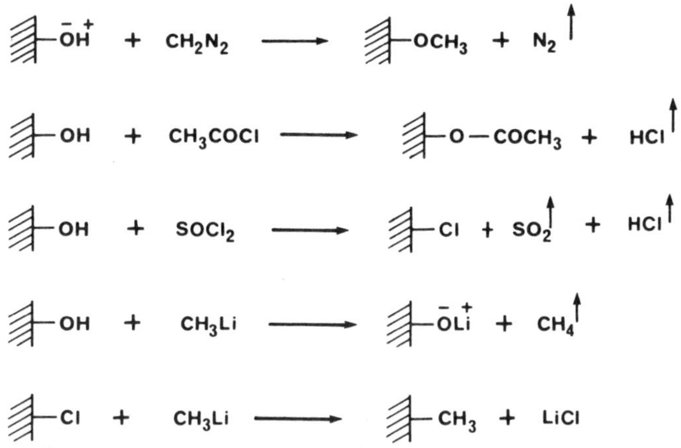

Figure 1.22   Some metathetical reactions of silicate surface hydroxyl groups.

mers [134] or, as in the case of the chrysotile derivatives (Figure 1.23) [135], they may retain the layered structure and morphology of the original minerals.

1.7.4   Ligand and Intercalate Reactions

Smectites can intercalate a wide variety of organic compounds, the adsorbed organic molecules being retained by dipolar interactions, by protonation and ionic interactions, or by coordination to interlamellar cations. These interactions will be considered in more detail in Chapter 4, but some examples which illustrate the reactions of organic and inorganic ligands are described in this section.

The intercalation of strongly basic amines, either as protonated or as coordinated species, has already been mentioned in Section 1.6. Pyridine can be intercalated into sodium montmorillonite to form complexes with basal spacings of 14.8 or 23.3 Å, corresponding to the respective Na^+ : pyridine:water ratios of 1:2:4 and 1:4:2. The structure of the 14.8-Å complex has been recently re-examined in detail. Contrary to some earlier reports, the Na^+ ions were found to be located in recesses in the basal oxygen sheet and their electrostatic potential was essentially unaltered by intercalation of pyridine molecules which are tenaciously bound instead to the silicate sheets [136]. The structure of this complex is in contrast to those formed by the intercalation of pyridine in Mg- or Ca-montmorillonites, in which the amine is present in its protonated form, or in transition-metal ion-exchanged minerals, in which the amine is coordinated to the metal ions.

The intercalation of transition metal ions bearing ammino-ligands may involve the formation of square-planar coordinated complexes instead of the octahedral complexes which would be favored in solution. For example,

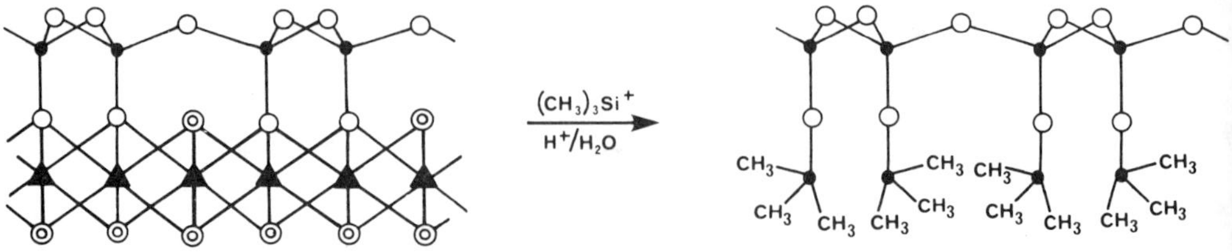

$(CH_3)_3Si^+$

$\xrightarrow{H^+/H_2O}$

CH_3 CH_3 CH_3 CH_3 CH_3 CH_3 CH_3 CH_3

Figure 1.23

the reaction of Cu^{2+} or Ni^{2+} ions with ethylene diamine (en) can form octahedral chelates having the structure $[M(en)_3]^{2+}$, or square-planar chelates $[M(en)_2]^{2+}$. In the latter, the ionic charge can be delocalized over the planar structure. This provides an optimum configuration for neutralization of the diffuse anionic surface charge of the montmorillonite exchange sites [137,138]. The increased stability of the square-planar complexes in montmorillonites and hectorites is apparently a linear function of the charge density of the silicate layers [138]. However, this conclusion must be qualified, as the lower-charged minerals used in the particular study were mostly montmorillonites whose charge had been reduced by partial Li-exchange and subsequent calcination. The extra stabilization may consequently have also depended on the degree of octahedral substitution and the effective residual octahedral charge in the lithiated smectites.

Although intercalated $[Cu(en)_2]^{2+}$ ions may be present as octahedral coordination complexes, with the opposing silicate sheets acting as weak ligands [139], $[Ni(en)_2]^{2+}$ ions do not interact with the mineral surfaces in this manner when intercalated in smectites [140]. The smectites can promote the disproportionation of $[Ni(en)_x(H_2O)_y]^{2+}$ ions in solution by preferential adsorption and ion exchange to produce complexes containing square-planar $[Ni(en)_2]^{2+}$ ions.

The state of coordination may be temperature-dependent in some systems. For example, dimethyl sulfoxide (DMSO) intercalated in Co-montmorillonite is present in both physisorbed and coordinated forms. At low temperatures or in the presence of moisture, the DMSO forms a tan-colored complex containing octahedral $[Co(DMSO)_4(H_2O)_2]^{2+}$ ions; this is reversibly converted to a purple tetrahedral complex on heating or on drying [141].

The increased ionization of residual water at higher temperatures can affect the stability of the coordination complexes. One interesting example is that of the stannic mesotetrapyridylporphyrin, Sn·Tp [142]. This can be intercalated in a hydrated Na-hectorite as the neutral molecule. However, on drying or on heating, the porphyrin complex is reversibly protonated and completely demetallated to form a mixture of intercalated Sn^{4+} ions and porphyrin dications. The Sn·TP reforms with increase in water content of the mineral complex and with decrease in interlamellar acidity. In the absence of the mineral, the Sn·TP is acid-resistant and is not completely demetallated in 100% sulfuric acid.

The acidity of interlamellar water can also catalyze the decomposition of intercalated $[Co(NH_3)_6]^{3+}$ ions to form unstable $[Co(NH_3)aq]^{3+}$ ions; these decompose forming cobaltous hydroxide, ammonium ions, and nitrogen [143]. The conversion of Prussian Brown (ferric ferricyanide) into Prussian Blue (a ferrous ferricyanide complex) is another reaction catalyzed by acidic surfaces. Kaolin was found to rapidly convert the Prussian Brown, whereas the reaction on silica gel was significantly slower [144].

REFERENCES

1 R. E. Grim, Clays—Uses, in *Kirk-Othmer Encyclopedia of Chemical Technology*, Vol. 6, 3rd ed., Wiley, New York, 1979.

2 W. D. Keller, Clays—Survey, in *Kirk-Othmer Encyclopedia of Chemical Technology*, Vol. 6. 3rd ed., Wiley, New York, 1979,

3 U. Hoffmann, *Angew. Chem. Int. Ed.*, 7, 681 (1968).

4 G. Brown, Ed., *X-Ray Identification and Crystal Structures of Clay Minerals*, Mineralogical Society, London, 1961.

5 G. W. Brindley and G. Brown, Eds., *Crystal Structures of Clay Minerals and Their X-Ray Identification*, Mineralogical Society, London, 1980.

6 W. D. Keller, *Clays Clay Miner.*, 25, 311 (1977); *Clays Clay Miner.*, 25, 347 (1977).

7 M. K. Lloyd and R. F. Conley, *Clays Clay Miner.*, 18, 37 (1970).

8 M. S. Camazano and S. G. Garcia, *An. Edafol. Agrobiol.*, 25, 9 (1966).

9 R. L. Ledoux and J. L. White, *J. Colloid Interface Sci.*, 21, 9 (1966).

10 P. M. Costanzo, C. V. Clemency, and R. F. Giese, Jr., *Clays Clay Miner.*, 28, 155 (1980).

11 H. van Olphen, *An Introduction to Clay Colloid Chemistry*, Wiley, New York, 1st ed., 1963; 2nd ed., 1977.

12 R. E. Grim, *Clay Mineralogy*, McGraw-Hill, New York, 1968.

13 S. L. Swartzen-Allen and E. Matijec, *Chem. Rev.*, 74, 385 (1974).

14 N. C. Lockhart, *J. Colloid Interface Sci.*, 74, 520 (1980).

15 P. A. Thiessen, *Z. Electrochem.*, 48, 675 (1942).

16 A. Weiss and J. Russow, *Proceedings of the International Clay Conference, 1963*, Vol. 1, Pergamon, New York, 1963, p. 203.

17 A. Weiss, *Z. Anorg. Allgem. Chem.*, 299, 92 (1959).

18 M. B. McBride, *Clays Clay Miner.*, 24, 88 (1976).

19 H. H. Murray, in *Paper Coating Pigments*, TAPPI Monograph No. 38, TAPPI, New York, 1976, p. 69.

20 L. E. Brooks and H. H. Morris, Aluminum Silicate-Kaolin, in T. C. Patton, Ed., *Pigment Handbook*, Wiley, New York, 1973.

21 P. S. Sennet, K. L. Turner, and H. H. Morris (Freeport Sulphur), U.S. Patent 3,476,576 (November 4, 1969).

22 N. H. Horton (Engelhart Minerals and Chemicals), U.S. Patent 3,520,719 (July 14, 1970).

23 A. Weiss and W. Thielepape (Erbsloh and Co.), U.S. Patent 3,309,211 (March 14, 1967).

24 A. Weiss, *Angew. Chem. Int. Ed.*, 2, 697 (1963).

25 W. D. Keller, H. Cheng, W. D. Johns, and C-S. Meng, *Clays Clay Miner.*, 28, 97 (1980).

26 J. H. Hodgkin and D. H. Solomon, *J. Macromol. Sci. Chem.*, A8, 635 (1974).

27 Georgia Kaolin Co., British Patent 1,228,538 (April 15, 1971).

28 J. D. Russell and A. R. Frazer, *Clays Clay Miner.*, 19, 55 (1971).

29 H. A. Stephens and A. N. Waterworth, *Brit. Found.*, **61**, 202 (1968).

30 G. Lagaly, *Clays Clay Miner.*, **27**, 1 (1979).

31 R. E. Grim and C. Kulbicki, *Amer. Miner.*, **46**, 1329 (1961).

32 K. F. Landgraf, *Chem. Erde*, **38**, 97 (1979).

33 N. Gueven, *Clays Clay Miner.*, **22**, 155 (1974).

34 P. Rengasamy, J. B. van Assche, and J. B. Uytterhoeven, *J. Chem. Soc., Farad. I*, **72**, 376 (1976).

35 W. Fertl and F. W. Jessen, *Clays Clay Miner.*, **17**, 1 (1969).

36 R. Levy and C. W. Francis, *Clays Clay Miner.*, **23**, 475 (1975).

37 G. Lagaly, *Clays Clay Miner.*, **27**, 1 (1979).

38 G. Lagaly, G. Schoen, and A. Weiss, *Kolloid-Z., Z. Polym.*, **250**, 667 (1972).

39 E. Frey and G. Lagaly, *J. Colloid Interface Sci.*, **70**, 46 (1979).

40 N. G. Vasilev, L. V. Golovko, A. G. Savkin, and F. D. Ovcharenko, *Colloid J. (USSR)*, **39**, 578 (1977).

41 C. C. Chou and J. L. McAtee, Jr., *Clays Clay Miner.*, **17**, 339 (1969).

42 G. Lagaly, H. Muller-Vonmoos, and G. Kak, *Proceeding of the Third European Clay Conference*, 1977, p. 97. Nordic Society for Clay Research, Oslo.

43 R. F. Giese, *Clays Clay Miner.*, **26**, 51 (1978).

44 R. I. Zlockevskaya and V. I. Divisilova, *Svyazannaya Voda Dispersnykh Sist.*, **43** 2, (1972); *Chem. Abs.*, **78**, 34317 (1973).

45 G. W. Brindley, K. Wiewiora, and A. Wiewiora, *Amer. Miner.*, **54**, 1635 (1969).

46 B. Rand, E. Pekenc, G. W. Goodwing, and R. W. Smith, *J. Chem. Soc., Farad. I*, **76**, 225 (1980).

47 H. Suquet, C. De la Calle, and H. Pezerat, *Clays Clay Miner.*, **23**, 1 (1975).

48 M. E. Harward and G. W. Brindley, *Clays Clay Miner.*, **13**, 209 (1964).

49 H. E. Doner and M. M. Mortland, *Soil Science Society American Proceedings*, **35**, 360 (1971).

50 T. D. Thompson and W. F. Moll, Jr., *Clays Clay Miner.*, **21**, 337 (1975).

51 G. F. Walker, *Clay Miner.*, **7**, 129 (1967).

52 G. F. Walker and W. G. Garrett, *Science*, **156**, 385 (1967).

53 E. T. Uskova, N. G. Vasilev, and I. A. Uskov, *Colloid J. (USSR)*, **30**, 118 (1968).

54 B. K. G. Theng, D. J. Greenland, and J. P. Quirk, *Clay. Miner.*, **7**, 1 (1967).

55 E. F. Vasant and G. Peeters, *Clays Clay Miner.*, **26**, 279 (1978).

56 A. Maes, P. Marynen, and A. Cremers, *Clays Clay Miner.*, **25**, 309 (1977).

57 J. R. Feldkamp and J. L. White, *Dev. Sedimentology*, **27**, 187 (1979); *Chem. Abs.*, **90**, 175345 (1979).

58 A. Weiss, *Angew. Chem. Int. Ed.*, **2**, 134 (1963); *Clays Clay Miner.*, **10**, 191 (1963).

59 J. A. Martin-Rubi, J. A. Rausell-Colom, and J. M. Serratosa, *Clays Clay Miner.*, **22**, 87 (1974).

60 G. Lagaly, St. Fitz, and A. Weiss, *Clays Clay Miner.*, **23**, 45 (1974).

61 G. Lagaly and A. Weiss, *Kolloid-Z., Z. Polym.*, **248**, 979 (1971); *Kolloid-Z., Z. Polym.*, **248**, 968 (1971).

62 A. Weiss, *Z. Anorg. Allgem. Chem.*, **297**, 257 (1958).

63 U. Kafkafi, *Isr. J. Chem.*, **6**, 367 (1968).

64 L. Bragg and G. F. Claringbell, *Crystal Structures of Minerals*, Bell, London, 1965.

65 W. C. Streib, Asbestos, in *Kirk-Othmer Encyclopedia of Chemical Technology*, Vol. 3, 3rd ed., Wiley, New York, 1978.

66 *Zeolite Molecular Sieves,* Advances in Chemistry Series, Nos. 101 and 102, American Chemical Society, Washington, 1971.

67 R. A. Rabo, Ed., *Zeolite Chemistry and Catalysis,* Advances in Chemistry Series, No. 171, American Chemical Society, Washington, 1976.

68 W. M. Meier and D. H. Olsen, in *Zeolite Molecular Sieves,* Advances in Chemistry Series, No. 101, American Chemical Society, Washington, 1971, p. 155.

69 D. M. Bibby, N. B. Milestone, and L. P. Aldridge, *Nature,* **285,** 30 (1980).

70 R. H. Daniels, G. T. Kerr, and L. D. Rollmann, *J. Amer. Chem. Soc.,* **100,** 3097 (1978).

71 H. A. Benesi and B. H. Winquist, *Adv. Catalysis,* **27,** 97 (1978).

72 N. G. Vasilev and F. D. Ovcharenko, *Russ. Chem. Rev.,* **46,** 775 (1977).

73 M. Sato, T. Kanbayoshi, N. Kobayashi, and Y. Shima, *J. Catalysis,* **7,** 342 (1967).

74 G. D. Chukin and B. V. Smirnov, *Proc. Acad. Sci. USSR, Dokl. Phys. Chem.,* **226,** 188 (1976).

75 J. J. Fripiat, A. Leonard, and J. B. Uytterhoeven, *J. Phys. Chem.,* **69,** 3274 (1965).

76 M. M. Mortland and K. V. Raman, *Clays Clay Miner.,* **16,** 393 (1968).

77 R. Touillaux, P. Salvador, C. Vandermeersche, and J. J. Fripiat, *Isr. J. Chem.,* **6,** 337 (1968).

78 Yu. I. Tarasevich and F. D. Ovcharenko, *Adsorbenty, Ikh Poluch., Svoistva Primen.,* Tr. Vses. Sovesch. Adsorbentam, 4th, 1976; *Chem. Abs.,* **89,** 118315 (1978).

79 V. P. Telichkun, Yu. I. Tarasevich, and F. D. Ovcharenko, *Proc. Acad. Sci. USSR, Dokl. Phys. Chem.,* **228,** 507 (1976).

80 D. H. Solomon and H. H. Murray, *Clays Clay Miner.,* **20,** 135 (1972).

81 J. W. Ward, *J. Catalysis,* **17,** 355 (1970).

82 R. H. Loeppert, L. W. Zelazny, and B. G. Volk, *Clays Clay Miner.,* **27,** 57 (1979).

83 C. Walling, *J. Amer. Chem. Soc.,* **72,** 1164 (1950).

84 H. Benesi, *J. Amer. Chem. Soc.,* **78,** 5490 (1956).

85 M. Frenkel, *Clays Clay Miner.,* **22,** 435 (1974).

86 D. H. Solomon, J. D. Swift, and A. J. Murphy, *J. Macromol. Sci. Chem.,* A5, 587 (1971).

87 H. Benesi, *J. Phys. Chem.,* **61,** 970 (1957).

88 M. Deeba and W. K. Hall, *J. Catalysis,* **60,** 417 (1979).

89 J. A. Helsen, R. Drieskens, and J. Chaussidon, *Clays Clay Miner.,* **23,** 334 (1975).

90 V. C. Farmer and J. D. Russell, *Trans. Farad. Soc.,* **67,** 2737 (1971).

91 H. Suquet, R. Prost, and H. Pezerat, *Clay Miner.,* **12,** 113 (1977).

92 J. D. Russell, *Trans. Farad. Soc.,* **61,** 2284 (1965).

93 V. C. Farmer and M. M. Mortland, *J. Chem. Soc.,* **A,** 344 (1966).

94 G. E. Striddle (National Lead), U.S. Patent 3,992,467 (November 16, 1976).

95 LaPorte Industries Ltd., French Patent 2,203,317 (November 10, 1974); *Chem. Abs.,* **83,** 35747 (1975).

96 I. M. Kolesnikov, *Russ. Chem. Rev.,* **44,** 306 (1975).

97 V. P. Telichkun, Yu. I. Tarasevich, V. V. Goncharuk, and F. D. Ovcharenko, *Proc. Acad. Sci. USSR, Dokl. Phys. Chem.,* **221,** 346 (1975).

98 S. H. C. Liang and I. D. Gray, *J. Catalysis,* **66,** 294 (1980).

99 D. Atkinson and G. Curthoys, *J. Phys. Chem.,* **84,** 1358 (1980).

100 D. H. Solomon, *Clays Clay Miner.,* **16,** 31 (1968).

101 T. Yamanaka and K. Tanabe, *J. Phys. Chem.,* **79,** 2409 (1975).

102 K. Fruehauf and U. Hoffmann, *Z. Anorg. Allgem. Chem.,* **307,** 187 (1961).

103 J. Kamil and I. Shainberg, *Soil Sci.,* **106,** 193 (1968).

104 D. G. Hawthorne and D. H. Solomon, *Clays Clay Miner.*, **20**, 75 (1972).

105 P. Bar-On and I. Shainberg, *Soil Sci.*, **109**, 241 (1970).

106 E. T. Uskova, V. B. Emelyanov, and I. A. Uskov, *Colloid J. (USSR)*, **29**, 646 (1967).

107 R. F. Conley and A. C. Althoff, *J. Colloid Interface Sci.*, **37**, 186 (1971).

108 A. P. Ferris and W. B. Jepson, *Proceedings of the Third European Clay Conference*, 1977, p. 54. Nordic Society for Clay Research, Oslo.

109 G. W. Brindley and C-C. Kao, *Clays Clay Miner.*, **28**, 435 (1980).

110 I. Novak and B. Cicel, *Clays Clay Miner.*, **26**, 341 (1978).

111 N. G. Vasilev, F. D. Ovcharenko, and M. A. Buntova, *Proc. Acad. Sci. USSR, Dokl. Chem.*, **216**, 407 (1974).

112 J. C. Davidtz, *J. Catalysis*, **43**, 260 (1976).

113 D. Riley and P. W. Arnold, *Soil Sci.*, **108**, 414 (1969).

114 I. Barshad and A. E. Foscolos, *Soil Sci.*, **110**, 52 (1975).

115 A. Toussaint, *14th. FATIPEC Congr.*, 657 (1978).

116 A. Torok. D. G. Hawthorne, and D. H. Solomon, (in preparation).

117 L. Heller-Kallai, *Clays Clay Miner.*, **23**, 462 (1975).

118 D. T. B. Tennakoon, J. M. Thomas, and M. J. Tricker, *J. Chem. Soc., Dalton*, 2211 (1974).

119 J. D. Russell, B. A. Goodman, and A. R. Fraser, *Clays Clay Miner.*, **27**, 63 (1979).

120 N. Malathi, S. P. Puri, and I. P. Saraswat, *J. Phys. Soc. Jap.*, **26**, 680 (1969).

121 G. J. Ross and C. I. Rich, *Clays Clay Miner.*, **22**, 335 (1974).

122 Z. Gerstl and A. Banin, *Clays Clay Miner.*, **28**, 335 (1980).

123 M. B. McBride, *Clays Clay Miner.*, **27**, 224 (1979).

124 J. H. Lunsdorf, *Catalysis Rev.*, **8**, 135 (1973).

125 F. R. Dollish and W. K. Hall, *J. Phys. Chem.*, **71**, 1005 (1967).

126 H. P. Boehm, *Adv. Catalysis*, **16**, 179 (1966).

127 H. Deuel, *Ber. Deutsch. Keram. Ges.*, **31**, 1 (1952).

128 H. Deuel and G. Huber, *Helv. Chim. Acta*, **34**, 1697 (1951); *Helv. Chim. Acta*, **35**, 1799 (1952).

129 R. Gentile and H. Deuel, *Helv. Chim. Acta*, **40**, 106 (1957).

130 J. Uytterhoeven, *Silicates Ind.*, **25**, 403 (1960).

131 R. M. Barrer and J. S. S. Reay, *Trans. Farad. Soc.*, **54**, 214 (1958).

132 D. J. Greenland and F. W. Russell, *Trans. Farad. Soc.*, **51**, 1300 (1955).

133 J. C. Bart, F. Cariati, L. Erre, C. Gessa, G. Micera, and P. Pui, *Clays Clay Miner.*, **27**, 429 (1979).

134 R. Atwal, B. R. Currell, C. B. Cook, H. G. Midgley, and J. R. Parsonage, *Organomet. Polym. Symp*, 231 (1977); *Chem. Abs.*, **89**, 25162 (1978).

135 J. J. Fripiat and E. Mendelovici, *Bull. Soc. Chim. Fr.*, 483 (1968).

136 J. M. Adams, J. M. Thomas, and M. J. Walters, *J. Chem. Soc., Dalton*, 1459 (1975).

137 M. I. Knudson, Jr. and J. L. McAtee, Jr., *Thermochim. Acta*, **4**, 411 (1972).

138 A. Maes and A. Cremers, *J. Chem. Soc., Farad I*, **75**, 513 (1979).

139 J. L. Burba III and J. L. McAtee, Jr., *Clays Clay Miner.*, **25**, 113 (1977).

140 R. A. Schoonheydt, F. Velghe, and J. B. Uytterhoeven, *Inorg. Chem.*, **18**, 1842 (1979).

141 G. A. Garwood, Jr. and R. A. Condrate, Sr., *Clays Clay Miner.*, **26**, 273 **(1978)**.

142 S. Abdo, M. I. Cruz, and J. J. Fripiat, *Clays Clay Miner.*, **28**, 125 (1980).

143 J. J. Fripiat and J. Helsen, *Clays Clay Miner.*, **14**, 163 (1966).

144 S. Yariv, *Isr. J. Chem.*, **453**, 7 (1969).

2

TITANIA PIGMENTS

Titanium dioxide (titania) and titania-based composites are the whitest and brightest of the commercial white pigments. This is a consequence of the high refractive index of titania and its relatively low and uniform absorption of visible light. Titanium dioxide may exist in one of several crystalline forms, but the commercial pigments are based on either the anatase or rutile forms of the mineral. The pigment particles are often modified by coating them with layers of other inorganic materials, for example, silica–alumina gels, to improve dispersion or reduce undesirable chemical activity.

This chapter briefly outlines the history of the early pigments and two of the basic processes currently used for the manufacture of titanias. This is followed by a description of the structure, surface chemistry, and photolytic activity of titania pigments. The chapter concludes with a discussion of some of the inorganic coating treatments used to modify the properties of the base pigments. The surface chemistry of these coated titanias is discussed in Chapter 3, as it largely resembles that of the oxides or mixed oxides used as the coating materials.

The chemistry of the titania pigments can be related to that of the hypothetical box structure (Figure 2.1). The surface contains basic terminal **(1)** and acidic bridged-hydroxyl groups **(2)**, which may be those of titania or of a hydrous oxide coating; labile Ti–O–Ti bonds **(3)**; water molecules adsorbed at Lewis acid sites **(4)** or bound to surface hydroxyl groups **(5)**;

and adsorbed anions such as sulfate or chloride process residues (X,6). The surface has potential electron donor and acceptor sites (7) and may contain adsorbed oxidants such as hydroxyl or hydroperoxyl radicals, or activated oxygen species (8) generated by photocatalytic processes described in Section 2.4.

2.1 HISTORICAL INTRODUCTION

All the titania pigments are of synthetic origin, unlike the clay mineral pigments and fillers described in the previous chapter. The first commercial titania pigments were manufactured in 1916 and consisted of titania–calcium/barium sulfate composites. Composite pigments were preferred because the early titanias were uneconomic and had inferior color and hiding power. The poor performance was largely due to the incomplete removal of Fe^{3+} and other ionic impurities and to an inadequate control of the dimensions of the pigment particles. During the 1920's, a range of titania-coated calcium and barium sulfates was manufactured, together with "reduced" (diluted) titanias, consisting of coprecipitates of anatase with other oxides such as alumina and silica or with colorless, insoluble metal salts. Manufacture of these materials continued through to the 1940's, when they were largely displaced by improved titania base pigments [1–3]. The reduced titanias by that stage had optical properties that approached those of the titania base pigments. Development of the reduced titanias and other composite pigments has continued [4,5] and may return to favor with the increasing scarcity and price of suitable titanium ores.

Anatase base pigments were introduced during the early 1930's. These pigments and the anatase-containing reduced titanias were optically superior to the other white pigments then in use, namely, zinc oxide, zinc

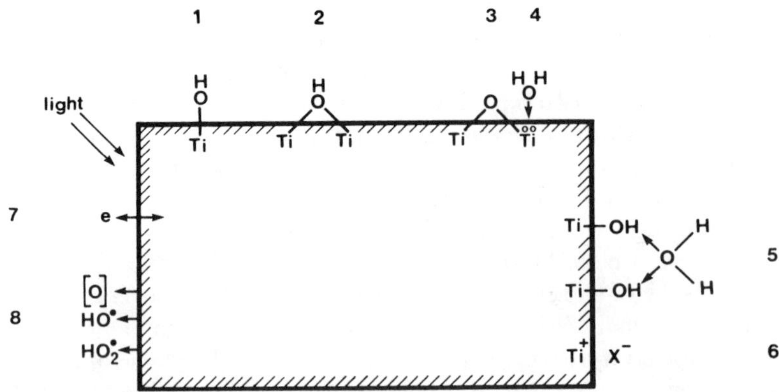

Figure 2.1 The principal surface features of a titania particle.

sulfide, white lead, and lithopone (a zinc sulfide–barium sulfate coprecipitate), but suffered from practical defects, the most noticeable of which were the poor "chalk resistance" and accelerated fading of organic colorants in titania-pigmented surface coatings. The chalking, or erosion of the organic binders, results from titania surface-catalyzed photodegradation reactions that can occur when the pigmented materials are exposed to sunlight or other sources of near-ultraviolet light.

The commercial development of the more chalk-resistant rutile-base pigments during the late 1930's and then of alumina–silica-coated rutile pigments having little residual photolytic activity, has resulted in titania becoming the premium white pigment for use in surface coatings, in many plastics applications, and in the production of high quality white papers.

2.2 PRODUCTION OF PIGMENTARY TITANIAS

Titanium derivatives occur in nature as minor components in many rocks. However, the only commercial sources are rutile ore, principally obtained from heavy sands, and ilmenite, an iron titanate, which is more widely distributed and has been the major source for pigment production. Ilmenite has an idealized composition $FeO \cdot TiO_2$, but natural weathering processes, particularly of ilmenite sands, can result in the oxidation and removal of ferrous iron [6]. The ores often contain appreciable amounts of chromium, niobium, and other heavy metals which can make them unsuitable for pigment production. Rutile and highly weathered ilmenites containing in excess of 65% TiO_2 are virtually insoluble in sulfuric acid and are unsuitable for use in the sulfate process described below.

There are two basic processes for the commercial production of pigmentary titania: the hydrolysis of titanyl sulfate solutions (the long-established sulfate process), and the high-temperature oxidation or hydrolysis of gaseous titanium tetrachloride (the more recent chloride processes). The choice of process has been partly dictated by the available raw materials, the sulfate process using ilmenite and the chloride process using rutile or beneficiated (reduced iron content) ilmenites. Each process has a large number of commercial variations, and it is only possible to give a broad outline of the basic processes in this book; more comprehensive surveys have been published elsewhere [1,7–9].

2.2.1 The Sulfate Process

The preparation of pigmentary titanias by the sulfate process involves several stages, namely, dissolution of the ilmenite ore in hot, concentrated sulfuric acid to form a solution of ferrous and titanyl sulfates; reduction of any ferric ions present; clarification to remove insoluble residues; cooling and removal of the bulk of the iron as crystalline, hydrated ferrous sulfate;

boiling and hydrolysis of the mother liquors to yield a titania precipitate; washing the precipitated hydrous titania to remove residual ferrous sulfate; and calcination to produce anatase or rutile pigments.

Hydrolysis and precipitation of the hydrous titania can be hastened by the addition of seed nuclei of anatase. The precipitated titania consists of aggregates of particles too small to be useful as pigments, and the hydrolysis product must be calcined at temperatures of 800 to 1000°C to allow the crystallization and growth of the anatase particles to a uniform and optimum size of 200 to 250 nm cross section. Titania particles of this size provide the greatest hiding power and the most uniform reflection of white light [10,11]. The hydrous anatase may be mixed with additives before calcination; these additives may include antimony trioxide, to reduce the photoactivity of the pigment, or phosphoric acid to inhibit rutile formation during calcination [6]. Anatase is unstable and can be converted to rutile by calcination, but complete conversion requires prolonged treatment at high temperatures unless the rutilization process is accelerated by use of a catalyst.

Commercial production of the rutile pigments is achieved by the addition of rutile seed nuclei during hydrolysis of the titanyl sulfate solution and by calcination of the resultant partially rutilized hydrated titania at temperatures of 900 to 1200°C. Formation of rutile pigments is also assisted by doping of the hydrous titania with Zn^{2+} or Al^{3+} ions, which can act as rutilization catalysts during the calcination process. These catalysts enable the rutilization to occur rapidly at lower calcination temperatures and thus avoid excessive particle growth and pigment discoloration which can occur on prolonged heating at higher temperatures.

The calcined products are ground in water containing silicate or phosphate dispersants to break up the sintered particle aggregates. The resultant dispersions are flocculated by the addition of acid or aluminum salts, and the filtered and washed solids dried at temperatures below 200°C to produce the uncoated rutile or anatase pigments. Alternatively, the pigment dispersions may be used for the production of coated titanias, which are described later in this chapter.

The sulfate process is technically the simpler of the two methods for production of titania pigments, and the process conditions for production of optimum particle sizes and crystalline form have been well established. Some of the disadvantages of the sulfate process are the difficulty in recycling the sulfuric acid component of the process, the limited market for the byproduct ferrous sulfate, and the environmental consequences of the disposal of large volumes of strongly acidic liquid wastes.

2.2.2 The Chloride Process

The chloride process is a relatively recent development and, unlike the sulfate process, can yield rutile directly. The raw material for the process is titanium tetrachloride, prepared by heating rutile ore or beneficiated ilmenites with carbon in a stream of chlorine at 900°C. After purification to remove residual ferric chloride and other volatile contaminants, the titanium tetrachloride is mixed with a small proportion of a rutilization catalyst, such as aluminum chloride, plus other additives, and the mixture vaporized and burned in a stream of oxygen at approximately 1000 to 1500°C; the chlorine byproduct is recycled. This process yields pigmentary titanias which only need to be neutralized to remove surface chloride ions, for example, by treatment with steam, and then milled and classified to remove particles having sizes outside the optimum range. Anatase can also be prepared by the chloride process.

Although the chloride process is superficially simple, proper design of the burners and control of the reaction conditions are critical for the production of pigments having an optimum particle size and free from excessive amounts of aggregates and oversized particles. Chloride process pigments have been claimed to have better brightness and chalk resistance than those prepared by the sulfate process. However, the differences are minor and are frequently disputed.

2.3 STRUCTURES OF TITANIA PIGMENTS

Titania is the mineral form of titanium dioxide, TiO_2. Most titania specimens are nonstoichiometric and are slightly deficient in oxygen. Titania can occur in three crystalline modifications, namely, the two different tetragonal forms, anatase and rutile (Figure 2.2), and a less common orthorhombic form, brookite [1]. Only rutile and anatase are of commercial importance as pigments.

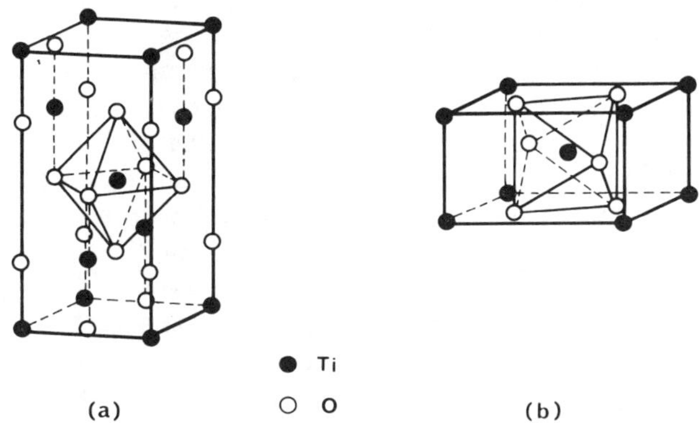

Figure 2.2 Titania unit cell structures. (a) Anatase. (b) Rutile. (From W. A. Kampfer, Titanium Dioxide, in *The Pigment Handbook*, T. C. Patton, Ed., Wiley-Interscience, New York, 1974; with permission.)

Table 2.1   Physical Properties of Titanias

Physical Properties	Anatase	Rutile
Specific gravity (g/ml)	3.9	4.2
Refractive index	2.52	2.76
Hardness (moh)	5 to 6	6 to 7

2.3.1   The Crystal Lattice and Bulk Properties

The rutile modification of titania has greater density, hardness, and a higher refractive index than the anatase form (Table 2.1) [12.13]. Both anatase and rutile absorb light in the near-ultraviolet (350 to 400 nm) region. Pure rutile has a yellowish cast resulting from the edge of this absorption band overlapping the visible spectrum. The apparent color of pigmentary grades is also influenced by the size of the particles, anatase pigments usually having a bluish cast which results from their smaller particle size.

Anatase is preferred for the pigmentation of artificial fibers as it is softer and less abrasive towards the spinning machinery, and the bluish cast of the pigment can compensate for yellow discoloration resulting from impurities in the fiber polymer. However, anatase is more photoactive than rutile and can catalyze the oxidative degradation of pigmented organic materials when these are exposed to sunlight or other sources of near-ultraviolet radiation. Although rutile is more expensive to produce, it provides greater reflectivity for visible light and, because of its lower photoactivity, it is generally preferred for the pigmentation of plastics and surface coatings which may be exposed to sunlight.

Pigmentary anatases and rutiles usually consist of irregular, but roughly spherical particles of uniform size, having diameters in the range 200 to 300 nm. Typical uncoated titania pigments have specific surface areas, calculated from vapor adsorption isotherms, of 5 to 8 m^2/g. Some anatase pigments may have needle-like, or acicular particles; acicular rutile-like "potassium titanate" pigments are also commercially available. Pigments give the maximum scattering of light of a particular wavelength, have the best brightness and hiding power when their particles have diameters equal to half that wavelength, and are evenly dispersed and separated by a similar distance [10,11,13].

The scattering of light, hiding power, and brightness of the pigment are functions of the difference in refractive indices of the pigment and its surrounding medium. The refractive indices of some typical white pigments and polymers are given in the Table 2.2. The clay minerals described in the previous chapter are suitable for use as pigments in paper, where light is reflected from a pigment–air interface. However, materials dispersed in organic media require higher refractive indices if they are to provide the efficient scattering of reflected light necessary for the pigment to have a

Table 2.2   Refractive Indices (n_D) of Some Common Pigments and Media

Rutile	2.76	Silica	1.55
Anatase	2.52	Polystyrene	1.59
Diamond	2.42	Polyesters and alkyds	1.53 to 1.57
Zinc sulfide	2.37	Poly(vinyl chloride)	1.52 to 1.55
Zinc oxide	1.99	Poly(methyl methacrylate)	1.50
Barium sulfate	1.64	Linseed oil	1.48
Calcium carbonate	1.57	Water	1.33
Kaolin	1.56	Air	1.00

good hiding power; in this respect, the titanias are virtually unmatched as white pigments [10].

Unlike the massive crystalline forms of titania, the pigments do not have well developed crystalline structures, but consist essentially of small domains of crystalline lattice with nonplanar surfaces, surrounded by disordered regions. The surface of massive rutile is formed from three crystal planes, the (110), (100), and (101). The (110) plane accounts for 60% of the surface, while the remainder is divided equally between the other two planes. The microcrystalline domains of the pigmentary rutile are believed to contain similar proportions of these surface planes [14], while the (110) would predominate in pulverized specimens as this is also the preferred cleavage plane [15]. The dominant plane in microcrystalline or pulverized anatase is frequently reported to be the (001) [e.g., 15,16], although the (111) plane may be more representative of the pigment surface [17]. The (110) plane of rutile contains equal numbers of four- and fivefold coordinated Ti^{4+} ions, while the (111) and (001) planes of anatase contain four- and fivefold coordinated cations, respectively. The preferred octahedral coordination of the surface cations of the crystalline materials can be achieved through the chemisorption of water or other ligands.

The pigmentary titanias contain lattice defects that include oxygen deficiencies associated with substituent ions of lower valence such as Fe^{3+}, Al^{3+}, Ti^{3+}, or electrons in lattice vacancies. The titanias can lose oxygen from lattice surface sites on heating, particularly in the presence of organic species or in an oxygen-free atmosphere [18]. Although the process can be reversed by cooling and exposure to oxygen, about 1 in 10^6 of the lattice titanium ions of the calcined pigments are frozen in the Ti^{III} state. Foreign cations in the lattice may include Zn^{2+}, Al^{3+}, Sb^{3+}, Si^{4+}, and P^{5+} ions derived from calciner or burner additives; adsorbed sulfate or chloride ions, or other ionic species derived from processing of the pigments; and Cu^{2+}, Cr^{3+}, Fe^{3+}, Mn^{4+}, and Nb^{5+} ions resulting from impurities in the ores [19]. A typical uncoated rutile pigment particle can be considered to have the composition shown in Table 2.3 [26], its surface having approximately 1.8×10^6 exposed lattice oxygens, one-third of which consist of hydroxylic groups resulting from the chemisorption of water.

Table 2.3   Composition of a Rutile Particle

Component	Number per Particle
Ti^{4+} (and Ti^{3+})	1.35×10^8
O^{2-}	2.70×10^8
Cr^{3+}	200
Fe^{3+}	2000
Al^{3+}	10^5
Zn^{2+}	10^6
P^{5+}, Nb^{5+}, and other ions	10^5
Oxygen vacancies	200

2.3.2   The Pigment Surface

Anhydrous titanias readily adsorb water. The resultant hydrated titania surfaces are amphoteric and contain a variety of hydroxylic species which include ionic and partially covalent hydroxyl groups and chemisorbed and physisorbed water molecules. The nature and surface locations of these adsorbed species have been extensively studied and the various conclusions have been frequently reviewed [e.g., 13,21,22]. Only a broad outline of the modes of formation, structures, and properties of the surface hydroxyls will be given here.

The bulk of the strongly bound surface hydroxyl groups result from the dissociative chemisorption of water and consist of approximately equal numbers of inherently basic and inherently acidic species [23]. On the basis of chemical and structural considerations [14,16,24], and theoretical calculations [25], the dissociative adsorption process is believed to involve the following sequence of events.

(i)   An initial adsorption of a water molecule at a five-coordinate surface Ti^{4+} site, preferably located on the (110) plane of rutile [14] or the (100) plane of anatase [16].

(ii)   Ionization of the water in the strong crystal surface field to give the Ti–OH species described as a terminal hydroxyl group (1; Figure 2.3); this dissociation is more extensive on rutile surfaces than on anatase.

(iii)   Migration of the liberated proton to a neighboring Ti–O–Ti site, with formation of a bridged-hydroxylic species (2) from the lattice O^{2-} anion.

This process is accompanied by the hydroxylation of other surface species such as Si^{4+} and P^{5+} through dissociative adsorption of water. Dissociative adsorption of water is less likely on Al^{3+} ions contained in the (110) plane of an Al-doped rutile than on the Ti^{4+} ions of a pure rutile [25]. Weakly bound water may be adsorbed by hydrogen bond formation with

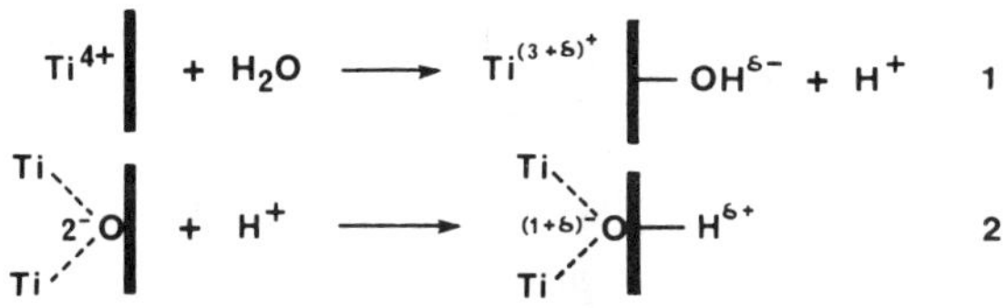

$$\text{Ti}^{4+}| + H_2O \longrightarrow \text{Ti}^{(3+\delta)+}| - OH^{\delta-} + H^+ \qquad 1$$

$$\underset{\text{Ti}}{\overset{\text{Ti}}{}}{}_2{}^{-}O| + H^+ \longrightarrow \underset{\text{Ti}}{\overset{\text{Ti}}{}}{}_{(1+\delta)}{}^{-}O| - H^{\delta+} \qquad 2$$

Figure 2.3

terminal hydroxyls, by non- or weakly dissociative adsorption on the surface cations contained in other crystal planes, and by condensation in micropores [21,26–28].

The various hydroxylic species can be differentiated by thermal analysis, by nuclear magnetic resonance and infrared spectroscopy, and by various chemical reactions. Considerable effort has been made to assign infrared absorptions to particular types of titania surface hydroxyl groups or adsorbed water. The heterogeneous nature of pigmentary titanias can introduce considerable complications in these interpretations because of the presence of crystalline and noncrystalline domains, and of hydroxylic species associated with surface impurities introduced during the manufacturing process. Despite the differences in their crystalline lattices, the hydroxylic populations on the surfaces of pigmentary anatase and rutile appear to be qualitatively similar [29].

The infrared spectrum of a fully hydrated rutile may contain eight distinct O–H adsorption bands in the 3 μm region. The structural assignments have been reviewed [30] and are now believed to be as follows.

3725 cm^{-1}:	SiOH impurity (or anatase TiO – H)
3700 cm^{-1}:	Terminal TiO – H on sites at lattice steps, edges, or apices
3680 cm^{-1}:	Terminal TiO – H on (100) or (101) planes
3655 cm^{-1}:	Terminal TiO – H on (110) plane
3610 and 3520 cm^{-1}:	Bridged TiO – H on (100) or (100) planes
3410 cm^{-1}:	Bridged TiO – H on (110) plane
3400 cm^{-1}:	Adsorbed and coordinated H$_2$O

Other absorption bands may be present in rutiles containing substituent cations or adsorbed anions. The assignments listed above are in accord with those given for the various types of hydroxyl groups on other hydrated metal oxides. They differ from those reported previously for rutile [21], principally for the peaks at 3600 to 3700 cm^{-1} which, on the basis of their apparent thermal stability and crystallographic considerations, were originally assigned to bridged TiOH groups [14].

The infrared spectrum of anatase contains four O–H adsorption bands [15,31]. Their probable assignments are:

$$3730 \text{ cm}^{-1}: \quad \text{Terminal } TiO-H \text{ on (001) or (111) plane}$$
$$3680 \text{ and } 3620 \text{ cm}^{-1}: \quad \text{Bridged (acidic) } TiO-H$$
$$3480 \text{ cm}^{-1}: \quad \text{Adsorbed and coordinated } H_2O$$

The thermal stability of the various hydroxylic species is markedly affected by the presence of adjacent defects or adsorbed anions and by the previous history of the particular samples. There is also no clear demarcation between the various stages in the loss of surface hydroxylic species. In general, though, condensed, weakly bound, physisorbed waters can be readily eliminated by heating hydrated rutiles or anatases at temperatures up to 150°C; dissociated water on rutile (100) and (101) planes is also eliminated at this stage. Hydrogen-bonded water molecules can be tenaciously retained. Some may only be desorbed with the elimination of the associated surface hydroxyl groups [32]. Heating above 200°C results in the loss of strongly bound, dissociated water from the rutile (110) and anatase (001) planes, leaving a small number of isolated terminal hydroxyls that are too widely separated on the titania surface for their diffusion together and subsequent condensation to occur at significant rates. The residual hydroxyl groups can be eliminated on heating at 300 to 400°C with the evolution of molecular hydrogen [33]; at these temperatures, lattice oxygens also begin to evaporate. Rehydration of the surface results initially in the rapid coordination of water followed by the slower dissociation process outlined above. In the case of anatase, the original hydroxylated state is not readily restored and a transitional hydrated anatase surface containing strongly bound, coordinated waters can persist for long periods at room temperature [31].

The terminal hydroxyls are ionic in character and weakly basic (anatase, pK_a 12.7) [34], and are capable of exchange with other anions. They can also react with CO_2 to form labile surface bicarbonate species that can be displaced by water [35]. Exchange of the hydroxyls with acid anions such as bisulfate or dihydrogen phosphate can result in a marked increase in the Bronsted acidity of the titania surface, dry titania with adsorbed $H_2PO_4^-$ ions, for example, having a Hammett acidity value (H_0) of -5.6 to -3.3 [36]. The terminal hydroxyl-bridged hydroxyl pairs formed by dissociation of water on the (100) and (101) planes can be displaced by the adsorption of organic bases such as pyridine, which are more strongly coordinated to the Lewis acidic Ti^{4+} ions than water. This displacement is shown by the loss of the 3690 and 3610 cm^{-1} bands in the spectrum of rutile treated with pyridine [14].

The bridged hydroxyls are partially covalent and weakly acidic (ana-

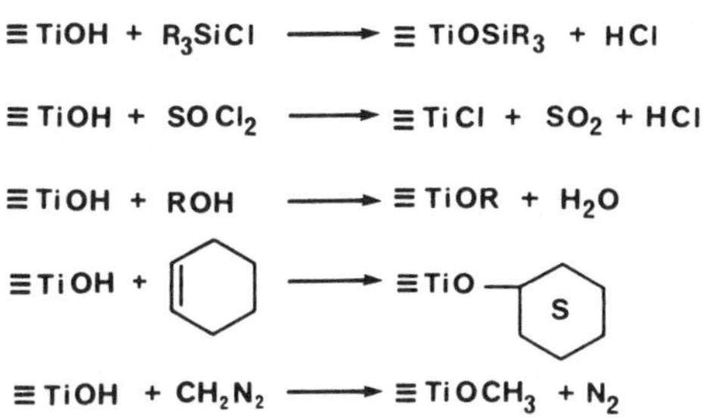

$\equiv$ TiOH + R$_3$SiCl $\longrightarrow$ $\equiv$ TiOSiR$_3$ + HCl

$\equiv$ TiOH + SOCl$_2$ $\longrightarrow$ $\equiv$ TiCl + SO$_2$ + HCl

$\equiv$ TiOH + ROH $\longrightarrow$ $\equiv$ TiOR + H$_2$O

$\equiv$ TiOH + [cyclohexene] $\longrightarrow$ $\equiv$ TiO—[cyclohexyl, S]

$\equiv$ TiOH + CH$_2$N$_2$ $\longrightarrow$ $\equiv$ TiOCH$_3$ + N$_2$

Figure 2.4 Some metathetical reactions of titania surface hydroxyl groups.

tase, pK_a 2.9) [34]. The Bronsted acidity of the anatase bridged hydroxyls is sufficient for the protonation of adsorbed trimethylamine or triethylamine, but not of weaker bases such as pyridine [37]. Ammonia may [23] or may not [37] be protonated by anatase hydroxyls. Protonation of triethylamine is not observed on rutile surfaces, the amine being adsorbed solely by coordination with the surface Ti^{4+} ions [37]. Ammonia is also adsorbed by coordination on rutile, even in the presence of water vapor [38]. The adsorption of amines can also occur by hydrogen bond formation with surface hydroxylic species.

The titania surface hydroxyl groups resemble those of alumina, described in Chapter 3, and they can undergo a variety of reactions which yield covalently bound surface substituent species [16]. For example, surface titanosiloxane species can be formed by the reaction of titania with chloro- or alkoxysilanes, titanyl alkoxides with alcohols [32], titanyl chloride species with thionyl chloride, and titanyl acylates with acyl chlorides [34]. Titanyl acylates can also be formed by the acid induced dissociation of surface Ti–O–Ti bonds [37]. Some of the hydroxyl groups are sufficiently acidic to add to cyclohexene, forming surface alkoxides [39], and to react with diazomethane forming surface methoxy groups. Whereas all the surface hydroxyls of rutile may be methylated using diazomethane, only about half of the anatase hydroxyls react in this manner [29]. Practical examples of these reactions are considered in more detail in Chapter 4. (See Figure 2.4.)

The Lewis acid sites on dehydroxylated titania surfaces resemble those of alumina, forming surface coordination (electron pair) complexes with a wide variety of organic bases. Unlike alumina, the dehydroxylated titanias have a shallow surface conduction band in the lattice electronic structure, and this facilitates single-electron transfer between the adsorbed species and other electron donors or acceptors. Lewis acidity of the titania surfaces can be enhanced by repeated cycles of hydration and dehydration; this treatment results in the gradual migration and condensation of isolated, thermally stable hydroxyl groups. Strongly acidic titanias prepared by this method can accept electrons from perylene with the formation of perylene radical-cations [15]. In contrast, dehydroxylated titanias containing surface Ti^{III} can reduce adsorbed nitrogen oxide (NO) to the NO$^-$ anion; oxidized titanias only form coordination complexes with NO [40].

Because of the amphoteric nature of the titania surfaces, dispersed titanias can buffer the pH of aqueous media at values close to neutrality [13]. The net charge which results from the pH-dependent protonation or deprotonation of the surface hydroxyl groups of titanias dispersed in aqueous or ionizing media can be measured by microelectrophoresis [41]. In terms of their respective electrochemistry, anatase is more basic than rutile; the zero point of charge (zpc) or, more correctly, the isoelectric point (iep) of anatase occurring at about pH 6.2, while that of rutile occurs at about pH 5.3. The actual values depend on the history of the samples, a calcined rutile having zpc at pH 4.8, while a hydrated rutile has a zpc at pH 5.6. These values refer to pure titanias; the electrochemistry of titania pigments is considerably affected by the presence of acidic or basic surface species such as P–OH, Si–OH, and Zn–OH, or other processing residues. The electrophoretic properties of a variety of commercial titanias have been reviewed [21].

2.4   PHOTOCHEMICAL ACTIVITY OF TITANIAS

The surface of titania is capable of oxidizing or reducing a wide variety of adsorbed or contacted inorganic and organic compounds when exposed to light having wavelengths of 300 to 400 nm. This activity can have a number of practical applications. The reversible photoreduction of adsorbed silver ions, for example, is the basis of the Itek photographic process [42], and other photographic processes have been developed on the oxidation or reduction of leuco-dyestuffs. The photodecomposition of water on titania surfaces can yield hydrogen, thus providing a potential means of solar energy storage, while a variety of other titania-catalyzed photochemical or photoassisted electrochemical techniques have been described in the scientific and patent literature.

It is well known that exposure of titania-pigmented plastics or surface coatings to sunlight results in oxidative degradation of the organic matrices, leading to embrittlement and to chalking, or erosion of the surface of pigmented materials; these processes are accelerated by the presence of moisture. Although the photoactivity of titania pigments is low compared to that of doped titanias and other useful photocatalysts, the contact times for reaction are usually very long, many thousands of hours in the case of exterior paint films, and considerable conversion of the contacted organic materials may occur. Many hypothetical mechanisms have been proposed for the chalking processes, but the details of these and many other simpler processes have not been established with certainty. The mechanisms for the formation of the various oxidizing species and their probable importance in the photodegradation processes will be discussed in this section; titania-catalyzed organic reactions will be discussed in detail in Chapter 5.

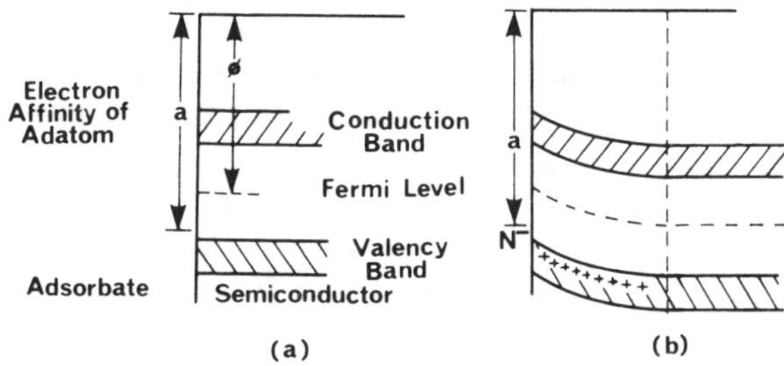

Figure 2.5 Electron band energy levels in the surface zone of titania. (a) Before irradiation. (b) After irradiation. [Adapted, with permission, from R. D. Murley, *J. Oil Colour Chem. Assoc.*, **45**, 16 (1962).]

2.4.1   Origins of the Photoactivity

The titania crystal lattice absorbs near-ultraviolet light, the absorption band for rutile having a maximum near 370 nm. This absorption results in the conversion of light quanta into excited lattice species, in which electrons are promoted from the valence band to the conduction band in the lattice electronic structure (see Figure 2.5). The energy gap between these bands in the rutile lattice is approximately 3.05 eV, and 3.2 eV in the anatase lattice. These energy levels correspond to light quanta having the respective wavelengths of 420 and 390 nm, which correspond in turn to the long wavelength edges of the rutile and anatase lattice absorption bands. Actinic light having wavelengths shorter than these critical values is absorbed with the formation of "excitons" or electron–hole pairs (h-e). These can dissociate into free, mobile holes (h^+) and electrons (e^-) that are capable of migration through the lattice to the surface or other reactive sites [43]. The formation of an exciton corresponds, in terms of discrete Ti^{4+} and O^{2-} ions, to the transfer of an electron from an oxygen $2p$ to a titanium $3d$ orbital [18,44].

The excitons are generated most effiently by adsorbed light having wavelengths near that of the near-visible band edge, which can penetrate deep into the titania lattice. The excitons are unstable and decay, transferring thermal energy to the lattice, unless they are trapped by some other process. Most titanias fluoresce when irradiated with ultraviolet light at low temperatures. This had been ascribed to exciton decay. However, pure rutile does not fluoresce and the fluorescence of other specimens may be due to the presence of impurity ions such as Cr^{3+} in the titania lattice [43].

Titania exposed to actinic light under vacuum or in an oxygen-free atmosphere can lose oxygen from the crystalline lattice. This results in the formation of a semiconducting surface layer containing Ti^{3+} ions [45] and causes the titania to become blue–grey in color. The deoxygenation is reversed when the titania is exposed to oxygen in the absence of the radiation. Irradiation of titania by near-ultraviolet light in the presence of oxygen results in the formation of one or more of the paramagnetic adsorbed oxygen species, O^-, O_2^-, and O_3^- [46–49]. Similar photoadsorption reactions have been observed on other semiconducting oxides, notably zinc and magnesium oxides, although the latter requires activation by more energetic radiation than that for zinc oxide or the titanias [47,50]. The formation of an oxygen radical-cation has been reported [51], but the species was subsequently found to be either adsorbed NO or NO_2 [52], formed presumably by oxidation of adsorbed ammonia [53].

The relationship between the photoadsorption of oxygen and the band structure of the semiconducting titania lattice can be illustrated by reference to Figure 2.5 [18,43]. The energy of chemisorption of the first molecule of oxygen has a value proportional to (a − ϕ), where (ϕ) is the ionization potential, that is, the extra energy above the Fermi level required for an electron to leave the titania surface, and (a) is the electron affinity of the adsorbed oxygen; the function (a − ϕ) must have a positive value for the absorption to occur.

The Fermi levels of unirradiated titanias are too low for this electron transfer to occur. However, on irradiation, the Fermi level of the crystal rises, reducing the value of (ϕ) and making the adsorption of oxygen energetically favorable. The adsorption of oxygen is limited by the development of a surface charge and the formation of an electric double layer between the chemisorbed anions and the positive holes remaining in the lattice. This double layer effectively lowers the Fermi level near the crystal surface until an adsorption equilibrium is attained [54]. Processes which can annihilate the holes will enable larger amounts of oxygen to be chemisorbed.

Although the photoadsorption of oxygen may occur on apparently anhydrous titania surfaces, kinetic studies have shown that the process is promoted by the presence of surface hydroxylic species, and retarded by dehydration [55] or by replacement of the surface hydroxyls by chloride [56] or by phosphate ions [57]. There have been various proposals that the adsorption of oxygen by hydrated titanias is accompanied by the formation of hydroxyl radicals, hydroperoxyl radicals and anions, hydrogen peroxide, solvated electrons, nascent atomic oxygen, and excited (singlet state) molecular oxygen, as well as the oxygen radical-anions observed on oxidized anhydrous titanias [43]. The principal products formed in the oxygen photoadsorption depend on whether the substrate is an outgassed, partially hydrated titania in a solid–gas system or titania dispersed in an aqueous (electrically conducting) medium. Fully hydrated titanias have an "aqueous-like" hydroxylic surface and their photoadsorption of oxygen can yield the same products as those obtained from dispersed materials. The photoadsorption of oxygen on hydrated titania surfaces is an example of the photooxidation of water on a semiconducting surface, a general process which has been extensively studied [58–61].

The currently favored hypotheses for the generation of oxidizing spe-

$$h\nu + \text{lattice} \rightleftharpoons (h-e) \rightleftharpoons h^+ + e^-$$

$$OH^- + h^+ \longrightarrow HO^{\bullet} \qquad\qquad 1$$

$$O_{2(ads)} + e^- \longrightarrow O_{2(ads)}^{-\bullet} \qquad\qquad 2$$

$$O_{2(ads)}^{-\bullet} + HO^{\bullet} \longrightarrow HO_2^{\bullet} + O_{ads}^{-\bullet} \qquad\qquad 3$$

$$O_2^{-\bullet} + H_2O \rightleftharpoons HO_2^{\bullet} + HO^- \qquad\qquad 4$$

$$HO_2^{\bullet} + e^- \longrightarrow HO_2^- \; ; \; HO_2^- + h^+ \longrightarrow HO_2^{\bullet} \qquad\qquad 5$$

$$HO_2^{\bullet} + OH^- \longrightarrow H_2O + O_{2(ads)}^{-\bullet} \qquad\qquad 6$$

$$HO_2^{\bullet} + HO_2^- \longrightarrow H_2O + O_{3(ads)}^{-\bullet} \qquad\qquad 7$$

Figure 2.6

cies on the irradiated pigment surface are essentially similar to the mechanism originally proposed by Bickley and Stone [43,55] for the adsorption of oxygen on a dehydrated rutile containing few residual hydroxyl groups. In this scheme (see Figure 2.6), the excitons or derived mobile holes generated by the adsorption of actinic light (Eq. 1) migrate to the titania surface and interact with surface hydroxyl groups, the holes becoming trapped in the form of "bound" hydroxyl radicals (Eq. 2). The conduction band electrons freed by this process can then transfer to adsorbed oxygen molecules forming superoxide radical-anions, O_2^- (Eq. 3). This process is apparently irreversible, as oxygen is not desorbed on cessation of the irradiation [55]; this is in contrast with the reversible photoadsorption of oxygen on anhydrous titania surfaces [43] and may reflect differences in the residual hydration of the titanias (see p. 67). The hydroperoxyl radicals $HO_2\bullet$ may be formed by the reaction of adsorbed O_2^- with bound hydroxyl radicals [55] or by protonation of the strongly basic radical-anion by titania surface hydroxylic species (Eqs. 4 and 5).

The rate of oxygen uptake decreases during irradiation, but may be enhanced on secondary irradiation after storage in darkness, indicating that some form of thermal dark-reaction can occur. This reaction was thought to involve the capture of lattice electrons by hydroperoxyl radicals to form hydroperoxide anions; these are more effective hole-traps than the surface hydroxyl groups [55] (Eq. 6).

Oxygen radical-anions may also be formed by reaction of the hydroperoxyl radicals with other adsorbed species (Eqs. 7 and 8) [55]. Alternative, speculative reactions of $HO\bullet$ and O_2^- species with adsorbed water have been variously proposed for the generation of oxygen radical-anions, other oxygenated radical species, and the atomic and singlet molecular oxygens which may be active intermediates in the titania-catalyzed photooxidations. Hydrogen peroxide has been detected in the photolysis products from aqueous titania dispersions [62] and has been assumed to be a product of the photoadsorption of oxygen by fully hydrated solids [17]; it is apparently not formed during the photooxidation of partially hydrated titanias [63]. The formation of hydroxyl and hydroperoxyl radicals [59] and of singlet (excited state) molecular oxygen [62] during the photooxidation of titania dispersions has been demonstrated by diagnostic trapping experiments.

Only a small proportion of the surface hydroxyl groups on hydrated titanias are apparently available for the trapping of holes formed by irradiation. The selection of hydroxyl groups appears to be on a random basis [64] and there is no concrete evidence to support an earlier hypothesis [13] that the hydroxyl radicals are derived largely from hydroxyl groups at the intersections of crystalline domains on the titania pigment surface.

Munuera and co-workers have re-examined in detail the kinetics and products of the photoadsorption of oxygen on anatase titanias having various degrees of hydration [31,46,65]; their conclusions are also relevant to the photooxidation of rutile. They found that the rate of photoadsorption depended on the concentration of basic (terminal) hydroxyls present and that coordinated molecular water or acidic hydroxyl groups do not greatly influence the photoactivity. They also found that the photoadsorption on a hydrated surface was accompanied by a slow photodesorption of oxygen and that the rate of photoadsorption changes from first-order to diffusion-controlled kinetics as the surface hydroxyl concentration declined. The photoadsorption and desorption processes were accompanied by the desorption of molecular water from the hydrated anatases. They proposed the scheme shown in Figure 2.7 for the photoadsorption on a hydrated anatase surface. Their model consisted of the (111) plane, in which the surface Ti^{4+} ions (9) are each coordinated with three lattice O^{2-} ions, one terminal hydroxyl (OH_b) and one bridged hydroxyl (OH_a) formed by dissociation of H_2O, and one other, undissociated, coordinated water molecule.

The kinetics indicated that excitons generated in the subsurface regions could only reach the surface through a limited number of "portholes"; they then diffuse across the surface to nearby Ti^{IV} sites bearing OH_b groups (HO^- anions), which act as hole-traps. The electrons freed in this process can then be trapped by oxygen or by $HO^{\bullet}$ radicals to reform HO^- anions. The reaction with oxygen requires that formation of the trapped exciton––TiOH complex (10) be accompanied by the loss of ligand water, as the oxidant has to be coordinated with the Ti^{III} center for electron transfer to occur. Diffusion of excitons across the hydrated surface is also accompanied by desorption of coordinated water molecules from the transient TiOH–exciton complexes; these photodesorbed waters could be rapidly readsorbed if the H_2O partial pressure in contact with the surface is not too low. The emerging excitons are rapidly trapped on surfaces having high HO^- concentrations, and a steady state with pseudo-first-order kinetics is quickly established. The hydroxyl radicals produced are highly mobile and

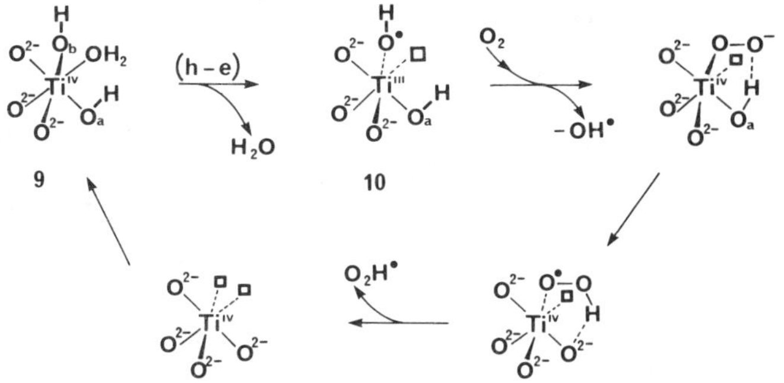

Figure 2.7   One mechanism for the photogeneration of oxidizing species on an oxygenated, hydrated titania (anatase) surface. [From A. R. Gonzalea-Elipe, G. Munuera, and J. Soria, *J. Chem. Soc., Farad. I*, 75, 754 (1979); with permission, Royal Society of Chemistry, London.]

can combine to form hydrogen peroxide. At low HO^- concentrations, the increased distance between portholes and the hole-traps results in a diffusion-controlled process and in the increased probability of exciton decay and consequent reduced photoactivity.

The photoadsorption of oxygen yields $HO_2{}^{\bullet}$ radicals which remain coordinated with the Ti^{4+} ions and which can be detected by esr spectroscopy on anatase irradiated at low temperatures. At room temperature or on continued irradiation, the $HO_2{}^{\bullet}$ radicals decompose to form the "photodesorbed" oxygen [46,65]. Hydrogen peroxide chemisorbed on a titania surface can be rapidly reduced by trapped excitons [65]; at low H_2O vapor pressures or on partially dehydroxylated surfaces, free H_2O_2 probably would not be formed during the photoadsorption process.

The initial processes in the photoadsorption of oxygen by anhydrous or dehydroxylated anatases are similar to those of the hydrated materials and depend on the presence of residual terminal hydroxyl groups for triggering the adsorption process by trapping surface excitons, with the formation of $HO^{\bullet}$ radicals [46]. However, as the dehydroxylated surface contains Ti^{4+} ions which are largely deficient in coordination (a Lewis acidic surface), the scarcity of acidic hydroxyls or coordinated waters results in the formation of isolated Ti^{3+} ions. These can either react with adsorbed oxygen, forming $O_2{}^-$ radical-anions, or else donate free electrons to the surface conduction band; these electrons can react with lattice O^{2-} and oxygen giving $O_3{}^{3-}$ radical-anions. The $HO^{\bullet}$ radicals can react with the lattice O^{2-} anions to regenerate the terminal hydroxyl "trigger" groups. The reactions proposed by Munuera for the photoadsorption of oxygen by hydrated and dehydroxylated anatase surfaces are summarized in the equations below [65].

The various photogenerated oxy-species are potential oxidants for adsorbed or contacted organic materials [50] and have been implicated in the photodegradation of a wide variety of polymers [66]. Direct electron transfer may also occur between an adsorbed organic species and the irra-

Hydrated surfaces	Dehydroxylated surfaces
$h \sim e + OH_b^- \longrightarrow e^- + OH^\bullet$	$h \sim e + OH_b \longrightarrow e^- \quad OH^\bullet$
$e^- + O_2 + H^+ \longrightarrow HO_2^\bullet$	$e^- + O_2 + O_{latt}^{2-} \longrightarrow O_3^{3-}$
$OH^\bullet + OH^\bullet \longrightarrow H_2O_2$	$OH^\bullet + O_{latt}^{2-} \longrightarrow O^- + OH_b^-$
$HO^\bullet + HO_2^\bullet \longrightarrow H_2O_2 + O_2$	$O^- + O_2 \longrightarrow O_3^-$
$HO_2^\bullet + OH^\bullet \longrightarrow H_2O + O_2$	$O_3^- + O^- \longrightarrow 2O_2^-$
$H_2O_2 + Ti^{\underline{iii}} \longrightarrow Ti^{\underline{iv}} + OH^\bullet + OH^-$	$O_3^- + O_3^- \longrightarrow O_2^- + O_2$
$OH^\bullet + H_2O_2 \longrightarrow H_2O + HO_2^\bullet$	$O_3^{3-} + O_3^{3-} \longrightarrow 3O_2^{2-}$

Figure 2.8

diated titania surface, as in the photoreduction of Methylene Blue; similar electron-transfer processes may occur when a titania–organic mixture is heated [67]. Consequently, in complex reactions such as the photodegradation of a titania pigmented paint film, the nature of the oxidizing species is uncertain and the observed oxidation products may be formed by more than one mechanism.

The products of the photolysis of adsorbed water or surface hydroxyls depend on whether electrons can flow between donor and acceptor centers via an external circuit [58]. In aqueous dispersions, the titania particles can behave like short-circuited photoelectrochemical cells for the photoassisted decomposition of water [59] (Figure 2.12), whereas the photocatalyzed reaction in a gaseous–solid system involve the decomposition of surface hydroxyls [58] (Figure 2.8). In aqueous media, the photogenerated holes are sufficiently energetic to oxidize water to $HO^\bullet$ radicals, which are intermediates in the production of both hydrogen peroxide and oxygen [59]. Adsorbed anions may also be oxidized; some examples are the oxidation of CN^- to CNO^- [68], of Br^- to Br_2 (in acid) or BrO^- (in alkali) [69], and of carboxylates to carboxyl radicals, which can undergo decarboxylation and dimerization to form hydrocarbons (a photo-Kolbe reaction) [70] (Figure 2.9).

$$CN^- + h^+ \longrightarrow CN_{ads}^\bullet \xrightarrow{OH^\bullet} H^+ + CNO^-$$

$$Br^- + h^+ \longrightarrow Br_{ads}^\bullet \xrightarrow{OH^\bullet} H^+ + BrO^-$$

$$BrO^- + Br^- + 2H^+ \longrightarrow Br_2 + H_2O$$

$$RCOO^- + h^+ \longrightarrow RCOO^\bullet \longrightarrow R^\bullet + CO_2\uparrow$$

$$2R^\bullet \longrightarrow R-R$$

Figure 2.9

$$HO^{\bullet} + O_{2(ads)}^{-\bullet} + 2\,Hg \longrightarrow H_2O + \left[Hg_2O\right] \xrightarrow{\;H^+\;} H_2O + 2\,HgO$$

Figure 2.10

Kaluza and Boehm proposed that $HO^{\bullet}$ radicals were the principal oxidants in the photooxidation of mercury [57] adsorbed on hydrated titanias. These radicals were also shown to be derived largely, but not exclusively, from terminal Ti–OH groups and to be accompanied by the formation of HO_2^- anions from adsorbed oxygen. The presence of both $HO^{\bullet}$ and O_2^- was found to be necessary for the oxidation of mercury, although organic species could be oxidized in anaerobic conditions on a hydrated titania surface (Figure 2.10).

Voelz and co-workers [71–73] have shown by spectroscopic and chemical methods that $HO^{\bullet}$ radicals are formed on irradiation of hydrated titania surfaces, and these and the coproduced $HO_2^{\bullet}$ radicals are the key intermediates in the oxidation of adsorbed organic molecules. They proposed that the radicals were formed by a process basically similar to the scheme shown in Figure 2.6.

In contrast, Bickley and co-workers have proposed that O_2^- radical-anions are the initial oxidants and that the bound hydroxyl radicals ($HO_s^{\bullet}$; Figure 2.11) serve mainly as sites for the reinjection of electrons into the titania lattice following the reactions between the superoxide radical-anions and the organic molecules [74,75]. This is illustrated in their hypothetical reaction sequence for the photooxidation of isopropanol (PrOH) to acetone (A) (Figure 2.11).

The ability of the various oxygen radical-anions to abstract hydrogen from, or undergo addition to organic molecules is well established, at least in homogeneous media. These active radical-anions are believed to be present on the surfaces of a variety of other heterogeneous oxidation catalysts [50]. Hydrogen abstraction by $HO^{\bullet}$ or other radical species can initiate autoxidative chain reactions between organic materials and

$$RO^- + h^+ \longrightarrow RO_s^{\bullet}$$

$$O_2 + e^- \longrightarrow O_2^{-\bullet}$$

$$O_2^{-\bullet} + PrOH \longrightarrow H_2O^{\bullet} + PrO^-$$

$$HO_2^{\bullet} + PrOH \longrightarrow H_2O_2 + PrO^{\bullet}$$

$$H_2O_2 + PrO^- \longrightarrow A + H_2O + OH^{\bullet} + e^-$$

$$OH^{\bullet} + PrO^{\bullet} \longrightarrow A + H_2O$$

$$e^- + RO_s^{\bullet} \longrightarrow RO^-$$

Figure 2.11

atmospheric oxygen. These and other surface-catalyzed photooxidation reactions will be discussed in Chapter 5.

There are some anomalies in these convenient, but relatively simple models for the initiation of photooxidation reactions. Most published mechanisms for the photoreactions between titania and organic species are based solely on the reactions of photogenerated oxidant (or reductant) species with supernatant organic molecules; they do not consider adsorbed organic complexes, the reactions of which are more likely to involve transfer of hydrogen, hydroxyl, or other groups with neighboring bound surface species.

The photoadsorption of oxygen is dependant on the availability of titania conduction-band electrons. Although the trapping of exciton-derived holes provides one source of these electrons, photoadsorption of oxygen to form O_2^- radical-anions can occur when titanias are irradiated with visible light having wavelengths of 400 to 560 nm, that is, below the threshold for exciton formation [48]; the alternative electron sources are believed to be donor impurities or other lattice defects. However, O_2^- radical-anions adsorbed on anatase are ineffective for the oxidation of isobutane, for example, unless the surface is also irradiated with light having wavelengths of 220 to 400 nm, that is, light which is sufficiently actinic for the generation of excitons or photoholes [54].

The failure of titania to oxidize alkanes when irradiated with visible light implies that O_2^- radical-anions alone have limited effectiveness as oxidants. This observation can be correlated with the chemistry of the superoxide (O_2^-) radical-anions in homogeneous media, which is predominantly that of a nucleophilic strong base, and only indirectly that of an oxidant. In protic media, the superoxide ions are rapidly converted to $HO_2^{\cdot}$ radicals, which can then react with additional superoxide ions to form nascent oxygen and hydrogen peroxide, both powerful oxidants. In aprotic media, the superoxide ions can undergo nucleophilic substitution with alkyl halides and carboxylic esters, for example, with the respective displacement of halide and alkoxide ions and the formation of organic peroxyl radicals; the latter may undergo dismutation with superoxide to form alkoxide or carboxylate anions and oxygen.

Partially hydrated titanias can catalyze the photoreduction of a number of inorganic or organic species which can be adsorbed at the surface reduction sites. These sites are probably the same as those involved in the reduction of oxygen, which are coordination deficient exciton–TiOH complexes shown in Figure 2.7, or their equivalents. Nitrogen, for example, can be reduced to ammonia or hydrazine via a hypothetical N_2H_2 intermediate (Figure 2.12) [64] and adsorbed Ag^+ ions reduced to Ag [42]. The generation of hydrogen from adsorbed water or other sources of protons can either involve electron transfer via a surface cathodic site (on hydrated TiO_2 or platinized TiO_2) or (in dehydroxylated samples) by the reduction of residual water by surface Ti^{III} sites [58].

$$[\text{Ti}^{IV}\text{OH}^-] \xrightarrow{h\nu} [\text{Ti}^{III}\text{OH}^\bullet]$$

$$\text{Ti}^{III}\text{OH}^\bullet + \text{H}_2\text{O}_{ads} \longrightarrow \text{Ti}^{IV}\text{OH} + \text{HO}^\bullet + \text{H}^+ + e \qquad (1)$$

$$2\text{HO}^\bullet \longrightarrow \text{H}_2\text{O}_2 \longrightarrow \text{H}_2\text{O} + \tfrac{1}{2}\text{O}_2\uparrow \qquad (2)$$

$$\text{H}^+ + e \longrightarrow \text{H}^* \longrightarrow \tfrac{1}{2}\text{H}_2\uparrow \qquad (3)$$

$$[\text{Ti}^{III}\text{OH}] \longrightarrow \text{Ti}^{III}_\square + \text{HO}^\bullet \qquad (4)$$

$$\text{Ti}^{III}_\square + \text{HO}^-_{surf} \longrightarrow \text{Ti}^{IV}\text{O}^{2-} + \text{H}^* \qquad (5)$$

$$\text{N}_2 + 2\text{H}^* \longrightarrow \text{HN} = \text{NH} \qquad (6)$$

Figure 2.12

Boonstra and Mutsaers [76] found that acetylene and ethylene could be hydrogenated and dimerized when irradiated in the presence of titania. Schrauzer and Guth subsequently found that oxygen was a byproduct of these reactions and that the photoreduction of nitrogen was inhibited by the presence of the more readily reduced (and more strongly chemisorbed) alkenes and alkynes [64]. Boonstra and Mutsaers proposed that, contrary to the generally accepted hypotheses, the photolysis of surface TiOH groups involved the formation of hydrogen atoms, which could either add to the unsaturated centers or else abstract hydrogen from the organic species. However, the organic reaction products point to the intermediacy of coordinated alkene or alkyne surface adducts which would not require "free" hydrogen atoms for their reduction. These reactions are described in greater detail in Chapter 5.

The photodegradation of a titania-pigmented polymer film by natural or accelerated weathering generally involves a system in which the surface zone exposed to actinic light is sufficiently permeable to oxygen and water for the titania to remain in a hydrated, oxidized state. Electron microscopy [11,72] of degraded specimens has shown that the oxidation reactions are not confined to the vicinity of the titania particles, indicating that the oxidizing species can diffuse from the pigment surface into the surrounding matrix. The formation and reactions of mobile hydroxyl and hydroperoxyl radical intermediates proposed by Voelz and co-workers can therefore provide a convenient, if not exclusively correct mechanism for the titania-catalyzed photooxidation of organic matrices. Other potential mobile oxidants include hydrogen peroxide, which is usually formed only in aqueous media, and singlet molecular oxygen. Atomic oxygen [77] is a less likely possibility.

The effects of the doping of titania pigments with foreign ions on the rate of degradation of pigmented polymer films has been studied by Torlaschi and co-workers. They found that Li^+, Na^+, K^+, Ba^{2+}, and Si^{4+} ions had no effect on the photoactivity of the pigment, while Ca^{2+}, Zn^{2+}, and Al^{3+} retarded the rate of photodegradation [78]; Al^{3+} substituent ions are known to act as lattice hole traps [79]. In contrast, variable valence ions like Sb^{5+} and Nb^{5+} caused serious photodegradation when present in titania at 150 ppm, while Cr^{3+}, Cu^{2+}, and Mn^{2+} were deleterious when present in only a few ppm [78]. The presence of Fe^{3+} ions in doped titanias also promotes the photoreduction of nitrogen [64].

The deleterious effects of the photoactivity of titania pigments can be reduced by a combination of techniques. The low, residual activity of a doped rutile, for example, could be overcome by encapsulating the pigment particles with an oxidation-resistant, nonporous inorganic coating that can neutralize or hinder the diffusion of oxidizing species from the titania surface; the coating would also prevent contact between the active surface and degradable organic materials. However, the titania would probably still not be completely insulated from its surroundings, as most coatings do not prevent the photoadsorption of oxygen [18] or the diffusion of molecular water to the underlying photoactive interface. The role of the inorganic coatings will be discussed in Section 2.5.

As the photoactive processes involve surface TiOH groups, suitable modification of these could be used to reduce the surface activity. Methods claimed to reduce activity include condensation with silanes or organosilicates, condensation with hydrous metal oxide components of an inorganic coating, and use of coatings which can act as an electron sink or increase the acidity of the surface TiOH groups. The photoactivity can be reduced by the adsorption of certain transition metal ions, most notably, those of manganese, on the titania surface; these deactivate the pigment by promoting the recombination of the mobile surface excitons. The photoactivity may also be reduced by replacing surface Ti^{4+} ions with other species such as the Zn^{2+} or Al^{3+} ions introduced as rutilization catalysts. These impurity ions tend to concentrate at the titania surface, and water coordinated with them is less likely to dissociate, and in turn provide HO radicals, than that coordinated with Ti^{4+} ions; hydroxyl groups attached to these impurity ions may also be less readily oxidized [73].

2.4.2   Comparison of the Photoactivity of Anatase and Rutile

The photoactivity of different titanias towards polymer substrates is often measured by the use of model systems; these include the photooxidation of isopropanol [80], leuco-dyestuffs [77], and mandelic acid [81].

Pigmentary anatase is considerably more active than pigmentary rutile in the photodegradation of titania-containing plastics and surface coating compositions, although there appear to be no qualitative differences in the photochemistry of the two forms of the mineral. The photoactivity of rutile is affected by the presence of residual domains of anatase, the increase in activity being out of all proportion to the percentage of anatase in the pigment [82], a 90% rutilized titania, for example, being twice as active as

"100%" rutile [83]. The reasons for the quantitative differences between the photochemical activities of rutile and anatase are not readily apparent, and many of the published explanations are facile or contradictory.

An early explanation was based on the observation that the Ti–O–Ti interatomic distances are smaller in the rutile lattice than those in anatase, and this was equated with an existence of "stronger" bonds in rutile. The rutile lattice has a greater energy of formation than that of anatase, and ultraviolet spectroscopy reportedly indicates that the Ti–O bonds of rutile have less ionic character [84], although they are still predominantly ionic [85]. The activation energy for loss of oxygen from rutile is actually less than that from anatase, and rutile is the more effective thermal oxidation catalyst [86]. There is no significant difference in the strength of the surface Ti–OH bonds of anatase and rutile, although the lower activity of rutile has been ascribed to the firmer bonding of its hydroxyl groups [73].

Boonstra and Mutsaers [56] found that there was an apparent linear relationship between the amount and rate of oxygen photoadsorption and the hydroxyl group concentration on the surface of anatase or rutile powders, and that anatase was capable of adsorbing up to twice as much oxygen as rutile containing the same number of surface hydroxyl groups. Murley [18] also noted a twofold difference in the amount of oxygen adsorbed per unit area of surface of anatase and rutile when these were irradiated for similar periods.

The differences in the photoadsorption of oxygen between anatase and rutile have been explained in terms of the energy levels of their crystalline band structures (Figure 2.5) [18]. The Fermi level of anatase is higher than that of rutile, as its lattice absorption band occurs at shorter wavelengths than that of rutile. This would allow a more extensive chemisorption of oxygen, although the difference may not be sufficient to explain the large observed differences in reactivity between the two crystalline forms of titania. The wider semiconduction band gap of anatase would also result in the formation of more energetic excitons for the generation of HO$^{\bullet}$ radicals. A nominally pure pigmentary rutile usually contains Al^{3+} or Zn^{2+} ions rutilization catalyst residues) as impurities, either within the crystal lattice or concentrated near the crystal surface. The light-absorbing layer in which excitons are generated has an approximate depth of 10 nm and the presence of foreign ions in this layer can hinder the migration of excitons to the pigment surface and promote their decay, thereby reducing the photo-oxidative efficiency of the rutile pigments.

Allen and co-workers have recently studied the luminescence of anatase and rutile pigmented polyolefins when irradiated at 77°K. They have attributed the difference in photoactivities of the two titanias to the difference in the energies of their excited states [87,88]. They also noted that the phosphorescence of the polyolefins, which probably arises from the decay of excited states of carbonyl impurities, is rapidly quenched by anatase but not by rutile. This was ascribed to one-electron transfer reactions between the triplet excited species and the anatase which could initiate radical-chain degradation processes [89].

The difference in photodegradative activities of rutile and anatase pigments may also result from differences in their physical structures. Pigmentary anatases, for example, normally have smaller sized particles with greater apparent porosity than rutile pigments, and consequently have greater specific surface areas for the generation of oxidizing species. The higher refractive index of rutile would enable more effective scattering of the incident radiation; this would provide better shielding of the deeper regions of a pigmented composite, thereby reducing the overall rate of photodegradation.

Bickley and Stone [55] found that their samples of rutile showed a greater rate of photoadsorption of oxygen than the anatase samples. They ascribed this to the greater absorption of near-ultraviolet light and the better crystallinity of the rutile sample, which resulted in increased efficiency in the conversion of light and migration of excitons to the titania surface. The increased photoactivity resulting from greater crystallinity would probably be partially offset by the reduction in activity caused by the annealing of surface defects.

The rate of photooxidation appears to be dependent on the concentration of anomalous sites located on lattice plane discontinuities; these may consist of the highly ionized Ti–O "portholes" [31] or their hydrated forms, the highly basic terminal TiOH species [26]. In the case of rutile, at least, the anomalous sites associated with the basic hydroxyl groups are irreversibly eliminated, with the development of more extensive crystalline domains during the annealing process [30].

2.5   INORGANIC COATED TITANIAS

The photoactivity of the titanias is generally detrimental for pigmented plastics, surface coatings, papers, and fibers that degrade when exposed to sunlight or other sources of near-ultraviolet radiation, particularly in the presence of moisture. The photodegradation can be useful in some systems, such as the "self-cleaning" anatase-pigmented exterior paints.

The unwanted degradation can be minimized by the use of rutile pigments, but these may still have sufficient activity to seriously limit the service life of the pigmented composites. Reduction of the residual photoactivity is particularly desirable in the case of pigments used in the newer surface-coating materials such as the acrylics or silicone polyesters. These polymers, unlike the commonly used alkyd resins, are transparent to near-ultraviolet light and are inherently stable in the absence of pigment- or impurity-initiated degradation [20]. The effects of residual pigment photoactivity are not so noticeable with the alkyd compositions because these degrade in sunlight, even in the absence of a photoactive pigment.

Most commercial titania pigments are manufactured with some form of treatment to reduce the surface photoactivity or to improve their dispersibility in various media [82]. This usually consists of coating the pigment particles with a layer of suitable inorganic material that can insulate the matrix materials from the reactive titania surfaces and surface derived oxidants, while also increasing the pigment–matrix compatibility.

Coating Techniques and Composition

There have been many hundreds of patents issued for methods for coating titanias. It is impractical to review all the various compositions and techniques that have claimed to improve the dispersion and reduce photoactivity of the pigments. Some of the earlier methods are described in a noncritical summary of British and United States patents relating to titania pigment production that were issued during the period 1946 to 1965 [9]; these methods are also described in reference 3. The scope of this section will be limited to a brief summary of general techniques and a description of some significant developments in coating technology.

Most of the early coated titanias were developed for the surface-coatings industry. Many of these coatings were subsequently adapted for use in the plastics, paper, and fiber industries. To some extent the development of the coatings paralleled that of the understanding of the chalking or photodegradation processes [3,11].

In the early 1930's, titania pigments were considered to be chemically inert. The nonbasic titanias showed poor affinity for the conventional oil-based resins, unlike the other widely used white pigments, most of which were basic salts or oxides capable of forming dispersant soaps with carboxylate-containing polymers. These titanias were difficult to disperse, or maintain in dispersion, unless used in conjunction with some other soap-forming pigment, such as zinc oxide, or with resins having high acidity.

The chalking process was originally believed to be due to an inherent incompatibility between the titania pigment and the matrix polymer, or its degradation products, which weakened the pigmented films and thus facilitated their erosion. The supposed release of adsorbed sulfate ions as sulfuric acid was also implicated in the chalking and fading of titania pigmented compositions. This "excess acidity" was overcome by treatment of the pigment with barium hydroxide or by coating with zinc oxide, thus forming a basic surface which was considered essential for good weather resistance of the pigmented compositions.

Coating of the pigments with insoluble silicates was also proposed for dispersion stabilization. These coatings were applied by the flocculation of silicate dispersed titanias with magnesium, calcium, or barium salts. Other coatings, introduced during the late 1930's for the improvement of dispersion characteristics and weather resistance of titania pigments, consisted of titanyl phthalates, insoluble fluorides, basic aluminum salts, silica, or antimony oxide. Alumina-coated anatases and zinc oxide- or antimony oxide-coated rutiles were commercially available in 1940, although the most frequently used titania pigments were still uncoated anatases and rutilized composites. The oxidative potential of the titania surfaces and its association with the chalking process had been recognized by this time.

In 1942, patents were issued for the use of insoluble silicates, including aluminum silicates or alumina–silica, for the precipitation coating of titanias [e.g.,90]. These coatings claimed to improve the dispersion properties (brightness) of the pigments in both aqueous and organic media, to reduce chalking and the fading of organic dyes, and they had greater stability than the earlier metal acylate coatings. The porous nature of these coatings was believed to provide "keying" of the binder resin to the pigment, thereby also improving the mechanical properties of the pigmented materials. These amorphous gel coatings comprised 2 to 5% by weight of the pigment, equivalent to a coating thickness of about 5 nm. The coatings were transparent to visible light and sufficiently thin to avoid degrading the reflective properties of the titania and yet, ideally, thick enough to encapsulate the particles and insulate the pigment surface from an organic matrix. One advantage of the use of a mixture of acidic and basic oxides lay in their ready coprecipitation and coating on to the pigment surface. Pure silica or alumina, when rapidly precipitated, tend to form separate colloidal phases which may not be adsorbed evenly or readily on the pigment surfaces [91].

The reasons why the alumina–silica coatings are so effective in reducing the photodegradative effects of titanias are not clear. The stabilization probably results from a combination of properties of the coatings. Some oxide coatings have been found to be singularly ineffective in reducing the photoactivity, so it is likely that the silica–aluminas provide more than just a physical barrier [20]. The coatings are generally permeable to water and do not prevent the photoadsorption of oxygen. However, they probably do limit the rate of diffusion of the oxidizing species from the pigment surface and may also promote their destruction [73].

The early silica–alumina coatings contained alumina-to-silica ratios of 1:1 to 2:1. They were usually applied by the addition of an aluminum sulfate solution to a dispersion of the titania in aqueous alkaline silicate at 60 to 80°C, the final pH of the mixture being adjusted to a value of 5 to 7; the batch coating times were 30 to 60 min. Silica–alumina coatings were also applied by the addition of a silicate solution to a slurry of the pigment in an aluminum salt solution. The first method was generally preferred as the alkaline silicate was a good dispersant for titanias and the product usually had better compatibility with alkyd resins and other synthetic polymers. The coated pigments prepared by either method need to be thoroughly washed to remove the byproduct salts prior to drying. Their removal can be assisted by the use of coating formulations containing counterions that form volatile salts capable of being eliminated during subsequent drying or calcination processes [92]. The relationship between the properties of the coatings and the silica–alumina component ratios will be discussed in Chapter 3.

Although silica, alumina, and titania were soon established as the standard components of the coatings, other coprecipitating species were included with, or substituted for these oxides [3]. The other species included phosphates or borates as the acidic oxide components and barium, magnesium, zinc, or zirconium as the basic oxide components. The addition of small amounts of variable valence ions, for example, cerium, chromium, manganese, cobalt, etc., to the alumina–titania–silica coatings has been variously claimed to improve the pigment brightness or the weather resistance of the pigmented materials. One frequently cited example is the use of manganese phosphate coatings to prevent the photodegradation of anatase-pigmented nylon fibers [93].

The inclusion of titanium or zirconium salts in the coating formulation results in the deposition of a titania– or zirconia–silica-enriched phase on the titania surface during the early stages of the precipitation process [94]. This is considered to improve the adhesion between the titania surface and the overlying coating. A typical titania–alumina–silica coating might contain these oxides in the respective ratios of 3:4:2. Zirconia has low photocatalytic activity and its adsorption on the surface of the titania lattice would help to mask the photoactive TiOH groups. The transition metal additives are believed to improve the chalk resistance of the coated pigment, either by inhibiting the formation of the various oxidizing species or by promoting their destruction as they diffuse from the pigment surface.

The alumina–silica coatings also alter the electrochemical behavior of the pigment surface from that of a rutile or anatase to that of an amorphous or pseudocrystalline aluminosilicate. The electrochemistry of the coated pigments is dependent not only on the relative proportions of silica, alumina, and other components of the coating, but on the concentrations of the soluble ionic precursors, the order of addition of the reagents, pH control, batch timing and temperature, washing procedures, and drying conditions used in their preparation.

These variables offer scope for optimizing coatings to suit particular applications, as they can affect not only the electrochemistry and photoactivity of the coated titanias, but also the porosity of the coatings, their adsorptive properties, and the brightness and hiding power of the pigments. The coated pigments may be further modified by secondary treatment with organic compounds, for example, long-chain alcohols or amines, polyols, or silicones, to reduce agglomeration during storage, to improve their compatibility with organic substrates, and to reduce absorption of water. Details of the coating techniques and formulations actually used in commercial pigment production are closely guarded trade secrets and the ultimate properties of the pigments can be very dependent on the attention to these details. Unfortunately, there are few guiding principles for the development of coatings, other than those contained in the "art," and scientific studies on the formation and properties of mixed oxide precipitates, such as those described in reference 91, are a recent development.

The coatings of mixed oxides seldom have a uniform composition, but may consist of a mixture of different phases [95,96]. For example, commercial alumina–silica coatings prepared by the classical methods, have been shown by electron microscopy to consist of silica-rich microparticles in porous alumina or silica–alumina matrices, the outer layers of which, being deposited last, tend to be alumina enriched.

Techniques have been developed for the application of dense, coherent, nonporous coatings which have good adhesion to the titania and which greatly reduce the diffusion of oxidants. These durable, densely coated pigments are often given an outer coating of a porous silica, alumina, or phosphate mixture to provide optimum dispersion qualities for particular applications. Multiple applications of a conventional coating may also be employed to ensure complete coverage of the pigment surface.

Dense, nonporous coatings of alumina or of an alumina–silica mixture can be applied by vapor phase treatment of the pigments, for example, by heating with aluminum chloride [97] or with a mixture of aluminum, titanium, and silicon chlorides in a fluidized bed at high temperatures [98]. Alumina and silica may also be deposited on the surface of chloride process pigments by introducing the respective chlorides into the reaction mixture [99]. Although low concentrations of Si^{4+} and Al^{3+} ions can be accommodated in the TiO_2 lattice, alumina and silica have low solubilites in titania. The addition of larger amounts of their precursors to the burner feed causes the oxides to concentrate in the surface zones of the particles.

Most of the coatings described in the literature are dried at temperatures below 200°C, and this preserves the coating's porous, hydrous oxide structure. Denser coatings having reduced porosity may be obtained by hydrothermal aging of the initially porous coatings, but some other methods are of interest. In several of these methods, the silica–alumina-coated pigments are calcined at 500 to 600°C to dehydrate and consolidate the coatings, thereby reducing their porosity and absorption of water [100,101]. The calcined coated pigment may then be given a second, conventional porous coat of silica–alumina to provide good compatibility and dispersion properties. Calcination has been used to improve the properties of other silicate coatings, including magnesium silicate [102].

Dense coatings of silica [103,104] and alumina [105,106] may also be formed by using a slow addition (or the *in situ* formation) of a precipitating agent to a dispersion of titania in hot aqueous silicate or aluminate solutions. These methods avoid the formation of high local concentrations of reacting ions which can occur in the older, conventional coating processes, and which may cause the oxides to precipitate as flocs in the dispersing media and on the pigment surface, thereby resulting in the formation of uneven and nonhomogeneous coatings. In contrast, the slow precipitation technique allows the nucleation and growth of the oxide phases on the titania surface, a process which favors the formation of a dense, homogeneous, uniform coating. The preferred methods include the precipitation of oxides by the addition of carbon dioxide to dispersions containing alkali silicates and/or aluminates and the precipitation of alumina from hot $AlCl_3$ solutions by hydroxide ions formed by the slow hydrolysis of urea. Basic, microcrystalline "boehmite" coatings can also be deposited on titanias from hot aluminate solutions at pH 9 [107]. Such coatings are reported to be apparently nonporous in adsorption studies, but still permeable to molecular water [108].

The silica–alumina coatings normally comprise 3 to 5% of the weight of the pigment, lesser amounts usually being insufficient to provide adequate surface coverage. Thicker, highly porous coatings equal to 10% of the weight of the titania are used in special low-binder applications where the optimum interparticle distance for maximum reflectivity is governed by the coating thickness rather than by the dilution in a matrix. Coatings comprising more than 12% of the weight of the titania may degrade the mechanical and optical properties of the pigment. These thick coatings also tend to separate from the pigment surface during the deposition process.

The weight ratios of alumina-to-silica in most conventional coatings and their approximate Al:Si atomic ratios range from 1:1 to 4:1. These proportions have been found empirically to provide the best balance of coating properties, an alumina-to-silica ratio of 2:1 generally being used as a standard composition. Silica-rich coatings have been proposed for special applications, for example, for improving the dispersion of titania in vinyl polymers or for use in aqueous systems while, according to some hypotheses [20,109], coatings having an alumina-to-silica ratio of 1:3 (Al:Si $\simeq$ 1:3) would ideally provide the greatest potential acidity and consequent reduction in photolytic activity. Patent examples [109] cite an optimum alumina-to-silica ratio of 1:2.5. This ratio also provides the maximum acidity for homogeneous silica–alumina compositions (see p. 92).

Coated titanias specifically intended for use in aqueous systems are often prepared using a final addition of silicate or phosphate to form a silicate- or phosphate-rich surface layer [110]. This treatment can mask the underlying reactive alumina-enriched layer [111] and can form coatings that provide good dispersion in aqueous media without the need for additional dispersants. However, pigments with silica-rich coatings have been reported to cause the instability and thickening of some vinyl emulsions, although high-silica coatings were found to result in the least thickening on storage of a wide variety of pigmented anionic polymer emulsions [112]. Titanias with high-silica coatings have also been recommended for use in poly(vinyl chloride) compositions [113]. Alumina-rich coatings may also cause thickening and gelation on storage of aqueous poly(vinyl acetate) emulsions, particularly those containing copolymerized acid [114]. This instability is believed to be due to the gradual conversion of the amorphous alumina to the more basic pseudo-boehmite, which can form polymeric salts with the dispersion-stabilizing acylate anions. Titanias having coatings with slightly to moderately enriched alumina content have been found to be the most suitable for use in electrodeposition paint systems [115]. The choice of the resin component, for example, an anionic charged polymer solution or emulsion in an alkaline medium, may have a greater influence on the deposition of the pigmented paint film than the pigment itself [116].

Conventional alumina–silica coatings may have a weakly basic, neutral, or weakly acidic character. However, strongly acidic pigments can be obtained by acid treatment of high-alumina coatings prior to drying [117,118]. These coatings are suitable for the polymerization of susceptible vinylic monomers and have a high affinity for the amines used in some secondary treatments. Calcined silica–alumina coatings, like the natural and calcined aluminosilicates described in Chapter 1, may also show strongly acidic properties in the absence of adsorbed water. In contrast, neutral, inherently oleophilic coatings can be formed using uncalcined magnesia–silica compositions; these coatings resemble in properties the natural magnesium silicate mineral, talc, and are claimed to provide improved dispersibility in organic media [119,120].

Most contemporary coated titanias use silica, alumina, or silica–alumina as the principal ingredients of the coating. However, recent patents have claimed that coatings of other colorless oxides, for example, stannic oxide [121], can give improved chalk resistance to pigmented films.

Although the various inorganic coatings discussed have been used mainly to improve dispersibility and to reduce the photodegradation of the pigmented composites, they can also prevent discoloration during the melt compounding of polyolefin or vinyl formulations. These formulations often contain phenolic antioxidants, ultraviolet screens, or phenate stabilizers that can react with uncoated titanias at the compounding temperatures of 150 to 200°C to form highly colored oxidation products [122].

REFERENCES

1   W. A. Kampfer, Titanium Dioxide, in T. C. Patton, Ed., *Pigment Handbook*, Vol. 1, Wiley-Interscience, New York, 1973.

2   L. Firing, *Chem. Ind.*, 250 (1974).

3   J. Barksdale, *Titanium*, 2nd ed., Ronald, New York, 1966.

4   L. H. Bixner (du Pont), U.S. Patent 3,528,838 (September 15, 1970).

5   J. J. Libera and C. R. Trompier (National Lead), U.S. Patent 3,832,206 (August 27, 1974).

6   D. M. Bell, *Chem. Ind.*, 547 (1977).

7   R. S. Darby and J. Leighton, Titanium Dioxide Pigments, in *Modern Inorganic Chemicals Industry*, Special Publication No. 31, Chemical Society, London, 1977.

8   G. H. J. Neville, Titanium Compounds (Inorganic), in *Kirk-Othmer Encyclopedia of Chemical Technology*, Vol. 20, Wiley, New York, 1969.

9   *Titanium Dioxide Production*, T.T.I.S. Publication No. 3, Translation and Technical Information Service, Braintree, 1969.

10   W. A. Kampfer and F. Stieg, Jr., *Color Eng.*, 35 (Jul/Aug, 1967).

11   G. Kaempf, *J. Coating Technol.*, 51 (655), 51 (Aug, 1979).

12   T. J. Wiseman, Inorganic White Pigments in G. D. Parfitt and K. S. W. Sing, Eds., *Characterisation of Powder Surfaces*, Academic, London, 1976.

13   H. S. Ritter, Surface Properties of Titanium Dioxide Pigments, in T. C. Patton, Ed., *Pigment Handbook*, Vol. 3, Wiley–Interscience, New York, 1974.

14   P. Jones and J. A. Hockey, *Trans. Farad. Soc.*, 67, 2679 (1971).

15   M. Primet, P. Pichat, and M-V. Mathieu, *J. Phys. Chem.*, 75, 1216 (1971).

16   H. P. Boehm, *Adv. Catalysis*, 16, 179 (1966).

17   G. Munuera, F. Moreno, and F. Gonzalez, *Proc. 7th Int. Symp. React. Solids*, Chapman and Hall, London, 1973, p. 681.

18   R. D. Murley, *J. Oil Colour Chem. Assoc.*, 45, 16 (1962).

19   H. Rechmann, *Ber. Bunsenges. Phys. Chem.*, 71, 277 (1967).

20   W. Hughes, *10th FATIPEC Congress Book*, Verlag Chemie, 1970, p. 67.

21   G. D. Parfitt, *Prog. Surf. Membr. Sci.*, 11, 181 (1976).

22   R. E. Day, in *Progress in Organic Coatings*, Vol. 2, Elsevier Sequoia, Lausanne, 1973/4, p. 269.

23   M. Herrmann and H. P. Boehm, *Z. Anorg. Allg. Chem.*, 368, 73 (1969).

24   G. Munuera and F. S. Stone, *Discuss. Farad. Soc.*, 52, 205 (1971).

25   M. J. Jaycock and J. C. R. Waldsax, *J. Chem. Soc., Farad. I*, 70, 1501 (1974).

26   A. C. Zettlemoyer and E. McCafferty, *Croat. Chem. Acta*, 45, 173 (1973).

27   P. Jones and J. A. Hockey, *J. Chem. Soc., Farad. I*, 68, 907 (1972).

28   V. S. Ienko and A. V. Uvarov, *Colloid J. USSR*, 37, 1051 (1975).

29   H. P. Boehm, *Discuss. Farad. Soc.*, 52, 264 (1971).

30   D. M. Griffiths and C. H. Rochester, *J. Chem. Soc., Farad. I*, 73, 1511 (1977).

31   G. Munuera, V. Rives-Arnau, and A. Saucedo, *J. Chem. Soc., Farad. I*, 75, 736 (1979).

32   P. Jackson and G. D. Parfitt, *J. Chem. Soc., Farad. I*, 68, 1443 (1972).

33   L. V. Denisova, *Colloid J. USSR*, 39, 293 (1976).

34   H. P. Boehm and M. Herrmann, *Z. Anorg. Allg. Chem.*, 352, 156 (1967).

35   P. Jackson and G. D. Parfitt, *J. Chem. Soc., Farad. I*, 68, 896 (1972).

36   J. Cornejo, J. Steinle, and H. P. Boehm, *Z. Naturforsch.*, 33B, 1238 (1978).

37   M. Primet, P. Pichat, and M-V. Mathieu, *J. Phys. Chem.*, 75, 1221 (1971).

38   G. D. Parfitt, J. Ramsbotham, and C. H. Rochester, *Trans. Farad. Soc.*, 67, 841 (1971).

39   H. P. Boehm, *Kolloid Z.*, 227, 17 (1968).

40   M. Primet, M. Che, C. Naccache, M-V. Mathieu, and B. Imelik, *J. Phys. Chim.*, 67, 1629 (1970).

41   A. L. Smith, in G. D. Parfitt, Ed., *Dispersion of Powders in Liquids*, Elsevier, New York, 1969, p. 39.

42   G. L. McLeod, *Photo. Sci. Eng.*, 13, 93 (1969).

43   R. I. Bickley, in M. W. Roberts and J. M. Thomas, Eds., *Chemical Physics of Solids and Surfaces*, Specialist Periodical Report No. 7, Chemical Society, London, 1978, p. 119.

44   A. L. Companion and R. E. Wyatt, *J. Phys. Chem. Solids*, **24**, 1025 (1963).

45   R. R. Addiss, Jr. and F. G. Watkins, *Photo. Sci. Eng.*, **13**, 111 (1969).

46   A. R. Gonzalez–Elipe, G. Munuera, and J. Soria, *J. Chem. Soc., Farad. I*, **75**, 748 (1979).

47   P. Meriaudeau and J. C. Vedrine, *J. Chem. Soc., Farad. II*, **72**, 472 (1976).

48   P. C. Gravelle, F. Juillet, P. Meriaudeau, and S. J. Teichner, *Discuss. Farad. Soc.*, **52**, 140 (1971).

49   M. Formenti, H. Courbon, F. Juillet, A. Lissatchenko, J. R. Martin, P. Meriaudeau, and S. J. Teichner, *J. Vac. Sci. Tech.*, **9**, 947 (1972).

50   J. H. Lunsford, *Catalysis Rev.*, **8**, 135 (1973).

51   R. D. Iyengar, M. Codell, J. S. Kava, and J. Turkevich, *J. Amer. Chem. Soc.*, **88**, 5055 (1966).

52   R. D. Iyengar and R. Kellermann, *Z. Phys. Chem.*, **64**, 345 (1969).

53   W. R. McLean and M. Richie, *J. Appl. Chem.*, **15**, 452 (1965).

54   E. N. Figurovskaya, V. F. Kiselev, and F. F. Volkenshtein, *Proc. Acad. Sci. USSR, Dokl. Phys. Chem.*, **161**, 339 (1965).

55   R. I. Bickley and F. S. Stone, *J. Catalysis*, **31**, 389 (1973).

56   A. H. Boonstra and C. A. H. A. Mutsaers, *J. Phys. Chem.*, **79**, 1694 (1975).

57   U. Kaluza and H. P. Boehm, *J. Catalysis*, **22**, 347 (1971).

58   H. van Damme and W. K. Hall, *J. Amer. Chem. Soc.*, **101**, 1501 (1979).

59   C. D. Jaeger and A. J. Bard, *J. Phys. Chem.*, **83**, 3146 (1979).

60   G. A. Korsunovskii, *Russ. J. Phys. Chem.*, **34**, 241 (1960).

61   T. Freund and W. P. Gomes, *Catalysis Rev.*, **3**, 1 (1969).

62   S. P. Pappas and R. M. Fischer, *J. Paint Technol.*, **46**, (599), 65 (Dec. 1974).

63   A. H. Boonstra and C. A. H. A. Mutsaers, *J. Phys. Chem.*, **79**, 1940 (1975).

64   G. N. Schrauzer and T. D. Guth, *J. Amer. Chem. Soc.*, **99**, 7189 (1977).

65   G. Munuera, *J. Chem. Soc., Farad. I*, **76**, 1535 (1980).

66   B. Ranby and J. F. Rabek, *Photodegradation, Photo-oxidation, and Photostabilization of Polymers*, Wiley–Interscience, New York, 1975.

67   A. V. Pamfilov, Ya. S. Mazurkevich, and E. P. Pakhomova, *Kinet. Catal.*, **10**, 752 (1969).

68   S. N. Frank and A. J. Bard, *J. Amer. Chem. Soc.*, **99**, 303 (1977).

69   J-M. Hermann and P. Pichat, *J. Chem. Soc., Farad. I*, **76**, 1138 (1980).

70   B. Kraeulter and A. J. Bard, *J. Amer. Chem. Soc.*, **99**, 7729 (1978).

71   H. G. Voelz, G. Kaempf, and H. G. Fitzky, *Farbe Lack*, **78**, 1037 (1972).

72   H. G. Voelz, G. Kaempf, and A. Klaerin, *Farbe Lack*, **82**, 805 (1976).

73   H. G. Voelz, G. Kaempf, and H. G. Fitzky, in *Progress in Organic Coatings*, Vol. 2, Elsevier Sequoia, Lausanne, 1973/4, p. 223.

74   R. I. Bickley, G. Munuera, and F. S. Stone, *J. Catalysis*, **31**, 398 (1973).

75   R. I. Bickley and R. K. H. Jayanty, *Farad. Discuss. Chem. Soc.*, **58**, 194 (1974).

76   A. H. Boonstra and C. A. H. A. Mutsaers, *J. Phys. Chem.*, **79**, 2025 (1975).

77   W. A. Weyl and T. Foerland, *Ind. Eng. Chem.*, **42**, 257 (1950).

78   S. Torlaschi, U. Barcucci, and G. Zorzella, *Ind. Vernice*, **24** (9), 10 (1970); *10th FATIPEC Congress Book*, Verlag Chemie, 1970, p. 25.

79   D. Zwingel, *Solid State Comm.*, **20**, 397 (1976).

80   R. B. Cundall, B. Hulme, R. Rudham, and M. S. Salim, *J. Oil Colour Chem. Assoc.*, **61**, 351 (1978).

81   A. E. Jacobsen, *Ind. Chem.*, **41**, 523 (1949).

82   J. Taylor, *J. Oil Colour Chem. Assoc.*, **38**, 233 (1955).

83   A. W. Evans and R. D. Murley, *Paint Technol.*, **26** (10), 16 (Oct. 1962).

84   Yu. Ya. Bobyrenko, A. B. Zholnin, and V. K. Konovalova, *Russ. J. Phys. Chem.*, **46**, 749 (1972).

85   J. Shanker, V. K. Jain, and S. C. Agrawal, *J. Chem. Phys.*, **72**, 4458 (1980).

86   Y. Oinishi and T. Hamamura, *Bull. Chem. Soc. Jap.*, **43**, 996 (1970).

87   N. S. Allen, D. J. Bullen, and J. F. McKellar, *Chem. Ind.*, 797 (1977).

88   N. S. Allen, J. F. McKellar, and D. G. M. Wood, *J. Polym. Sci., Polym. Chem. Ed.*, **13**, 2319 (1975).

89   N. S. Allen, J. F. McKellar, G. O. Phillips, and D. G. M. Wood, *J. Polym. Sci., Polym. Lett. Ed.*, **12**, 241 (1974).

90   British Titan Products, British Patent 545,604 (June 4, 1942).

91   P. B. Howard and G. D. Parfitt, *Croat. Chim. Acta*, **50**, 15 (1977).

92   D. P. Fields (Du Pont), U.S. Patent 3,418,147 (December 24, 1968).

93   O. B. Edgar, C. F. Richie, D. R. Lawrance, W. Hughes, and G. Lederer (I.C.I. and British Titan), British Patent 1,162,799 (1969).

94   T. J. Wiseman and P. B. Howard (British Titan), Ger. Offen. 2,313,544 (September 27, 1973).

95   R. M. Cornell, A. M. Posner, and J. P. Quirk, *J. Chem. Tech. Biotechnol.*, **30**, 187 (1980).

96   G. Kaempf and H-G. Voelz, *Farbe Lack*, **74**, 37 (1968).

97   A. Pechukas (Pittsburgh Plate Glass), U.S. Patent 2,441,225 (January 18, 1944).

98   British Titan, French Patent 1,157,699 (August 29, 1969).

99   British Titan, French Demande 2,008,555 (January 23, 1970).

100   W. T. Siuta (Titangesellschaft), U.S. Patent 3,409,501; French Patent 1,504,211 (December 1, 1967).

101   National Lead, British Patent 1,164,849 (October 24, 1969).

102   Fabriques de Produits Chimiques de Thann et de Mulhouse, British Patent 1,116,590 (June 6, 1968).

103   A. J. Werner (Du Pont), U.S. Patent 3,437,502; French Patent 1,508,441 (January 5, 1968).

104   R. K. Iler (Du Pont), U.S. Patent 2,885,366 (May 5, 1959).

105   H. Lott, Jr. and R. H. Walsh (P.P.G. Industries), U.S. Patent 3,545,944 (December 8, 1970).

106   T. J. Wiseman (British Titan), British Patent 1,134,249 (November 20, 1968).

107   T. F. Swank (Cabot Corp.), U.S. Patent 3,770,470 (November 6, 1973).

108   A. C. Zettlemoyer, R. D. Iyengar, and P. Scheidt, *J. Colloid Interface Sci.*, **22**, 172 (1966).

109   G. M. Sheehan, G. L. Roberts, Jr., and P. M. Dupree (American Cyanamid), British Patent 1,202,540 (1970); reported in Ref. [12]. Also related U.S. Patent 3,658,566 (August 7, 1969).

110   Cabot Corp., Belgian Patent 705,375 (March 1, 1968).

111   T. L. O'Connor and J. P. Bourgault (Cabot Corp.), U.S. Patent 3,449,271 (June 10, 1969).

112   F. D. Robinson and B. J. Tear, *Polym. Paint Col. J.*, 390 (1972).

113   G. M. Sheehan, G. L. Roberts, Jr., and P. M. Dupree (American Cyanamid), U.S. Patent 3,769,255 (October 30, 1973).

114   F. D. Robinson, *J. Oil Colour Chem. Assoc.*, **53**, 691 (1970).

115   F. D. Robinson and B. J. Tear, *J. Oil Colour Chem. Assoc.*, **53**, 265 (1970).

116   T. Entwistle, *J. Oil Colour Chem. Assoc.*, **55**, 480 (1972).

117   D. H. Solomon, D. G. Hawthorne, J. H. Hodgkin, and J. D. Swift (C.S.I.R.O.), U.S. Patent 3,957,526; Ger. Offen. 2,206,371 (September 2, 1972).

118   J. H. Hodgkin and D. H. Solomon, *J. Macromol. Sci. Chem.*, **A8**, 621 (1974).

119   D. H. Solomon and J. H. Hodgkin (C.S.I.R.O), Sth. Afr. Pat. 72/8955 (October 18, 1974).

120   J. H. Hodgkin and D. H. Solomon, *J. Macromol. Sci. Chem.*, **A8**, 635 (1969).

121   M. A. Claridge and P. D. Howden (La Porte), British Patent 1,365,999 (September 4, 1974).

122   R. Dyer and H. W. Boch, *Soc. Plast. Eng. Tech. Pap.*, **17**, 175 (1971).

3

INORGANIC PIGMENTS

The clay minerals and titanias described in the previous chapters represent only a few of the many types of inorganic fillers and pigments used in plastics, elastomer, and surface coatings industries. Other important white pigments and fillers, which will be the subjects of this chapter, include various forms of silica, alumina, calcium carbonate, and zinc oxide, while colored inorganic pigments covered include the various chromates and iron oxides. It is not practical to describe the many other inorganic pigments and extenders. Readers are referred to comprehensive treatises such as the *Pigment Handbook,* edited by T. C. Patton [1], for information on these materials. The materials described in this chapter can serve as models for the reactive surface chemistry of many of the other oxides and salts used as pigments. Details of the industrial applications of other fillers and reinforcements used in plastics and elastomers can be found in technological manuals such as reference 2.

3.1 SILICA

Silica and its derivatives are used as reinforcing fillers in plastics and elastomerics, as desiccants, and as thickening agents in a variety of organic liquid media. Silica is available in three widely different forms: small particle, usually highly porous forms obtained by precipitation and drying of hydrated silica gels or by the hydrolysis at high temperatures of silicon tetrachloride; nonporous particulates obtained by crushing sand; and silica or silicate glass fibers.

Silica gel is formed by acidification of aqueous silicate solutions. The nascent silicic acid undergoes rapid polymerization to form a sol, which consists of cross-linked, hydrated silica particles having diameters of the order of 1 nm. Condensation between particles occurs as the sol ages, forming a macroscopic gel and causing the mixture to solidify. Further aging of the gel results in spontaneous shrinkage and extrusion of the trapped liquid. This shrinkage can reach a metastable state, forming a product that can be titurated to a fine, apparently dry powder; this may contain more than 60% water trapped in the silica network and in the spaces between the partially coalesced primary (sol) particles. These hydrogels can be dehydrated at low temperatures to form highly porous xerogels, or at higher temperatures to form desiccant gels. The specific surface area of the gels can range from 400 to 800 m^2/g, with pore volumes of 0.9 to 1.2 ml/g. The water in the hydrogels can be displaced by organic liquids. This provides a useful means of storing liquid reagents in a "solid" form. The porosity of the xerogels is usually well defined and is dependent on the methods of manufacture of the hydrogel and on the dehydration conditions; these aspects have been reviewed elsewhere [3,4]. The hydrogels and xerogels usually contain residual salts or metal ions, so pure gels for scientific study are prepared by the spontaneous hydrolysis of silicate esters in water instead.

Hydrated silicas can be obtained as dense precipitates by the use of low concentrations of silicate and near-neutral precipitation conditions which allow the nucleation and slow growth of compact particles or aggregates. The precipitated silicas consist of aggregates of primary particles that have diameters of 1 to 2 nm, the average diameter of the aggregates being 10 to 20 μm. The precipitated silicas are porous, having specific surface areas of 100 to 300 m^2/g [3], although the pore volume is probably much less than that of the xerogels. The precipitated silicas are frequently used as reinforcing agents in plastics and elastomers.

The pyrogenic or fumed silicas are prepared by flame-hydrolysis of silicon tetrachloride or by volatilization of sand in a plasma arc. These materials contain little processing residues and are frequently used as thickening agents for organic media. The primary particles contained in the commercial materials have diameters of 10 to 20 nm, although these readily aggregate to form micron-sized agglomerates. In liquid media, the primary particles can form three-dimensional chain structures. These chain structures are responsible for the thixotropic character and gellation of silica dispersions.

Apart from considerations of particle size and porosity, the surfaces of the fine-particle silicas are similar and equivalent to that of a hydrated silica bearing surface silanol groups. These may consist of isolated silanols or various types of paired and hydrogen-bonded silanols that can be identified by means of infrared spectroscopy, by their behavior on dehydration, or on chemical reaction with boron trichloride [5]. The adsorbed water can be eliminated by heating the hydrated gels at temperatures below 200°C.

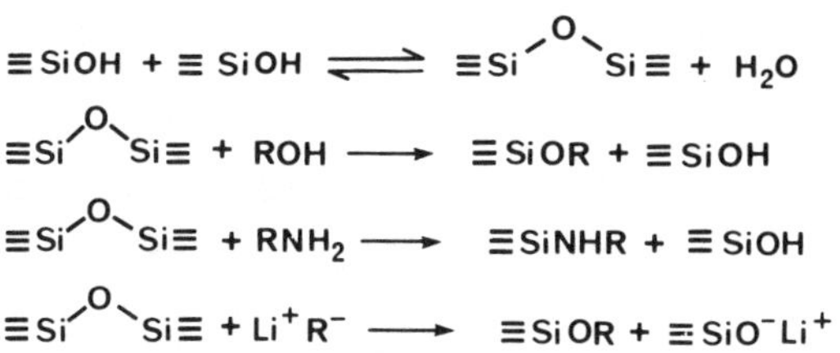

$$\equiv SiOH + \equiv SiOH \rightleftharpoons \equiv Si{\overset{O}{\diagup}}{\diagdown}Si \equiv + H_2O$$

$$\equiv Si{\overset{O}{\diagup}}{\diagdown}Si \equiv + ROH \longrightarrow \equiv SiOR + \equiv SiOH$$

$$\equiv Si{\overset{O}{\diagup}}{\diagdown}Si \equiv + RNH_2 \longrightarrow \equiv SiNHR + \equiv SiOH$$

$$\equiv Si{\overset{O}{\diagup}}{\diagdown}Si \equiv + Li^+R^- \longrightarrow \equiv SiOR + \equiv SiO^-Li^+$$

Figure 3.1

This can also result in the condensation of many of the paired silanols. Other, less favorably oriented pairs of silanols are eliminated at higher temperatures, with the formation of strained, highly reactive siloxane groups (see Figure 3.1). Isolated silanol groups are resistant to dehydration, and their elimination may require calcination at temperatures over 600°C.

Dehydrated silicas slowly rehydrate on exposure to moisture by hydrolysis of the surface siloxane bonds, although the dehydration of silica gels is partially irreversible. The strained siloxane groups can also react with alcohols and with amines to form alkoxides (esters) and silylamines [6,7] or with lithium aryls to form phenoxysilicates [8].

Fine-particle silicas are extensively used as adsorbents for the chromatographic separation of organic compounds. The adsorption process on silicas that have not been modified by silylation or esterification largely depends on formation of hydrogen bonds between surface silanol groups and the adsorbed molecules (see Chapter 4). The surface silanol groups differ in activity, the most reactive comprising adjacent (separated by less than 5 Å) hydrogen-bonded silanol groups [9]; these paired silanols are sufficiently acidic to form hydrogen bonds with aromatic hydrocarbons. Gels having specific adsorptive properties can be prepared by precipitation in the presence of an organic activator, which is adsorbed on the nascent polysilicic acid sol, and "prints" a particular spatial arrangement of silanol groups; this allows the subsequent preferential adsorption of the activator from otherwise intractable mixtures. Silica printed with dimethylaniline, for example, can be used for the separation of this amine from a mixture with diethylaniline [10].

Crystalline polysilicic acids can be prepared by controlled hydrolysis of silicate solutions or by leaching of certain minerals. These hydrated silicas can consist of layers of two-dimensional sheets which, like the layer silicates described in Chapter 1, can intercalate water and other polar molecules including long-chain alcohols and amines, which also may form ordered arrays in the interlamellar spaces [11].

The silanol groups of the hydrated silicas can be deprotonated by strong inorganic bases. Pure hydrated silicas are only weakly acidic and cannot protonate weak bases such as pyridine; they can protonate ammonia or alkylamines, but the ionic adducts can dissociate on evacuation or on heating at 100°C. However, the presence of coprecipitated or adsorbed metal ions, particularly those of Al^{3+}, can polarize the adjacent silanol groups and enable them to act as strong Bronsted acids, capable of forming stable salts with ammonia alkylamines. Alternatively, the presence of Al^{3+} ions in a highly condensed silica can result in the formation of acidic bridged-hydroxylic species. These Bronsted acid species are catalytic centers for the dehydration of alcohols and isomerization of alkenes. Pure silica prepared by hydrolysis of ethyl silicate does not catalyze these reactions, but activity can be induced by the presence of 0.01% of Al^{3+} ions [12]; the activity of commercial silicas is probably due to trace impurities of this nature.

The surface silanols can be esterified by reaction with diazomethane or by condensation with alcohols, and can undergo metathetical reactions with thionyl chloride, acyl halides, and metal alkyls; these reactions have been outlined in Chapter 1 (see Figure 1.22) [13–15]. The surface silanols may also be utilized in the bonding of the organic surface modifying reagents described in Chapter 4 and are responsible for the hydrophilic character of silica; anhydrous silica, which lacks these groups, is hydrophobic.

The surfaces of the nonporous silica particles and fibers have reactive silanol groups that are broadly similiar to those of the fine-particle silicas. The surface siloxane groups of the massive silicas are also susceptible to hydrolysis by adsorbed water. This hydrolysis can weaken the bonding between organic matrices and the silica surface, and can seriously affect the reinforcement of plastics by silica fibers. The problem is particularly serious for materials reinforced with silicate glass fibers; these materials are more readily hydrolyzed, yielding soluble silica salts, which, through osmosis, can accelerate the imbibition of water and failure of the composite at the resin–glass interfaces. The alkaline silicate salts can also cause the hydrolysis of the vinyl acetate polymers that are frequently used to size the fibers. This reaction results in the formation of acetic acid and a polyvinyl alcohol interfacial layer that is permeable and swollen by water, thereby compounding the water sensitivity of the composites. In some cases involving fiberglass-reinforced polyester boat hulls and water tanks, this osmotic imbibition has resulted in the delamination of the composite and massive "blistering" of the structures. The water sensitivity of fiberglass-reinforced composites can be reduced by surface modifications that render the fiber surfaces hydrophobic and hydrolysis-resistant, or by pretreatment of the fibers with multifunctional silanes that can form a nondesorbable polymeric interfacial layer with the matrix polymer in the composite [16].

3.2 ALUMINA

Alumina, hydrated alumina, and aluminum hydroxides are used as fillers or flame retardants in plastics media and are important as components of inorganic coatings used for the surface modification of other particulates, most notably those of titania pigments. The term "alumina" is often loosely applied to crystalline or amorphous aluminum hydroxide and aluminum hydroxide oxide (hydrated alumina), as well as to aluminum oxide.

Aluminum hydroxide, $Al(OH)_3$, can form two crystalline phases, bayerite and gibbsite. Bayerite can be formed by precipitation from cold alkali aluminate solutions by addition of acid, while gibbsite is formed by hydrolysis of hot aluminum salt solutions. Bayerite is converted to the more stable gibbsite on aging, and both can be dehydrated to the aluminum hydroxide oxide, boehmite, $Al(OH)O$, on heating at 160°C; boehmite can also be formed directly from hot aluminate solutions. The mineral diaspore consists of another crystalline form of $Al(OH)O$ and only occurs in nature.

Commercial hydrated aluminas are usually mixtures of gibbsite and boehmite, although a pure microcrystalline boehmite is available (DuPont, Baymal® colloidal alumina). The aluminum hydroxides and boehmite are converted to various transitional, "activated" forms of alumina that are used as catalysts and desiccants [17–19] by heating above 400°C. Rehydration of the surface of a calcined alumina can be very rapid; as with the titanias and most other metallic oxides [20], the surfaces of the hydrated alumina fillers and commercial aluminas that have been exposed to moist air are covered with reactive hydroxyl groups. The adsorption of water by anhydrous alumina involves an initial coordination at surface Al^{3+} sites, followed by its dissociation into terminal and bridged hydroxyl groups; the process is similar to that described in Chapter 2 for the hydration of titanias.

The alumina surface hydroxyl populations are more complex than those of silica, but contain readily desorbed paired-hydroxyl groups and thermolysis-resistant isolated hydroxyl groups, as well as water molecules bound as ligands to surface Al^{3+} ions. These ligand waters, the bridged hydroxyls, and most of the terminal hydroxyl groups, can be eliminated on heating, leaving coordinately unsaturated Al^{3+} ions; these are Lewis acids and can form adsorption complexes with a wide variety of donor molecules.

The surface of a hydrated alumina is amphoteric and can form salts with acids and strong bases. The surface hydroxyls of crystalline boehmite, or of the partially crystalline "pseudoboehmite" which can form from hydrated alumina gels, are more basic than those of aluminum hydroxide. The Bronsted acidity of a pure hydrated alumina is very low, although alumina can readily adsorb and form complexes with polybasic anions such as sulfate or phosphate. These adsorbed anions can form strong Bronsted acidic centers, if they are present in their protonated forms, namely, bisulfate or hydrogen phosphate ions.

The hydrated alumina surfaces, like those of titania, contain approximately equal numbers of "acidic" hydroxyl groups, which can be esterified with alcohols or with diazomethane, and "basic" hydroxyls, which can

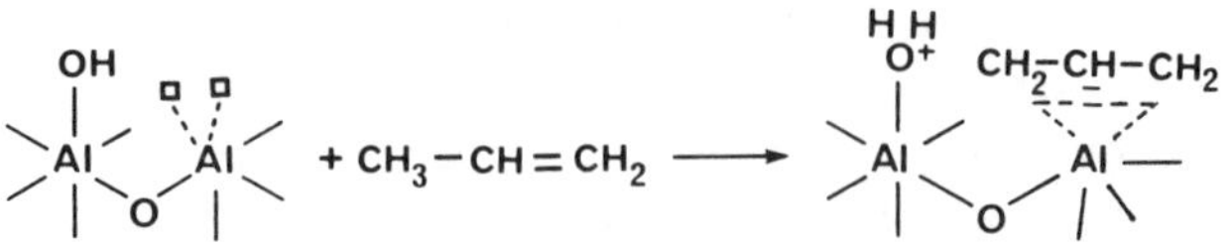

Figure 3.2 The probable mechanism for the dissociative adsorption of propene on an alumina surface.

react with carbon dioxide and nitrogen dioxide to form surface bicarbonates and nitrates [13,20,21]. The surface hydroxyls can also react with acyl chlorides and anhydrides to form surface acylate species; these reactions are similar to those of titania (see Figure 2.4). The relative stability of these surface species differs from those on silica, alumina surface alkoxide groups, for example, being more susceptible to hydrolysis than those of the alkoxysilicas.

Although pure hydrated alumina has almost negligible Bronsted acidity, the intrinsic Lewis acidity of dehydrated alumina surfaces and their ability to catalyze a wide range of organic reactions have long been recognized [22]. The anhydrous alumina surface has a number of potential acidic sites. The catalytic activity does not arise from the coordination deficient surface Al^{3+} ions alone, but from Al–O ionic pairs consisting of acidic Al^{3+} cations and adjacent strongly basic O^{2-} or HO^- anions [23]. Recent studies have indicated that the Lewis site probably involved in most organic reactions consists of an Al^{3+} ion that bears a basic, terminal hydroxyl group and that is linked, via the underlying lattice oxygens, to an adjacent coordination deficient Al^{3+} ion (Figure 3.2) [19]. The terminal hydroxyl group is a highly efficient electron donor and can accept protons from adsorbed organic molecules. Compounds such as alcohols, amines, and unsaturated hydrocarbons are chemisorbed on these active sites with dissociation to form alkoxides, alkamides, and allylic carbanions.

Anhydrous alumina can also catalyze electron-transfer reactions and can convert adsorbed polycyclic aromatic hydrocarbons to radical-cationic species, the electron acceptor probably being coadsorbed oxygen [24]. Some of the reactions catalyzed by active aluminas are discussed in greater detail in Chapter 5, and more comprehensive reviews are contained in references 19 and 25.

3.3 SILICA–ALUMINAS

Silica and alumina can coprecipitate to form a mixed oxide gel that has substantially different chemical properties from those of either silica or alumina. The most important difference is the development of strong

Bronsted acidity which can result from the incorporation of aluminum ions into the silica network. Acidic silica–aluminas are widely used as cracking catalysts in the petrochemical and other industries; these materials are silica-rich, containing silica-to-alumina ratios greater than 2:1, and are calcined to remove the bulk of the hydroxyl groups of the original gel structures.

Silica-alumina compositions are also used as surface coatings on a variety of pigments, including titanias, to improve their dispersibility in various media and, in some cases, to insulate a reactive pigment surface from degradable organic matrices. Except for some special examples described in Section 2.5.1, the pigment coating compositions are usually alumina-rich, containing alumina-to-silica ratios greater than 1:1, and nonacidic; the pigment coatings are frequently not calcined and may largely retain a hydrous gel structure. The properties of the hydrated gels differ from those of the calcined materials and from those of the crystalline aluminosilicates discussed in Chapter 1. There is some evidence that the properties of relatively thin layers of an oxide gel deposited on a foreign oxide lattice may differ from those of the bulk material.

The strong Bronsted acidity of some silica–alumina compositions is in marked contrast to the lack of acidity of the pure alumina or silica gels. The acidic centers are bridged-hydroxyl groups formed by the protonation of tetrahedral aluminate anions held in the silicate network (Figure 3.3); the acidity is similar in origin to that of the decationized acidic zeolites described in Chapter 1. Silica gel containing adsorbed aluminum ions may also be acidic, and acidic mixed oxide gels can be formed from silica and magnesia [26], titania [27], and some transition metal oxides. In contrast, the adsorption of excess silica on magnesium hydroxide can result in the formation of a hydrated magnesium silicate having a talc-like 2:1-layer structure [26]; mixed hydrogels of silica (70%) and alumina (30%) can form montmorillonite-like products [38].

X-ray fluorescence studies [29,30] of "homogeneous" bulk alumina-silica compositions, prepared by the cohydrolysis of alkyl aluminate and silicate ester solutions, have shown that high-alumina compositions contain Al^{3+} ions in both octahedral and tetrahedral environments, the octahedral Al^{3+} ions being converted to tetrahedral coordinately unsaturated species on dehydration. A high-alumina gel particle prepared in this manner consists of an anionic silica–alumina core, surrounded by a cationic hydrated alumina shell [30]. The Al^{3+} ions in a high-silica composition, containing less than 30% alumina, are all in a tetrahedral environment that is stabilized by the surrounding silica tetrahedra (Figure 3.3). Theoretical studies [31] have confirmed that the Bronsted acidity of the silica–aluminas is most probably associated with aluminum ions in tetrahedral coordination with silicate ions and that the acidic species is a bridged-hydroxyl group.

The structure and acidity of a series of calcined "homogeneous" sili-

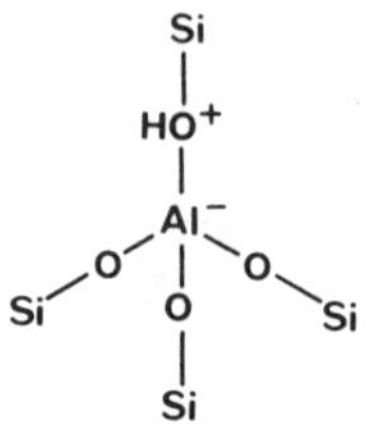

$$Si$$
$$|$$
$$HO^+$$
$$|$$
$$Al^-$$
$$O \quad | \quad O$$
$$Si \qquad O \qquad Si$$
$$|$$
$$Si$$

Figure 3.3   The Bronsted acid site of a silica–alumina.

ca–aluminas containing 0 to 100% silica have been examined using both infrared and photoelectron (xps) spectroscopy [32–34]. (The calcination employed would be unlikely to change the distribution of Al and Si atoms present in the hydrated gel.) These studies also indicated that the acid sites are associated with hydroxyls adjacent to tetrahedral Al^{3+} ions, which could be distinguished from those associated with octahedral Al^{3+} ions; however, only about 1 in 40 of these acidic sites is capable of adsorbing pyridine at 100°C. In this instance, xps could not distinguish amine adsorbed on Lewis sites from that adsorbed on Bronsted sites, although the evolution of Bronsted and Lewis acidity could be followed by infrared spectroscopy.

The Bronsted acidity of the silica–alumina develops as the alumina content is increased from 0 to 25%, but is decreased for samples containing greater than 50% alumina, due to the dilution of the acidic alumina–silica mixed oxide phase by an alumina phase; the Bronsted acidity vanishes for samples containing greater than 80% alumina. The acidity of samples containing less than 30% alumina is not affected by atmospheric moisture, indicating its Bronsted nature. The Lewis acidity of the dehydrated silica–aluminas increases with increase in the alumina content from 0 to 70% [33]. Other workers have also reported that Lewis and Bronsted acid sites may occur together in calcined silica–aluminas having alumina contents of 10 to 70%, but that only Lewis sites occur in materials containing more than 70% alumina; these Lewis acid sites are deactivated by water [34].

Only tetrahedral Al^{3+} ions are present in silica–alumina compositions containing less than 30% alumina content (Al:Si < 1:2). The xps response of the surfaces does not change appreciably for compositions containing 50 to 70% alumina (Al:Si, 1:1 to 2:1); however, the porosity of the gels increases with an increase in their alumina content. In gels containing more than 50% alumina, the silica–alumina phase is progressively diluted with alumina, which can appear as a separate, distinct phase when the alumina content exceeds 70% (Al:Si, > 2.5:1). The mixed-oxide phase is not observed in compositions containing more than 85% alumina [33,34].

There are two broad populations of hydroxyl groups in the "anhydrous" catalytic silica–aluminas; these are nonacidic or weakly acidic isolated silica hydroxyls and the acidic bridged-hydroxyls. Because the protons of the bridged-hydroxyls are required for charge neutralization, these groups are not readily eliminated on calcination and can rapidly reform when the calcined silica–alumina is exposed to water vapor. Partial dehydroxylation forms Lewis acid sites which can promote nearby residual bridged-hydroxyls to superacid centers, with Hammett acidity values, $H_0 < -12$ [35]. Most silica–alumina compositions contain regions rich in either silica or alumina, and these will have their own distinctive hydroxyl populations. Silica–aluminas equilibrated with the atmosphere will also contain physically adsorbed water. The infrared assignments for the various hydroxylic species present in the silica–alumina coatings have been summarized elsewhere [36–38].

The hydrated and "anhydrous" silica–alumina surface hydroxyl groups, like those of silica, alumina, and other hydrous oxides, may be esterified by treatment with alcohols or with diazomethane and can undergo similar metathetical reactions with thionyl and acyl halides or metal alkyls [13].

Mixed Oxide Coatings

The structures and chemical properties of the silica–alumina cracking catalysts have been extensively investigated and reviewed [e.g., 39,40]. This section will be largely devoted to mixed-oxide coatings, the most important of which are the compositions used for coating titania pigments.

The mixed-oxide coatings usually consist of a mixture of different phases, rather than a homogeneous composition. Consequently, a coated pigment may have chemical properties that differ significantly from those of other pigments having coatings with the same empirical composition [41]. An extensive variety of techniques have been used in the study of the chemical and physical properties of coatings to provide fine details of the structural and chemical nature of the effective pigment surface. Prior to the development of techniques, such as scanning electron microscopy and photoelectron spectroscopy, which can examine the composition of selected areas of the silica–alumina surface, the structure of the coatings had to be deduced using transmission electron microscopy for the estimation of coating form and homogeneity from a variety of chemical, electrochemical, spectroscopic, and adsorption studies. The probable structure of the coating would then be deduced from a comparison between its properties and those of various reference samples of silica, alumina, and alumina-silicas having known structures [38]. A recent comprehensive review of methods for the characterization of pigment surfaces has been published [42].

Published work on the chemistry and structures of pigment coatings and of mixed-oxide systems, in general, often contains indefinite or contradictory conclusions. These ambiguities are probably due to unreported differences in the preparation of nominally similar materials by different workers and to the often inhomogeneous nature of the coatings. This latter aspect has been demonstrated through the electron microscopic examination of a range of coated titanias [43]; by this method, the apparent texture and crystallinity of the visible surfaces could also be used to differentiate between the various regions of silica, alumina, and silica–alumina. While some of the conclusions of this study have been questioned, the electron micrographs did show that grossly inhomogeneous coatings could be produced by the use of improper coating techniques. The conventional coatings usually appear to be amorphous, and although crystalline phases may develop in bulk mixed-oxide systems, they are either absent or are often too poorly developed to be identified by x-ray diffraction examination; crystallinity may, however, be apparent in their electron microdiffraction patterns [44].

Silica coatings on titania surfaces, formed by acidification of heated titania–silicate dispersions, often consist of coagulated, porous layers of 4-nm-diameter silica particles which had been precipitated in solution and subsequently deposited on the pigment surface. The coagulation of the silica is accelerated by the presence of multivalent cations, such as Al^{3+} [45], and these ions may also limit the size of the colloidal particles that are initially formed in the liquid phase. The silica gel coatings, like the bulk gels, are amorphous. Coherent pseudocrystalline coatings of alumina can be produced by the hydrolysis of hot aluminate solutions; while the crystalline phases of hydrated aluminas are well established in bulk specimens, there is little evidence that they exist as such in pigment coatings. Hydrated alumina coatings that have been aged, or heated above 150°C, are often described as containing "pseudoboehmite"; like boehmite, they are appreciably more basic than amorphous hydrated aluminia.

The precipitation of alumina, silica, and combinations of alumina and silica on titania surfaces has been extensively studied by Parfitt and co-workers [45–47], while the initial adsorption of hydrolyzable metal ions on titania and silica has also been examined in detail by James and Healy [48]. The structure of the precipitated product depends on many factors, and true aluminosilicates are only formed in alkaline conditions and at low concentrations of free Al^{3+} ions [49].

Silica–alumina coatings are often applied to titania by addition of an aluminum sulfate solution to the pigment dispersed in sodium silicate solution, the pH of the mixture being adjusted to values of 6 to 8 to complete the precipitation process. The alkaline silicate solutions usually contain, or readily form, microcolloidal polysilicate anions when neutralized or acidified. On addition of the aluminum salt, the Al^{3+} ions are adsorbed on, and cause the rapid flocculation of, the polysilicate anions; the remaining Al^{3+} ions will be hydrolyzed and coprecipitate as hydrated alumina. The precipitated product consists of domains of silica enmeshed in an alumina gel. Dehydration of the hydrous oxide coatings at temperatures below 200°C results in the elimination of adsorbed water and the condensation of some of the hydroxyl groups to form additional oxide linkages with adjacent hydroxyls and with the titania surface.

The nature of the final surface is dependent on the ratio of silica to alumina in the coating composition. A silica–alumina ratio of 4.4:1 gives a silica-rich surface, whereas a silica–alumina ratio of 1.1:1 results in a mixed silica–alumina surface; coatings containing higher proportions of alumina have an alumina surface. In each of these compositions, a silica-rich layer is deposited first. Alumina precipitated in the presence of sulfate ions will tend to retain these ions; the amount of adsorbed sulfate is pH-dependent and is greatest when the pH of the supernatant liquid is between 3 and 5 [46].

The electrokinetic properties of coated pigments and the adsorption of organic species have been used extensively to determine the nature of the silica–alumina coatings; key aspects of this work have been reviewed by Day [38] and Wiseman [50]. The relevance of some of the early work is limited by lack of details of the coating compositions and by the presence of zinc oxide, derived from rutilization catalyst residues, in many of the coatings [51].

The electrokinetic [52] and adsorption [53] properties of a series of coated zinc-free rutile pigments have been studied to determine their dependence on the coating composition; the coatings constituted 2 to 4% of the weight of the pigments and contained 23 to 94% alumina, the silica–alumina ratios ranging from 3:1 to 1:19 [54]. The zero point of charge (zpc), or pH of the suspending medium at which the surface of the pigment is electrically neutral, for coatings containing less than 56% alumina (Al:Si = 1.5:1), was found to decrease from a value of 4.4 to an extrapolated value of 1.5, the zpc of a pure hydrated silica, with increasing silica content. The zpc of coatings containing more than 56% alumina occurred at an approximately uniform pH of 4.4, close to that of the rutile base pigment, instead of increasing with increase in alumina content; the zpc of a hydrated alumina occurs at pH 9. The uniformity of the zpc of the high-alumina coatings shows that the character of the surface hydroxyls remained unchanged in these samples, but that the electrophoretic behavior was governed by the silica content of the surface layer or the presence of residual adsorbed sulfate ions [52]. In other studies, the amount of stearic acid adsorbed as the ionized carboxylate was found to have a linear dependence on the alumina content of the coating. In contrast, the amount of stearylamine adsorbed increased with increasing silica content, the adsorbed amine reportedly being bound as the protonated form on coatings containing more than 50% silica (silica-rich); the amine may also be coordinately bound on alumina-rich coatings [53].

These results are broadly in agreement with those noted above for the "homogeneous" bulk alumina–silicas. However, there are significant differences: the conventional coatings tend to consist of separate domains of silica and alumina, and the chemical and electrophoretic properties of the coated pigments show that these largely govern the character of the surface. Significant numbers of strongly acidic species are not present in the precipitated coatings, even in those containing an alumina–silica ratio of 1:2.5, although strongly acidic, homogeneous coatings can be prepared by alternative techniques.

Coatings of one oxide on another usually show the surface chemical and electrokinetic properties of the outer oxide layer, if this outer layer is substantially complete. Ideally, these properties should be largely independent of the method of application, although the choice of method used will probably affect the porosity of the coating, itself an important physical property. Exceptions are known; for example, silica deposited on titania has zpc at the same pH as bulk silica, but silica deposited on alumina from an alkaline silicate solution, while showing a zpc that occurs at a pH value close to that of pure silica, also shows an inflexion in the electrophoteric mobility curve at about pH 5, indicating the presence of alumina in the outer coating.

Multiple coating techniques are complicated by the possibility that a component of an inner coating may dissolve and be leached out by the reagent solutions, and subsequently be redeposited as a mixed phase with the outer coating. This process may be adventitious, as in the case of zinc rutilization catalyst residues. These residues are concentrated in the surface layers of the rutile base pigment, dissolve during the pigment dispersion process, and then are reprecipitated as a component of the alumina–silica coatings. The presence of zinc in the coatings can have a marked effect on the adsorption of organic materials, as it is more basic than hydrated alumina and can, for example, cause the rapid physical adsorption, followed by the slow chemisorption, of oil-based resins [51]. Alternatively, the deliberate leaching and redeposition of one component of the pigment base or of an inner coating may form part of a patented process; one example is the partial leaching of an alumina-coated titania by a titanyl sulfate solution and the subsequent deposition of a mixed titania–alumina or titania–alumina–silica coating.

3.4 CALCIUM CARBONATE

Calcium carbonate, in its various forms, is the most widely used of the mineral fillers and pigments, over 1.5 million tons being used annually in plastics formulations alone. Calcium carbonate is also used extensively in the paper and paint industries.

Calcium carbonate can exist in two crystalline modifications, calcite and aragonite. Aragonite is harder, and has greater density and refractive index (s.g., 2.93; n_D, 1.53) than calcite (s.g., 2.71; n_D, 1.48). Calcite is formed by crystallization of calcium carbonate at temperatures below 30°C and occurs in limestone, marble, and chalk. Aragonite is obtained by crystalliza-

$$Ca^{++} \Big| CO_3^{--} + H_2O \rightleftharpoons Ca^{++} \Big| (O\bar{H})_2 + CO_2 \uparrow$$

Figure 3.4

tion at higher temperatures and is a major constituent of sea shells. Precipitated calcium carbonate may consist of a mixture of the two crystalline forms, the relative proportions being dependent on the processing conditions. In the following discussion, reference will only be made to calcite, although it would be equally applicable to the other forms of calcium carbonate.

The commercial grades of calcium carbonate may consist of crushed or beneficiated limestone, chalk, marble, or seashells, or may be prepared by precipitation, for example, by carbonation of calcium hydroxide suspensions. The finely divided material used as a filler and extender in the paints and plastics industries is usually obtained by the crushing and fluid-energy milling of natural forms of calcite; their preparation and beneficiation have been reviewed elsewhere [55,56]. Dolomite, a naturally occurring calcium magnesium carbonate, is also used as a filler; its chemical properties resemble those of calcite.

Calcite dissociates on heating to form calcium oxide and carbon dioxide, the equilibrium pressure of the latter reaching one atmosphere at about 900°C. Significant dissociation does occur at lower temperatures, and CO_2 is desorbed from freshly cleaved crystalline surfaces at room temperature. A considerable proportion of the surface CO_3^{2-} ions of calcite can be displaced by the chemisorption of atmospheric water (Figure 3.4) [57]. As a consequence of this hydrolysis, the mineral particles have a basic, oleophobic, hydroxylic surface.

The weak basicity of the calcite surface is usually not detrimental to its use as a filler in plastics, although it is usually pretreated or compounded with additives such as carboxylic acids to render the surface hydrophobic (as a processing aid) and to improve the mechanical properties of the product; these additives tend to neutralize the surface. Calcite, unlike the other commonly used fillers, talc and kaolinite, is susceptible to attack by acids. This is disadvantageous in applications that require an acid-resistant composite, but can be advantageous in polyvinyl chloride (PVC) formulations, for example, as the calcite can neutralize the HCl produced by decomposition of the polymer during melt-compounding or on exposure to sunlight. As HCl promotes the degradation and discoloration of PVC, its absorption by the calcite improves the thermal and photolytic stability of the filled polymer, provided the calcite has a low Fe^{3+} content.

The basic, hydroxylic surface of calcite facilitates the preparation of oleophilic grades, which can be readily obtained by surface modification of the mineral by reaction with long-chain fatty acids such as stearic acid, with alkyl sulfonic acids, or with organotitanates; these treatments are described in Chapter 4. Insoluble oleophilic calcium acylate coatings are also formed by reaction between calcite and soluble salts of long-chain fatty acids.

Predispersed calcium carbonates for the preparation of paper coatings can be prepared by treatment of precipitated calcium carbonate with a phosphate dispersant [58]. Alumina-coated calcite, prepared by treatment of an aluminate containing suspension of the mineral with carbon dioxide [59], and calcium aluminate-coated calcite, prepared by calcination of an alumina-coated calcite [60], have been claimed to have improved dispersion in organic media. Calcium carbonates with an acidic surface can be prepared by coating with silica–alumina gels [61].

3.5   ZINC OXIDE

Zinc oxide was once an important white pigment in paint formulations but has been supplanted by the titanias in these applications. Zinc oxide is an important component in vulcanized rubber formulations and is used extensively as a photoconductive pigment for xerographic paper coatings. The oxide is still used in paint formulations, but primarily as a nontoxic fungistat and as a reactive modifier for oil and latex-based paints [62].

Zinc oxide pigments are prepared by oxidation of zinc vapor. Anhydrous zinc oxide particles, like those of most other oxides, rapidly develop a hydroxylic surface on exposure to the atmosphere, the oxide readily adsorbing water by dissociative chemisorption and hydrogen bonding processes similar to those occurring on anhydrous titania or alumina.

Zinc oxide is amphoteric, but more basic than either alumina or titania; it can, for example, react with ester groups of oil-based film-forming polymers to form zinc salts, thereby introducing ionic cross-links which can increase the viscosity or even result in phase separation. Partial saponification of vinyl ester polymers by zinc oxide pigments may also result in cross-linking through interactions between the polymer hydroxyl groups and Zn^{2+} ions or the oxide surface. This reactivity can be deleterious to the storage qualities of some types of paint, particularly latex formulations, but can also improve the mechanical qualities of the dried film. Saponification of linseed-oil-based paint media and the formation of soluble zinc salts can result in moisture-induced blistering of paint films, although the zinc soaps formed by the reaction between zinc oxide and oil-based media can assist in the dispersion of other, more inert pigments such as titania. The surface hydroxyl groups of the hydrated oxide can condense with alcohols and alkoxysilanes and undergo other reactions typical of a hydrated metal oxide surface; the hydroxyl groups can also form strong hydrogen bonds with alcohols [85] and other acceptor species.

Anhydrous zinc oxide, like alumina, is a Lewis acid. It can form coordination complexes with ammonia [63] and amines, and can chemisorb a wide

$$h\nu + lattice \rightleftharpoons (h - e) \rightleftharpoons h^+ + e^-$$

$$-ZnOH + h^+ \longrightarrow Zn^{2+} + HO^{\bullet}$$

$$O_2 + e \longrightarrow (O_2^{-\bullet})_s$$

$$(O_2^{-\bullet})_s + H_2O \rightleftharpoons (HO_2^{\bullet})_s + HO^-$$

$$2HO^{\bullet} \longrightarrow H_2O_2$$

$$(HO_2^{\bullet})_s + e^- \longrightarrow (HO_2^-)_s$$

$$(HO_2^-)_s + H_2O \rightleftharpoons H_2O_2$$

Figure 3.5

range of alcohols and carbonyl compounds as alkoxides and enolates [62], and alkenes such as propylene, as allylic carbanions [64]. These anionic surface adducts are the intermediates in a variety of zinc-oxide-catalyzed oxidation and disproportionation reactions, for example, the oxidation of isopropanol to acetone [65]. The catalytic properties of zinc oxide have been reviewed [66].

Crystalline zinc oxide, like titania, is normally deficient in oxygen and is an $n-$ type semiconductor with an electronic band gap of about 3.2 eV. Zinc oxide can oxidize organic compounds when irradiated with near-ultraviolet light having wavelengths shorter than 380 nm. The oxidation processes are similar to those that can occur on titania. The intermediate oxidants formed on hydrated surfaces include $HO^{\bullet}$ and $HO_2^{\bullet}$ radicals, HO_2^-, and hydrogen peroxide (Figure 3.5) [67]. Dehydroxylated zinc oxide can adsorb oxygen as O_2^- radical-anions; these are the oxidants in the conversion of adsorbed allylic anions to acrolein [66], for example, or adsorbed ammonia to nitrogen oxides and nitrate [63].

Despite its moderate photocatalytic activity, zinc oxide is often used to improve the weather resistance of some plastics. The oxide strongly absorbs near-ultraviolet light and can screen the underlying susceptible polymer layers from sunlight-induced degradation.

3.6 COLORED PIGMENTS

The colored inorganic pigments used in paints and plastics comprise a diverse group of oxides, hydrated oxides, sulfides, and salts; in many instances, these may contain mixtures of anions or cations in order to obtain a composite material of the required color. The technology of these pigments is amply described in *Pigment Handbook,* edited by Patton [1]. Some of the more common pigments, notably the iron oxides and chromates, have also been reviewed by Hermann [68].

The hues of most of the colored inorganic pigments are determined not only by their molecular compositions, but also on particle size and crystalline form. Many of the commercial synthetic pigments are coprecipitated or coated with other inorganic materials to stabilize colors by retarding recrystallization or dehydration, or by acting as a barrier between a labile pigment surface and oxidizable organic matrices or reactive contaminants such as sulfur dioxide, hydrogen sulfide, acids, or bases. The inorganic coatings may also be used to improve dispersion and to prevent the adsorption of driers or other formulation additives. The coatings frequently consist of silica, alumina, silica–alumina, silica–magnesia, or insoluble phosphates. Organic modification by long-chain acylates or sulfonates or by silylation may also be used to improve color stability and dispersion in organic media.

In contrast to the titanias, the clay minerals, and the other white pigments described in this book, the surface reactions of the colored pigments have not been extensively studied. Most of the pigments contain hydroxylic surfaces like those of the respective metal oxides and may also contain labile anions such as chromate or sulfide. In many respects, their surface chemistry would be qualitatively similar to that of alumina, titania, or other oxides. However, the pigments usually contain variable valence species, and electron transfer reactions with adsorbed species could have significant effects, both on the pigment color and on the properties of organic materials in which the pigment might be dispersed.

3.6.1   Iron Oxides

The synthetic iron oxide pigments are usually based on the ferric oxides hematite (red, α-Fe_2O_3), maghemite (brown, γ-Fe_2O_3), the ferric hydroxide oxides goethite [yellow, α-$Fe(O)OH$], lepidocrocite [orange, γ-$Fe(O)OH$], or the spinel, magnetite (black, $FeO\cdot Fe_2O_3$). Hematite and goethite are the most common pigmentary forms, while magnetite, maghemite, and their derivatives are used in magnetic recording media. The synthetic pigments are often doped with other metals, for example, manganese, to modify their colors. The color dependence on average particle size is illustrated by the dominant wavelength of magnetite, which changes from 600 to 635 nm as the particles increase from 0.2 to 1.2 μm in size [69]; iron oxides with particles of 0.1 μm or less are used as transparent colorants.

The hydrated oxides are heat sensitive, lepidocrocite being transformed to goethite or dehydrated to maghemite, while goethite is dehydrated to hematite. The dehydration occurs at temperatures of 160 to 200°C, but the thermal stability of the hydrated oxides may be improved by coating with alumina–silica. Temperature-stable iron oxide derivatives having spinel-like structures can be formed by calcining hematite or acicular goethite with divalent metal oxides such as zinc or cobalt oxides, which produce, respectively, yellow and deep brown pigments.

A number of natural iron earths such as the ochers, umbers, and siennas are also used as pigments. These contain a mixture of iron oxides, silica, alumina, and iron-containing clays. The natural and synthetic iron oxide pigments are inexpensive and can be used for the preparation of a range of tinted colors. The pure iron oxides have relatively high refractive indices and coefficients of absorption of ultraviolet light and are valuable as protective pigments for exterior paints.

Goethite, lepidocrocite, and hematite are the respective structural analogues of diaspore, boehmite, and corundum (α-alumina). The pigment surfaces are hydroxylated and contain hydrogen-bonded water molecules. The hydrated oxide surfaces are amphoteric and, like those of alumina, contain basic "isolated" hydroxyl groups and acidic "bridged" hydroxyl groups [20,70]. These can form hydrogen bonds with Lewis bases and are readily displaced by heating with amines or by treatment with phosphate ions, which can form anionic chelation complexes with the surface Fe^{3+} ions [71]. The hydration of hematite is not as strong as that of alumina. The surface hydroxyl groups can be eliminated by heating at 300 to 400°C, leaving a coordination-deficient Lewis acidic surface that can chemisorb ammonia [63] and other Lewis bases. Heating at higher temperatures results in the elimination of lattice oxygen and the formation of maghemite.

The character of the surface hydroxyl groups is dependent on the method of preparation of the pigment. Goethite in aqueous media is weakly basic; hematite is essentially neutral, the acidity of the hydroxyls being intermediate to those of alumina and titania. The surface hydroxyl groups of hematite can be esterified by heating with aliphatic alcohols in the presence of an amine catalyst [72] and are both sufficiently acidic for the chemisorption of quaternary ammonium ions [73] and sufficiently basic for the chemisorption of stearate ions [74] or sulfonate ions [75]; these various reactions have been used for the organic modification of iron oxide pigment surfaces.

Anhydrous hematite, like alumina, can catalyze a variety of organic reactions. These include the isomerization of butene at low temperatures [76] and the ketonization of acetic acid at high temperatures [77]. Iron oxides containing adsorbed sulfate ions can develop an acid surface, having Hammett acidity values of $+1.5$ to -3.0, that is capable of polymerizing vinyl ethers [78].

Iron oxide pigments are generally regarded as being unreactive towards organic matrices, although this is not strictly correct. The iron oxides can catalyze transesterification reactions and can react with carboxylate-containing paint media; they can also catalyze a variety of radical-mediated reactions in aqueous and organic media.

The iron oxides are efficient combustion catalysts at temperatures above 250°C and are used as one of the active agents in some self-cleaning coatings for domestic ovens. Hematite appears to have two types of catalytic oxidation sites, both associated with adsorbed O_2^- radical-anions and/or lattice oxygens [76]. One is involved in selective oxidation, for example, the conversion of butene to butadiene, and the second is involved in combustion, that is, the nonselective oxidation of adsorbed organic molecules to carboxylates and eventually to carbon dioxide; the latter sites in anhydrous hematite are active at temperatures as low as 20°C [79]. Ammonia adsorbed on iron oxides is also partially converted at low temperatures to chemisorbed nitrogen oxides [63]. Combustion catalysis at higher temperatures involves the interconversion of Fe_2O_3 and Fe_3O_4 and the migration of Fe ions between the interior and the surface of the lattice [80]. Catalyzed combustion proceeds more readily with maghemite than with hematite, as the former has an oxygen lattice similar to that of magnetite. Hematite is converted to maghemite on heating with organic materials such as glycerol, presumably via Fe_3O_4 domains. Iron oxides have appreciable photocatalytic properties and can catalyze the photooxidation of sulfites [81] and the deamination of aqueous aminoacids.

The iron oxides can initiate radical reactions by redox processes with reductants or hydroperoxides, as well as by electron transfer with adsorbed cocatalytic species. The oxides can also catalyze the decomposition of hydrogen peroxide and hydroperoxides via radical intermediates [82]; the decomposition mechanisms are probably similar to those of the Fe^{3+} ion catalyzed reactions of hydrogen peroxide (Fenton's reagent) [83]. Iron oxides can also promote the autoxidation of cumene to cumene hydroperoxide (see Chapter 5), although the catalysis appears to involve reactions between chemisorbed oxygen species and cumene, in addition to the Fe^{II}/Fe^{III} redox-promoted autoxidation processes [84] (see Figure 3.6).

The redox reaction between ferric oxide and sulfur dioxide can initiate the polymerization of acrylic monomers [85], while the reactions with oxygen or trace peroxidic impurities are probably involved in the polymerization of methyl methacrylate by the oxide [86] and in the photodegradation of polyethylene [87]. Ferric oxide can partially oxidize and cross-link low-density polyethylene during melt-compounding at 120 to 180°C [88], probably via radical induced processes. Ferric oxides can catalyze the radical depolymerization of polystyrene and poly(methyl methacrylate) [89], and the pigments can promote the titania-catalyzed photodegradation of fluorescent "brighteners" [90]. The oxides can also catalyze the dehydrochlorination of poly(vinyl chlorides) by a dual mechanism, consisting of an ionic process, catalyzed by the surface Fe^{3+} ions or $FeCl_3$ formed in the reaction, and a radical process which involves surface oxygen species [91]. The surface reactivity of the pigments can be reduced by coating with inorganic oxides and mixed oxides [92–94] and by treatment with organosilanes.

3.6.2 Chromates

Common chromate pigments include the yellow and orange lead chromate and chromate–molybdate derivatives, the less toxic yellow barium

$$Fe^{3+} + H_2O_2 \longrightarrow Fe^{2+} + H^+ + HO_2^\bullet$$

$$Fe^{2+} + H_2O_2 \longrightarrow Fe^{3+} + HO^\bullet + HO^-$$

Figure 3.6

chromate, and the corrosion-inhibiting pigment, zinc chromate. The chromate pigments, like the iron oxides, are often coprecipitates or are encapsulated with organic or inorganic coatings, variously, to improve dispersion, to reduce leaching of potentially toxic chromate ions, to improve light and weather resistance, and to prevent discoloration reactions with organic matrices. The inorganic coatings may, for example, be based on alumina [95], silica [96], or antimony oxide–silica–alumina compositions [97,98].

Chromic acid is a weak acid, and chromate ions can be displaced from the pigment surfaces by treatment with solutions containing phosphate, sulfonate, or long-chain acylate anions [e.g., 74] to yield surface-modified products. The rust-proofing action of organic coatings containing sparingly soluble chromates such as zinc chromate is believed to be due to the polarization of ferrous metal surfaces by chromate ions leached from the pigment and to the sealing of defects in the coating by insoluble products [99]. The leaching of surface chromate ions results in the normal and basic chromate pigments having hydroxylic surfaces that can undergo surface modifying reactions with alkoxy- or chloro-silanes or with titanate esters.

The chromate pigments can oxidize a variety of organic matrices, hydrogen sulfide, and sulfur dioxide which might be adsorbed from the atmosphere. These reactions involve the reduction of the chromate Cr^{VI} to Cr^{III} state and to a resultant darkening of the pigment color. The chromates, including lead chromates, are thermal oxidation catalysts; the pigments can oxidize surfactants and adsorbed species such as stearic acid and stearates, waxes, and polyolefins at 160°C [100]. The oxidation of adsorbed organic materials is accelerated by actinic light, and related reactions are utilized in photoengraving resists based on poly(vinyl alcohol)-chromate compositions. Chromate pigments can catalyze the esterification and cross-linking of carboxylic acid–aminoplast resin formulations [101], probably by ionic processes. Soluble chromates have been used as hydrogen atom scavengers to prevent homopolymer formation during the radiochemical grafting of polyester fibers with acrylic acid [102].

REFERENCES

1 T. C. Patton, Ed., *Pigment Handbook*, 3 Vols., Wiley–Interscience, New York, 1973.

2 H. S. Katz and J. V. Milewski, Eds., *Handbook of Fillers and Reinforcements for Plastics*, Van Nostrand–Reinhold, New York, 1978.

3 D. Barby, Silicas, in G. D. Parfitt and K. S. W. Sing, Eds., *Characterization of Pigment Surfaces*, Academic, London, 1976.

4   T. D. Coyle, Silica (Introduction); P. K. Mather, Silica (Amorphous); in *Kirk-Othmer Encyclopedia of Chemical Technology*, Vol. 18, 2nd ed., Wiley, New York, 1969.

5   V. M. Bermudez, *J. Phys. Chem.*, **75**, 3249 (1971).

6   B. A. Morrow and I. A. Cody, *J. Phys. Chem.*, **79**, 761 (1975).

7   J. B. Peri, *J. Phys. Chem.*, **70**, 2937 (1966).

8   H. P. Boehm, M. Schneider, and H. Wistuba, *Angew. Chem., Int. Ed.*, **4**, 600 (1965).

9   L. R. Snyder and J. W. Ward, *J. Phys. Chem.*, **70**, 3941 (1966).

10   H. Bartels and B. Prijs, in *Advances in Chromatography*, Vol. 11, Dekker, New York, 1974, p. 115.

11   G. Lagaly, *Adv. Colloid Interface Sci.*, **11**, 105 (1979).

12   P. B. West, G. L. Haller, and R. L. Burwell, *J. Catalysis*, **29**, 489 (1973).

13   H. P. Boehm, *Adv. Catalysis*, **16**, 179 (1966).

14   H. Deuel, J. Wartmann, K. Hutschneker, U. Schobinger, and C. Guedel, *Helv. Chim. Acta*, **62**, 1160 (1959).

15   J. Wartmann and H. Deuel, *Chimia*, **12**, 83 (1958).

16   H. Ishida and J. L. Koenig, *J. Polym. Sci., Polym. Phys. Ed.*, **18**, 1931 (1980).

17   G. MacZurar, Aluminum Oxide, in *Kirk–Othmer Encyclopedia of Chemical Technology*, Vol. 2, 3rd ed., Wiley, New York, 1978.

18   T. Saito, *Z. Anorg. Allg. Chem.*, **391**, 167 (1972).

19   H. Knoezinger and P. Ratnasamy, *Catal. Rev. Sci. Eng.*, **17**, 31 (1978).

20   H. P. Boehm, *Discuss. Farad. Soc.*, **52**, 264 (1971).

21   A. C. Zettlemoyer and E. McCafferty, *Croat. Chem. Acta*, **45**, 173 (1973).

22   H. Pines and W. O. Haag, *J. Amer. Chem. Soc.*, **82**, 2471 (1960).

23   H. Knoezinger, *Adv. Catalysis*, **25**, 184 (1975).

24   B. D. Flockhart, I. R. Leith, and R. C. Pink, *J. Catalysis*, **9**, 45 (1967).

25   H. Knoerzinger, H. Krientenbrink, H-D. Mueller, and W. Schulz, *Proceedings of the 6th International Congress on Catalysis*, Vol. 1. (London, 1976), Chemical Society, London, 1977, p. 183.

26   H. Muraishi and S. Kitahara, *Nippon Kaguka Kaishi*, 1457 (1978).

27   K. Tanabe, M. Itoh, K. Morishige, and H. Hattori, *Prep. Catal. Proc. Int. Symp.* 1975, 65 (Elsevier, 1976); *Chem. Abs.*, **91**, 63236 (1979).

28   A. Toussaint, *14th FATIPEC Congr.*, 1978, p. 657.

29   A. Leonard, S. Suzuki, J. J. Fripiat, and C. De Kimpe, *J. Phys. Chem.*, **68**, 2608 (1964).

30   P. Cloos, A. J. Leonard, J. P. Moreau, A. Herbillon, and J. J. Fripiat, *Clays Clay Miner.*, **17**, 279 (1969).

31   I. D. Mikheikin, A. I. Lumpov, G. M. Zhidomirov, and V. B. Kazonskii, *Kinet. Catal.*, **19**, 850 (1978).

32   C. Defosse, P. Canesson, P. G. Rouxhet, and B. Delmon, *J. Catalysis*, **51**, 269 (1978).

33   R. E. Sempels and P. G. Rouxhet, *J. Colloid Interface Sci.*, **55**, 263 (1976).

34   P. O. Scokart, F. D. Declerck, R. E. Sempels, and P. G. Rouxhet, *J. Chem. Soc., Farad. I*, **73**, 359 (1976).

35   I. D. Mikheikin, A. I. Lumpov, G. M. Zhidomirov, and V. B. Kazonskii, *Kinet. Catal.*, **20**, 410 (1978).

36   P. G. Rouxhet and R. E. Semples, *J. Chem. Soc., Farad. I*, **70**, 2021 (1974).

37   G. D. Parfitt, *J. Oil Colour Chem. Assoc.*, **54**, 717 (1971).

38   R. E. Day, in *Progress in Organic Coatings*, Vol. 3, Elsevier Sequoia, Lausanne, 1976, p. 269.

39 H. A. Benesi and B. H. C. Winquist, *Adv. Catalysis,* **27**, 97 (1978).

40 K. Tanabe, *Solid Acids and Bases,* Academic, New York, 1970.

41 G. D. Parfitt, *Pure Appl. Chem.,* **48**, 415 (1976).

42 G. D. Parfitt and C. H. Rochester, Surface Characterization: Chemical, in G. D. Parfitt and K. S. W. Sing, Eds., *Characterization of Pigment Surfaces,* Academic, London, 1976.

43 G. Kaempf and H. G. Voelz, *Farbe Lack,* **74**, 37 (1968).

44 V. S. Sazhin, N. N. Stremilova, L. A. Gavrilova, A. S. Kostenko, V. F. Shabanov, and N. F. Ryabchikova, *J. Appl. Chem. USSR,* **50**, 395 (1977).

45 G. D. Parfitt, *Croat, Chem. Acta,* **45**, 189 (1973).

46 P. B. Howard and G. D. Parfitt, *Croat. Chem. Acta,* **50**, 15 (1977).

47 D. N. Furlong, K. S. W. Sing, and G. D. Parfitt, *J. Colloid Interface Sci.,* **69**, 409 (1979).

48 R. O. James and T. W. Healy, *J. Colloid Interface Sci.,* **40**, 65 (1972).

49 E. Matijevic, F. J. Mangravite Jr., and E. A. Cassell, *J. Colloid Interface Sci.,* **35**, 560 (1971).

50 T. J. Wiseman, Inorganic White Pigments, in G. D. Parfitt and K. S. W. Sing, Eds., *Characterization of Pigment Surfaces,* Academic, London, 1976.

51 A. F. Sherwood and S. M. Rybicka, *J. Oil Colour Chem. Assoc.,* **49**, 648 (1966).

52 G. D. Parfitt and J. Ramsbotham, *J. Oil Colour Chem. Assoc.,* **54**, 356 (1971).

53 A. F. Sherwood, S. M. Rybicka, J. Milne, M. G. White, and V. T. Crowl, Paint Research Station (Teddington) Memo No. 372; reported in ref. [52].

54 H. Rechmann, *J. Oil Colour Chem. Assoc.,* **48**, 837 (1965).

55 R. H. Lepley, Calcium Carbonate, in *Kirk-Othmer Encyclopedia of Chemical Technology,* Vol. 4, 3rd ed., Wiley-Interscience, New York, 1978.

56 R. F. Hall, Calcium Carbonate, Natural; P. F. Worner, Calcium Carbonate, Synthetic; in T. C. Patton, Ed., *Pigment Handbook,* Vol. 1, Wiley-Interscience, New York, 1973.

57 T. Morimoto, J. Kishi, O. Okada, and T. Kadota, *Bull. Chem. Soc. Jpn.,* **53**, 1918 (1980).

58 D. Kaufman (Wyanotte Chemicals), U.S. Patent 3,598,624 (August 10, 1971).

59 K. Shibazaki, K. Miwa, H. Hasegawa, Y. Furazawa, and Y. Takahashi (Shiraishi Kogyo), Jpn. Kokai Tokkyo Koho 79,131,599 (October 12, 1979); *Chem. Abs.,* **92**, 95702 (1980).

60 K. Shibazaki, S. Edagawa, K. Miwa, and Y. Takahashi (Shiraishi Kogyo) Jpn. Kokai Tokkyo Koho 79,131,598 (October 12, 1979); *Chem. Abs.,* **92**, 95703 (1980).

61 J. D. Swift, D. G. Hawthorne, B. C. Loft, and D. H. Solomon (C.S.I.R.O.), U.S. Patent 3,963,512 (January 15, 1976).

62 M. Nagao and T. Morimoto, *J. Phys. Chem.,* **84**, 2054 (1980).

63 Yu. B. Belokopytov, K. M. Kholyavenko, and S. V. Gerei, *J. Catalysis,* **60**, 1 (1979).

64 T. T. Nguyen and N. Sheppard, *J. Chem. Soc., Chem. Comm.,* 868 (1978).

65 E. Akiba, M. Soma, T. Onishi, and K. Tamura, *Z. Phys. Chem., Neue Folge,* **119**, 103 (1980).

66 C. S. John, in C. Kemball and D. A. Dowden, Eds., *Catalysis,* Vol. 2, Specialist Publication, Chemical Society, London, 1980, p. 169.

67 J. R. Harbour and M. L. Hair, *J. Phys. Chem.,* **83**, 652 (1979).

68 E. Herrmann, Inorganic Coloured Pigments, in G. D. Parfitt and K. S. W. Sing, Eds., *Characterization of Pigment Surfaces,* Academic, London, 1976.

69 G. Kaempf, *J. Coatings Technol.,* **51** (655), 51 (Aug. 1979).

70 C. H. Rochester and S. A. Ropham, *J. Chem. Soc., Farad. I,* **75**, 591 (1979); *J. Chem. Soc., Farad. I,* **75**, 1073 (1979).

71   Y. D'Ydewalle, V. Biermans, and L. Baert, *Meded. Fac. Landbouwwet, Rijksuniv. Gent*, 4, 1519 (1978); *Chem. Abs.*, 91, 44972 (1979).

72   S. Watatani and H. Hosoda (Hitachi Maxwell), Japan Kohai 77 50,599 (April 22, 1977); *Chem. Abs.*, 87, 77651 (1977).

73   V. A. Salmanov and R. M. Rasulova, *Tr. Gos. Nauchno-Issled. Proektn. Inst. "Gipromorneft"*, 13, 36 (1977); *Chem. Abs.*, 91, 159091 (1979).

74   F. Korkay and F. Szanto, *Farbe Lack*, 85, 258 (1979).

75   A. Miyake and S. Watatani (Hitachi Maxwell), Jpn. Kokai Tokkyo Koho, 78,134,796 (November 24, 1978); *Chem. Abs.*, 90, 114159 (1979).

76   W. H. Cheng and H. H. Kung, *J. Phys. Chem.*, 83, 1737 (1979).

77   J. C. Kuriacose and S. S. Jewur, *J. Res. Inst. Catal. Hokkaido Univ.*, 24, 73 (1976).

78   M. Hino and K. Arata, *J. Polym. Sci., Polym. Chem. Ed.*, 18, 235 (1980).

79   V. A. Kuznetsov, S. V. Gerai, Ya. B. Gorokhovatskii, and E. V. Rozhkova, *Kinet. Catal.*, 18, 361 (1977).

80   K. Sakata, F. Ueda, M. Misono, and Y. Yoneda, *Bull. Chem. Soc. Jap.*, 53, 324 (1980).

81   S. N. Frank and A. J. Bard, *J. Phys. Chem.*, 81, 1484 (1977).

82   M. D. Irmatov, Yu. D. Norikov, M. G. Bulygin, and E. A. Blyumberg, Deposited Doc. 1973, VINITI 7384-73; *Chem. Abs.*, 86, 154986 (1977); VINITI 7385-73; *Chem. Abs.*, 87, 5142 (1977).

83   W. G. Barb, J. H. Baxendale, P. George, and K. R. Hargrave, *Trans. Farad. Soc.*, 4, 591 (1951).

84   N. P. Evmenenko, Ya. B. Gorokhovatskii, and M. V. Kost, *Kinet. Catal.*, 12, 1275 (1971).

85   T. Yamaguchi, H. Tanaka, T. Ono, M. Endo, H. Ito, and O. Itabashi, *Kobunshi Ronbunshu*, 32, 120 (1975).

86   T. Yamaguchi, H. Tanaka, A. B. Moustafa, T. Ono, H. Ito, O. Itabashi, M. Endo, M. Ohuchi, and L. Saito, *Chem. Ind.*, 619 (1974).

87   M. Z. Borodulina, V. S. Bugorkova, N. P. Provorova, M. S. Kurzhenkova, and T. N. Zelenkova, *Plast. Massy*, (7), 72 (1978).

88   H. J. Schneider, W. Reicherdt, and K. Thinius, *Plaste Kaut.*, 17, 310 (1970).

89   T. Takahashi, K. Suzuki, and M. Abe, *Shikizai Kyokaishi*, 39, 55 (1966); *Chem. Abs.*, 68, 79018 (1968).

90   I. N. Bykova, L. A. Kaminskaya, G. A. Davydova, A. B. Pakshver, and L. I. Belenkii, *Izv. Vyssh. Uchebn. Zaved, Tekhnol. Tekst. Prom-sti.*, 78 (1978); *Chem. Abs.*, 89, 60901 (1978).

91   Y. Uegaki and T. Nakagawa, *J. Appl. Polym. Sci.*, 21, 965 (1977).

92   S. Nobuoko, Y. Omote, and H. Serikawa (Agency Industrial Science Technology), Jpn. Kokai Tokkyo Koho 78,136,038 (November 28, 1978); *Chem. Abs.*, 90, 123213 (1979).

93   K. Shimazawa, H. Sawai, and Y. Mura (Toda Kogyo Corp.), Japan Kokai 78 36,538 (April 4, 1978); *Chem. Abs.*, 89, 60564 (1978).

94   N. Abe, S. Kubo, M. Kanemaru, and S. Kozu (Nippon Chemical Industry), Japan Kokai 78 43,729 (April 20, 1978); *Chem. Abs.*, 89, 44678 (1978).

95   K. Kubo, M. Horiguchi, M. Hasegawa, and T. Anzai (Dainichiseika Color), Jap. Pat. 74 16,531 (April 23, 1974); *Chem. Abs.*, 82, 45215 (1975).

96   C. H. Elstermann and F. Hund (Farbenfabricken–Bayer), Ger. Offen. 1,952,537 (November 4, 1971); *Chem. Abs.*, 76, 87306 (1972).

97   Colores Hispana S. A., Spanish Patent 394,522 (February 16, 1974); *Chem. Abs.*, 82, 60052 (1975).

98   D. D. Dobson and J. Mitchell (I.C.I.Ltd), Ger. Offen. 2,115,450 (November 11, 1971).

99   M. Piens, *J. Coatings Technol.,* **51**, (655), 66 (Aug. 1979).

100   G. N. Gorelik, N. A. Novoselova, and L. N. Ioffe, *Lakokras. Mater. Ikh Primen,* 17 (1977).

101   L. M. Vinogradova and A. T. Sanzharovskii, *Lakokras. Mater. Ikh Primen.,* **5**, 43 (1972). (1972).

102   K. Rao, M. H. Rao, P. N. Moorthy, and A. Charlesby, *Proc. Chem. Symp. 1972,* **1**, 327 (1972); *Chem. Abs.,* **81**, 171042 (1974).

SURFACE MODIFICATION OF PIGMENTS AND FILLERS

4.1   INTRODUCTION

Many of the inorganic fillers and pigments manufactured for incorporation in paints, plastics, and other organic media are treated with organic reagents to modify their surface characteristics or are used with formulation additives that effectively modify the surfaces in the compounding processes. Surface modification may be used for a variety of reasons that include improvement of the dispersion of the mineral in organic media, modification of the rheology of a mineral dispersion, and improvement of the mechanical properties of filled plastics compositions. In this chapter we propose to confine our discussions largely to surface chemistry and adsorption processes that are relevant to the modification of inorganic particulates and to the dispersion, flocculation, and reinforcement of mineral-containing compositions. Organic reactions catalyzed by minerals are discussed in the following chapters.

The various surface modifying reactions and interactions have been divided into five groups for the purposes of discussion.

(i) Modification by adsorption of acids, bases, salts, and neutral compounds. These processes include examples of chemisorption in which coordination, hydrogen bond formation, or proton transfer occurs between the mineral surface and the adsorbed species.

(ii) Modification by ion exchange. This process is technically different from that of modification by the adsorption of salts, as any displaced ions and associated counterions are not retained on the mineral surface after ion exchange treatment.

(iii) Modification involving covalent bond formation between the adsorbed species and reactive groups present on the mineral surface. The principal covalent bond forming reactions discussed in this section are those between surface hydroxyl groups, or chemisorbed water molecules, and reactive silanes or metal alkoxides, or the condensation of these hydroxylic species with epoxides and isocyanates.

(iv) Modification by the adsorption of polymers and the dispersion or flocculation of minerals.

(v) Modification by encapsulation polymerization, or processes in which the modifying polymer is formed by the *in situ* polymerization of adsorbed monomer.

Discussion of these topics will be preceded by a brief outline of some aspects of the physical chemistry of surfaces, and of the dispersion and reinforcement processes that are affected by the surface modification of pigments and fillers.

4.1.1 Adsorption of Low-Molecular-Weight Compounds

The process of adsorption is fundamental to the modification of minerals by treatment with chemicals and to most other reactions occurring on, or adjacent to, the mineral surfaces. Provided no subsequent irreversible chemical reactions occur in the adsorbed phase, the adsorption is a reversible process and can be described by Eqs. 1 and 2, in which M represents the mineral surface, A and B, the adsorbates, and MA or MB, the interfacial adsorbed phase.

$$M + A \rightleftharpoons MA \tag{1}$$
$$MA + B \rightleftharpoons A + MB \tag{2}$$

The rates of the adsorption and desorption reactions and the equilibrium constant for the overall reaction are functions of the free energy change (ΔG) of the reactants resulting from the adsorption, desorption, or overall process. The free energy of the system has two components, enthalpy and entropy (Eq. 3), the free energy change at temperature T being the function of the enthalpy change ΔH, which can be related to the heat of reaction (adsorption) and the entropy change ΔS, a thermodynamic concept which describes the change in degree of "orderliness" of the system.

$$\Delta G = \Delta H - T\Delta S \tag{3}$$

Adsorption or other processes can only occur spontaneously if there is a decrease in the free energy of the system, that is, if ΔG has a negative value. In favorable cases, the enthalpic and entropic contributions to free energy of adsorption can be determined from the effects of changes in temperature on the adsorption equilibrium. The enthalpic contribution is usually dominant in the adsorption of vapors on to mineral surfaces, but the entropic contribution can be more important in adsorption from solution. In the latter process, the enthalpy changes are often small and the adsorption is governed by changes in entropy of the adsorbate or solvent molecules; the entropic contribution is particularly important in the adsorption of polymers.

The relationship between the quantity of adsorbate in the interfacial phase and the activity, that is, the concentration in solution or the vapor pressure of the adsorbate in the surrounding medium, is described by an adsorption isotherm. The shape of the isotherm for the adsorption of nonionic species depends on a number of factors that include the binding energy distribution for the surface sites, the relative affinities of the adsorbate molecules for the surface and for each other, the presence and binding energies of other adsorbate species, the solubility of the adsorbate (in cases of adsorption from solution), and the porosity of the surface. The adsorption and desorption isotherms for a given system are ideally identical, but pronounced hysteresis is often observed in the adsorption of polymers or low-molecular-weight species on porous surfaces, where adsorption may occur more readily than desorption. Three frequently observed isotherms are illustrated in Figure 4.1; values M and S correspond, respectively, to monolayer coverage of the surface and saturation pressure or concentration of the adsorbate in the fluid phase.

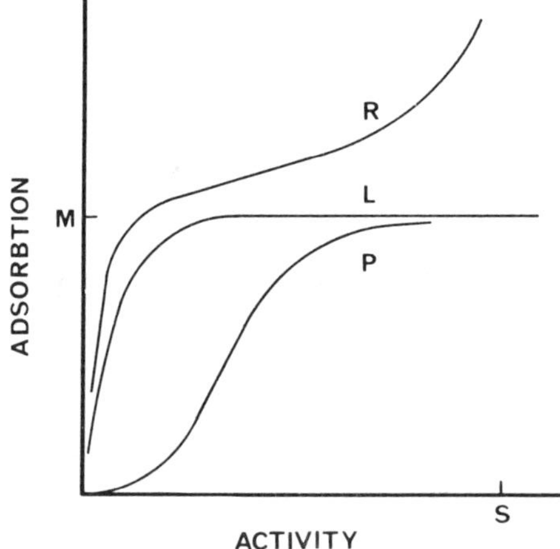

Figure 4.1 Some typical adsorption isotherms. See text for explanation.

The simplest isotherm (curve L) describes the physical adsorption of a vapor, the molecules of which have negligible interaction with each other, on a nonporous surface that has binding sites of equal energy. This idealized adsorption is described by the Langmuir isotherm (Eq. 4), where ϕ represents the fraction of surface covered by the adsorbate when in equilibrium with vapor at pressure p. In an alternative form of the Langmuir isotherm (Eq. 5), the amount of material adsorbed (x) by a definite mass of adsorbent (m) is expressed in terms of the concentration (c) of sorbate in the surrounding phase; A and B are constants, the nature of which is described in the standard texts [e.g., 1]. The Langmuir isotherm is characterized by an adsorption plateau corresponding to monolayer surface coverage and by an approximately linear relationship (at low values) between the degree of adsorption and activity of the adsorbate.

$$\frac{\phi}{1 - \phi} \propto p \tag{4}$$

$$\frac{x}{m} = \frac{Ac}{1 + Bc} \tag{5}$$

In most real systems, the adsorbed molecules have some affinity for each other. Long-range interactions between the surface and the sorbent can result in multilayer adsorption, shown by the absence of an adsorption plateau. The "knee" in the nonideal adsorption curve R corresponds only approximately to a monolayer coverage. In gaseous–solid systems, the isotherm may be described by the BET equation (Eq. 6), in which f is the partial pressure of adsorbate, v is the volume adsorbed, and v_m the volume of adsorbate corresponding to monolayer coverage of the solid surface; c is a function of temperature and the difference in the heats of liquefaction and monolayer adsorption.

$$\frac{f}{v(1-f)} = \frac{1}{v_m c} + \frac{c-1}{v_m c} \qquad (6)$$

Porous solids may show an adsorption plateau at higher adsorbate activities. The "knee" in such cases corresponds to filling of the pores or interlamellar spaces, with little further adsorption occurring until the fluid phase is saturated with adsorbate. The adsorption sites on mineral surfaces often have a broad distribution of binding energies and adsorption on these solids may more accurately be expressed, over a limited range of adsorbate concentrations, by the classical, but empirical Freundlich isotherm (Eq. 7), in which the constant $n \gg 1$.

$$\frac{x}{m} \propto c^{1/n} \qquad (7)$$

Often the adsorbate must displace a strongly adsorbed species or the adsorbate molecules may have large energies of association in the interfacial phase relative to the energy of adsorption. These factors can result in an S-shaped isotherm curve (Figure 4.1; P). The adsorption isotherms for long-chain alkyl species on solids are frequently of this type, and the adsorption of a sparingly soluble material from solution can often be described by a similarly shaped isotherm. Numerous other isotherms have been described for the adsorption of one or more nonelectrolytes at solid–liquid or solid–vapor interfaces and for the adsorption in binary liquid systems [1,2].

The competitive adsorption of nonelectrolytes containing single functional groups can be described by the alternative semiquantitative rules [1]. "A polar surface will preferentially adsorb the more-polar component from a nonpolar solution" and "A nonpolar surface will preferentially adsorb the less-polar component from a polar solution." These rules can be extended to give Traube's rule: "The adsorption of organic substances from aqueous solutions increases strongly and regularly with the ascent of a homologous series." This is seen, for example, in the adsorption of aliphatic acids, alcohols, and amines on minerals in aqueous suspension, where the adsorption coefficient of the organic compound increases with chain length. The reverse sequence occurs for the adsorption of a homologous series of compounds on a polar surface from nonpolar solutions.

The adsorption of an organic electrolyte on a mineral can follow a complex isotherm which may include contributions from the ionic interactions of the anions and cations with the electric double layer at the surface, and the nonionic interactions which may also occur between the adsorbed species. Often the adsorption isotherms may resemble the Langmuir isotherm, the "knee" in the curve corresponding to the neutralization of surface charge [3]. The Stern equation, which describes the adsorption of ions in the diffuse ionic layer surrounding the charged surface, can be expressed in a form similar to the Langmuir equation.

The adsorption of electrolytes by clay minerals has been analyzed and described by Van Olphen [4]. Other descriptions may be found in standard texts on colloid chemistry, although the isotherms discussed often only relate to adsorption on ionic crystals, which have a constant surface potential; these isotherms differ quantitatively from those of systems containing, for example, clay minerals or charged polymer lattices, the particles of which have a constant surface charge. The kinetics of the adsorption and diffusion of small molecules on zeolites, swelling clay minerals and organoclays, and other molecular sieves are reviewed by Barrer [5].

4.1.2 Adsorption of Polymers

When polymers are adsorbed on solid surfaces from solution, the molecular chains frequently assume coil-like configurations in the interfacial phase. The chains will contain sequences of polymer segments ("trains") adsorbed in extended configurations on the surface, unadsorbed segments in the form of a series of "loops" which, together with the polymer chain ends ("tails"), are capable of extension into the liquid phase [6,7]. The average number of polymer segments contained in each train, loop and tail, the segment density, and the mean maximum thickness of the adsorbed polymer layers in model systems can be calculated using statistical mechanics, although practical difficulties and the need for simplifying assumptions and approximations have so far prevented uniquely correct solutions for real systems.

Recent work by Lal and colleagues has shown that the adsorption configuration can be described for a hypothetical polymer containing N segments, each having an adsorption energy (E), the energy difference between a segment–surface and a solvent–surface bond. Monte Carlo calculations [7,8], which take into account the excluded volume effect, show that the fraction of adsorbed polymer increases and the thickness of the adsorbed layer decreases with increase in adsorption energy. The polymer can only be adsorbed if the adsorption energy is negative, but if the energy is less than a critical value, $-E/kT = 0.4$, the polymer is virtually nonadsorbed, that is, the mean fraction of adsorbed segments approaches zero. At high-adsorption energy, $-E/kT/ \geqslant 0.9$, the polymer is adsorbed in an extended configuration on the surface, and the thickness of the layer, which is determined by the mean number of segments in the loops, decreases slightly with increasing chain length of the polymer. The mean number of segments contained in trains and loops changes little with chain length or with adsorption energies in the range $-E/kT = 0.5$ to 0.9; however, the number of segments in the tails increases markedly with increase in chain length and decrease in adsorption energy. For weakly adsorbed polymers, the thickness of the layer is dominated by the lengths of the polymer tails, as the loops remain short and only contain an average of 4 to 20 segments [7]. The effective thickness of the adsorbed layers can be much larger and may be of the order of 100 to 800 Å for typical homopolymers adsorbed on a metal surface [9]. Thick adsorbed layers can also form on pigment surfaces [10,11]. The interactions between polymer layers, attached to particles in suspension, can provide an important means for dispersion stabilization.

The adsorption of polymers differs in many respects from the adsorption of low-molecular-weight species [12]. The adsorption can be very slow in reaching an equilibrium state, often requiring contact times of the order of 100 to 1000 hours. The slow attainment of equilibrium is due to the large number of configurations that the adsorbing polymer chain can adopt. Numerous theoretical isotherms have been developed, but most experimental data would fit a Langmuir isotherm as well as any other [1]. The adsorption of polymers can occur to a considerable extent, even at very low equilibrium concentrations, and the amount adsorbed quickly reaches a limiting value with increase in concentration. The amount of polymer adsorbed on oxide surfaces from toluene solution is close to the value calculated using the mean molecular cross sections, obtained by interfacial tension measurements of the polymers adsorbed at a toluene–water interface; the latter phase provides a convenient approximation for the hydroxylic oxide surfaces [13].

Larger amounts of polymer are adsorbed from a poor solvent, that is, one close to the θ-point, than from a thermodynamically good solvent; chemical modification of the mineral surfaces can strongly affect the adsorption coefficients [14]. The adsorption coefficients of polymers may increase or decrease slightly with increase in temperature, the variation usually being greatest for θ-solvent systems [6]. (A θ-solvent is one in which the polymer is just at the point of precipitation, that is, one in which the interaction between solvent molecules and the polymer segments is equal to the interaction between the segments themselves. For most solvent compositions, this state will occur at one critical temperature, θ-point.) An increase in adsorption with increase in temperature indicates that the adsorption is endothermic and involves an increase in entropy of the system, most probably that of the solvent molecules desorbed from the surface or released from the solvated polymer chains on adsorption. (The adsorbed chains have greater mean segmental densities in the adsorbed state.) The polymers usually are not readily desorbed on dilution of the liquid medium; they can be displaced by other polymers or by treatment with a better solvent than that used for the original adsorption [1,12].

4.1.3 Hydrogen Bond Formation

The adsorption of molecules on minerals involves some form of binding between the adsorbate and the mineral. This may include one or more of the specific forms of chemisorption introduced at the beginning of this chapter, but will also include nonspecific physisorption interactions such

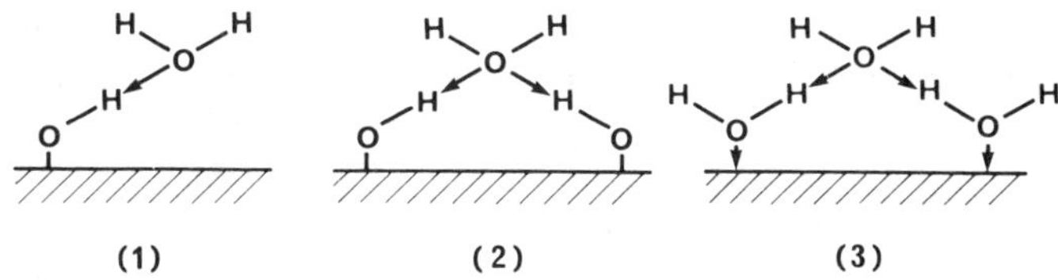

Figure 4.2 Hydrogen-bonded structures of water on silica surfaces.

as London–van der Waals attractions. Another specific interaction that is common to most practical mineral organic systems is the formation of hydrogen bonds. Because of their ubiquitous nature, hydrogen bonds will be discussed separately from the other forms of chemisorption. Most forms of chemisorption rely on hydrogen-bonded adsorption as a prelude for reaction with the surface or as a supplementary means of maintaining adsorption of the "chemisorbed" species. Hydrogen bond formation in the adsorption process has been reviewed [e.g., 15,16] and is the subject of a recent book [17].

In the previous chapters we have indicated that most mineral particulate surfaces bear surface hydroxylic species, unless the material has been dehydrated by rigorous outgassing at elevated temperatures. These surface species may be structural (lattice) hydroxyls, those formed by adsorption of hydrated oxides or by hydrolysis of the lattice, or those of chemisorbed water. Adsorbed water molecules are also important in the interactions of neutral organic molecules with charged silicate surfaces, as they can act as hydrogen-bonding bridges between the organic molecules and the charged species or sites.

Organic species capable of forming hydrogen bonds with mineral surfaces include proton donors such as alcohols, phenols, and acids, and proton acceptors which can act as Lewis bases, that is, species which contain uncoordinated electron pairs or π-electron clouds; these include alcohols, carbonyl and carboxylic species, amines, ethers, and unsaturated and aromatic hydrocarbons. The hydrogen-bonding affinity of a large number of simple organic molecules for hydrated silica surfaces has been analyzed and tabulated [18]; the strength of the bond can be estimated from the perturbation of the infrared SiO–H adsorption bands. The adsorbed organic molecules may contain more than one hydrogen-bonding center, and these may form multiple hydrogen bonds with the mineral surface.

The external surfaces of most common mineral fillers and pigments sufficiently resemble those of hydrated oxides for silica and alumina to serve as models. Silica that has been outgassed at elevated temperatures contains few residual surface hydroxyl groups and does not readily adsorb Lewis bases. Dehydroxylated silica does not form strong hydrogen bonds with water or other proton donors. This is due to the covalent nature of the siloxane bonds and weak basicity of the surface oxygens. The hydroxylated surface can be regenerated by hydrolysis of the siloxane bonds on prolonged contact of the anhydrous silica with water or alcohols. Adsorbed water and alcohol molecules may be bound via single surface hydroxyls (Figure 4.2; 1), but bonding through two adjacent hydroxyl groups (2) is a preferred arrangement [18]. The protons of the adsorbed water molecules are available for hydrogen bond formation with other water molecules of basic species (3). This process enables most oxide surfaces to retain one or two layers of water molecules in a hydrogen-bonded network that often can only be desorbed by drying the mineral at temperatures above 100°C.

The interaction of water molecules with the surfaces of layer silicates can involve weak hydrogen bonds to charged centers on the oxygen sheets, in addition to the stronger interactions with the adsorbed countercations [19]. Hydrogen bond formation is more pronounced with silicates in which the anionic charge originates in the tetrahedral layers; uncharged silicate surfaces, like those of talc, are essentially hydrophobic.

Hydrogen bond formation with the surface of metal oxides can follow a more complex pattern than that on silica. The hydrated oxide surfaces are usually virtually saturated with hydroxylic groups which, like those of silica, can form hydrogen bonds with acceptor molecules. The dehydroxylated surfaces of most metal oxides rehydrate rapidly but, because of their Lewis acidity, they may also form coordination complexes with hydrogen bond acceptors. In the case of adsorbed water or alcohols, coordination with Lewis acid centers on the oxide surface may enhance the acidity of their hydroxyl groups sufficiently for hydrogen bond formation to occur with other adsorbed species. The surface oxygens of "anhydrous" alumina, unlike those of silica, are also sufficiently basic to form hydrogen bonds with donor molecules [19].

4.2 DISPERSION AND REINFORCEMENT

The efficiency of most fillers, pigments, or other mineral additives is dependent on the degree of dispersion of these materials in the suspending medium. The maximum optical effectiveness, for example, of a pigmented polymer composition is only achieved if the pigment is uniformly and finely dispersed in the surrounding medium. The process of obtaining and maintaining a good dispersion involves at least two forms of surface–medium interaction, namely, the wetting of the mineral ingredients by the liquid or molten medium and the stabilization, or retardation of flocculation, of the dispersed particles after their disagglomeration by mechanical shearing of the wetted mineral suspensions [10,20,21].

4.2.1 Surface Wetting

The mineral fillers and pigments are usually supplied in the form of dry powders, consisting of air-filled porous agglomerates of particles that form

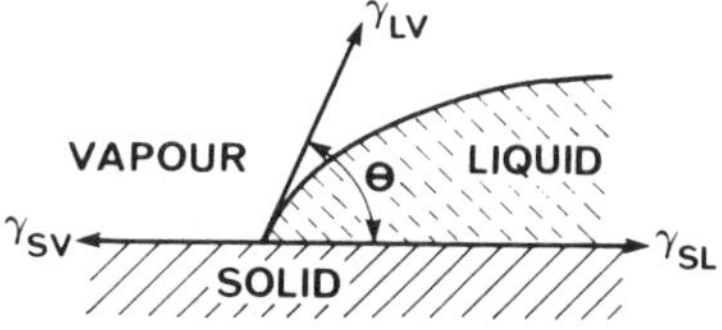

Figure 4.3 Surface tension diagram for the solid–liquid–vapor interface formed by a drop of liquid on a planar surface.

during the packing, storage, or transport of the solids. In the initial stage of the dispersion process, the liquid medium has to wet the mineral in order to displace air from the surfaces and interstices of the agglomerates before the particles can be dispersed. Displacement of adsorbed moisture may also be necessary for the effective dispersion of minerals in water-immiscible media. The disruption of agglomerates is usually assisted by the use of surface-active agents or grinding aids, either as mineral treatments or as formulation additives. The use of wetting agents not only reduces the mixing time and energy requirements, but also helps to avoid overmilling and fracture of particles which, in the case of colored pigments, can result in changes in their hue.

The formal treatment of the wetting process can be found in numerous reference works and reviews [e.g., 1,21,22]. The quantitative aspects will not be described in this book. Reference 21 also includes a survey of surface active agents and dispersion technology.

The wetting of a solid by a liquid is often described in terms of the equilibrium contact angle (θ) formed at the solid–liquid–vapor(air) triple interface (Eq. 8) or the spreading coefficient (S; Eq. 9); both are functions of the surface tensions of the three paired interfaces (Figure 4.3).

$$\gamma_{SV} = \gamma_{SL} + \gamma_{LV}(\cos\theta) \tag{8}$$

$$S = \gamma_{SV} - (\gamma_{LV} + \gamma_{SL}) \tag{9}$$

The maximum spontaneous wetting required for displacement of air from interstices is obtained when the contact angle (θ) is zero or when the spreading coefficient has a large, positive value. The wetting agent should reduce the liquid–solid surface tension (γ_{LS}) as well as the liquid–vapor tension (γ_{LV}). The wetting of particulate minerals is usually determined indirectly, using microcalorimetry, for example, and may be expressed in terms of the heat of immersion of the mineral in the liquid. In technical practice, the wetting of pigments and fillers is often gauged by their "oil adsorption values" [21].

Most of the aluminosilicate minerals and titanias have highly polar surfaces that are hydrophilic rather than oleophilic and are not readily wetted by organic materials such as hydrocarbon solvents, oils, and polymers, which have only low-to-moderate polarity. These minerals can be readily rendered oleophilic and wettable by the chemisorption of organic molecules, preferably those having structural similarities with the dispersing medium.

4.2.2   Stabilization of Dispersions

The agglomerated minerals do not spontaneously disperse in organic media, but must be disrupted by shearing forces that accompany the mixing process. The dispersed mineral particles will begin to flocculate immediately after the shearing ceases, unless the viscosity of the medium is very high or unless some stabilizing process is present to hinder the close approach and mutual attraction of the particles. This attraction results from long-range London–van der Waals forces that can operate when the particles are brought into close proximity by thermal (Brownian) motion or sedimentation. This close approach can be restrained by two types of repulsive interaction, namely, the Coulombic interactions between similarly charged particles and the steric or entropic interactions between long-chain molecules adsorbed on the particulate surfaces. Coulombic repulsions are most effective in highly polar media, for example, aqueous systems, whereas dispersion stabilization through steric interactions is required in media of low polarity.

When a mineral is suspended in an aqueous medium, the ionization of surface groups or adsorbed species or the adsorption of ions can result in the particles developing an electric charge, which is neutralized by the formation of a diffuse layer of oppositely charged ions in the medium adjacent to the mineral surface [23]. When two particles collide, the overlapping of their diffuse ionic layers results in a repulsive force between them which is proportional to their radii, their surface potentials (ζ-potentials), and the dielectric constant of the medium. As the attractive and repulsive forces increase at different rates with decreasing separation of the particles, the net interaction over a range of interparticle distances may be repulsive (Figure 4.4). If the potential energy of this repulsive interaction exceeds the kinetic energy of the particles due to their thermal (Brownian) motion, the dispersions will resist flocculation. Sedimentation of the individual particles in the gravitational field will still occur. Repulsive interactions between the settling particles may be insufficient alone to prevent the eventual formation of a dense, hard-caked sediment that may not be easy to redisperse.

Ionic stabilization processes have been extensively studied, and the underlying concepts of ionic dispersion and flocculation are embodied in the DLVO (Deryaguin–Landau–Verwey–Overbeek) theory. A more precise and detailed explanation of the process can be found in references 23–26 or in most books on physical or colloid chemistry. The dispersion of pigments in aqueous systems and the mode of action of some practical dispersants have also been reviewed [27].

The ionic stabilization of dispersions of clays, or of other solids with ionizable surface species, is usually less effective in organic media, as the stabilizing extended diffuse ionic layers do not form in media of low dielectric constant. Instead, the adsorbed counterions are tightly paired with

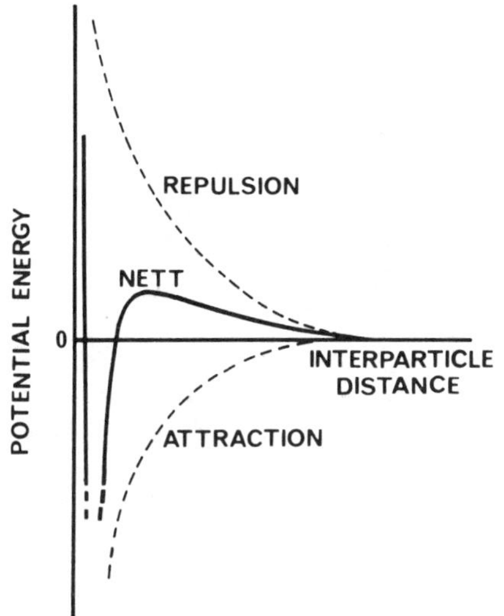

Figure 4.4 Potential energy curves for the interaction of two ionic stabilized, dispersed particles.

surface ionic sites or are constrained to the immediate vicinity of the mineral surface. Adsorbed polar molecules can enhance the ionic interactions, and they can, for example, promote the disaggregation of organophilic smectites used as gellants for nonpolar or weakly polar oganic liquids (see Section 4.4.2).

Mineral dispersions can be stabilized in nonaqueous media by the adsorption of surfactants or polymers. The basic concepts of the nonionic or steric stabilization processes have only recently been developed, although some forms of steric stabilization have been used since antiquity; an example is the use of protective colloids such as egg albumin or gelatin in the preparation of aqueous paints and inks. Steric stabilization can operate in both aqueous and nonaqueous media and can stabilize dispersions in saline media which would flocculate ionic-stabilized systems. Some reviews of the modern theories of steric stabilization can be found in references 6,10,28–30. We only attempt to give an elementary outline of the process here.

Molecules that are effective as steric stabilizers require two properties: (1) that they contain groups or segment blocks which are strongly adsorbed on the mineral surface and (2) that the remainder of the molecule be solvated and able to extend into the surrounding medium. The stabilizers are frequently polymeric molecules, which can form a thick, solvent-swollen, adsorbed layer around each particle. Nonadsorbed polymers can also hinder the close approach of particles [7], and steric stabilization can operate in polymer melts as well as in solvent-based systems [31].

The stabilization process can be illustrated by consideration of two dispersant-coated particles in collision (Figure 4.5). As the particles approach, they are mutually attracted by London–van der Waals interac-

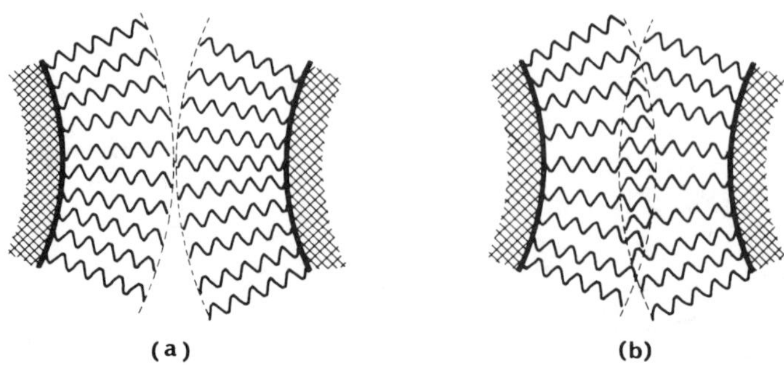

(a)   (b)

Figure 4.5   Steric stabilization by adsorbed long-chain organic species. (a) Protective shells of solvated chains surrounding particles in collision. (b) Zone of repulsion formed by the intermingling and compression of these shells.

tions, the energy of interaction having a minimum value when the interparticle distance is slightly less than twice the mean thickness of the dispersant layers. Closer approach of the particles results in interpenetration of the layers of adsorbed polymer and the development of a net repulsive force between the particles (Figure 4.6). These repulsive forces arise from two principal sources: (1) the restriction of polymer segmental motion in the interpenetration zone, which is equivalent to a decrease in entropy of the system; and (2) the osmotic forces which counter the increase in segmental densities of the "compressed" polymer loops or tails and which are equivalent to an increase in enthalpy [32]. Displacement of solvent from strongly solvated polymer chains during interpenetration can also result in an increase in enthalpy. Stabilization in polymer melts is largely due to the elastic contributions from the compressed polymer chains to the net repulsion [31].

The restriction of segmental motion, particularly of polymer tails [29], is believed to be the dominant stabilizing force in most media of low or moderate polarity. Such steric stabilization is described as entropic stabilization. In favorable cases, entropic-stabilized dispersions can be recognized by the increase in their stability with increasing temperature.

The thickness of the adsorbed layer necessary for the effective stabilization of the dispersed particles increases with increasing particle size. Small particles such as those of carbon black or fumed silica require stabilizing layers with thickness of the order of 10 to 20 Å; these can be provided by the adsorption of low-molecular-weight surface active species such as the long-chain quaternary ammonium ions. However, dispersions containing pigment or filler particles, with equivalent spherical diameters of 0.2 to 1.0 μm, may require adsorbed layers of 100- to 200-Å thickness for effective stabilization; such layers can only be provided by high-molecular-weight polymers [10,22]. As the polymer must be solvated and swollen, the dispersing medium must be a better than θ-solvent for the segments of the polymer not attached to the particle surface. Weaker solvents may be suit-

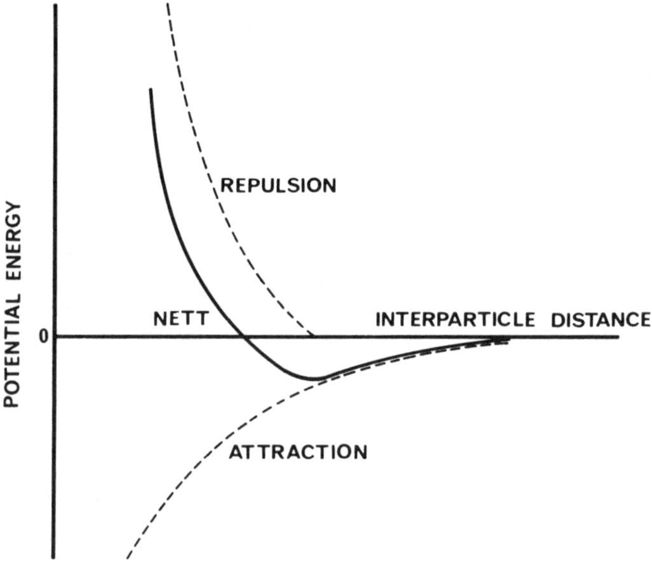

Figure 4.6 Potential energy curves for the interaction of two steric stabilized, dispersed particles.

able for dispersion of minerals, provided ionization of substituent groups in the coating can occur to assist in the steric stabilization process. A considerable variety of polymeric dispersants has been reported in the scientific, technical, and patent literature, and some examples will be discussed later in this chapter.

The interactions shown in Figure 4.4 may be qualitatively true for small particles. However, attractive interactions between large particles decay less rapidly than those between small particles. (The standard equations showing an inverse sixth power dependence of the London–van der Waals forces on interparticle distance are only applicable for interactions between small molecules, that is, between virtual point species.) As a result, the relatively large (0.1 to 1.0 μm diameter) particles, that constitute many fillers and pigments may have a weak net attractive interaction at intermediate interparticle distances, as in Figure 4.6 [10]. This may result in partial flocculation of the dispersion, although the weakly bound aggregates would also be easily disrupted by the Brownian motion of the pigment particles; the weak flocculation phenomenon is useful in control of the rheology of paint systems.

4.2.3 Rheology and Reinforcement

An ideal particulate dispersion should show Newtonian flow behavior (Figure 4.7; curve A), in which the rheological shear stress (τ) is directly

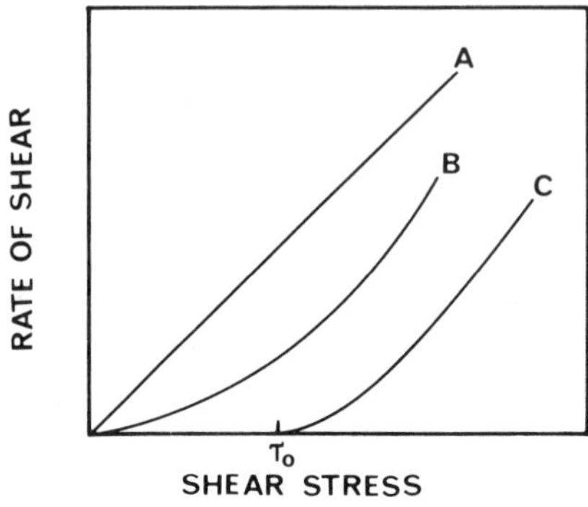

Figure 4.7 Three types rheological behavior exhibited by particulate suspensions. Curve A, Newtonian flow. Curve B, pseudoplastic flow. Curve C, pseudoplastic flow with yield.

proportional to the viscosity (η) and shear rate (D) (Eq. 10), with the viscosity independent of the shear rate.

$$\tau = \eta D \tag{10}$$

Suspensions of mineral particles in organic liquids commonly show non-Newtonian plastic or pseudoplastic rheological behavior, in which the apparent viscosity of the suspension and shear stress is markedly dependent on the shear rate (curve B). In some systems (curve C) a certain minimum shear stress, the yield stress (τ_0), must be applied for the suspension to flow. Suspensions of titania pigments or clay mineral fillers usually show pseudoplastic behavior that can be described by Casson's equation (Eq. 11) [33]. The values of yield stress and viscosity at infinite shear rate (η_∞) are characteristics of the particular suspension system. The rheological behavior of other liquid–solid systems may be more accurately described by alternative equations [34].

$$\tau^{1/2} = \tau_0^{1/2} + \eta_\infty^{1/2} D^{1/2} \tag{11}$$

The yield stress reflects the shear stress required for breakdown of the "structures," or agglomerates of mineral particles. These structures usually reform when the shearing ceases, but this is a time-dependent process, with time scales of seconds to hours for noticeably thixotropic systems. This can result in marked hysteresis in shear stress with a change in shear rate. The interparticle forces that hold the structures together may arise from several sources: from London–van der Waals attractions; from ionic interactions, as in the case of the edge–face interactions of the layer silicates; and from the formation of water bridges between particles, if these carry multilayers of adsorbed water and are suspended in a water-immiscible medium [34,35]. These interactions may be reduced by the use of wetting agents, dispersants, or by drying the mineral before use [35].

The mechanical properties of a mineral filled or pigmented polymer composition are also dependent on the properties of the mineral–polymer interface [36]. In most plastics, the modulus of the filler or pigment is considerably greater than that of the polymer matrix. As a result, unless the surface of the mineral has been modified, the particulate-filled compos-

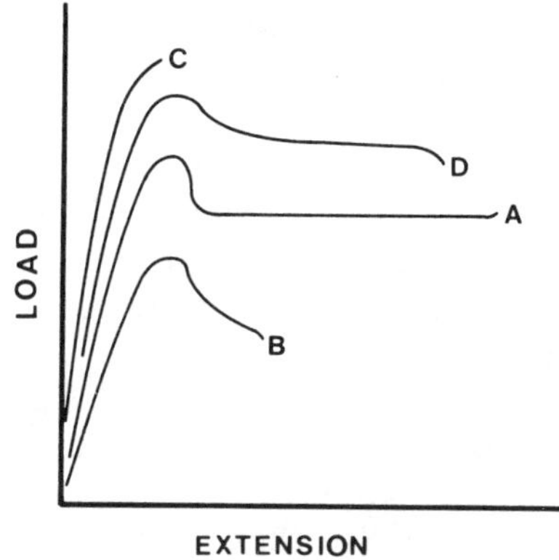

Figure 4.8 The tensile behavior of some polymers and filled polymer systems. See text for explanation.

ite will be harder and less flexible than the unfilled polymer. It will also be more brittle, because interfacial stresses, which develop when the composite is subject to mechanical stress, can result either in adhesive failure at the interface or in cohesive failure of the matrix polymer adjacent to the interfaces. The qualitative tensile properties of some classes of mineral–polymer composites containing nonfibrous fillers are shown in Figure 4.8.

Unfilled, semicrystalline, flexible polymers such as high-density polyethylene or polypropylene usually have a maximum tensile strength that is approximately equal to the yield stress at which irreversible deformation occurs; these polymers may undergo considerable subsequent deformation or extension before breaking (curve A). The incorporation of filler particles that have low interfacial adhesion with the polymer results in the reduction of both yield stress and extension-at-break (curve B). In contrast, the inclusion of a similar amount of a filler that is capable of forming strong interfacial bonds with the polymer may result in an increase in the elastic modulus and yield stress for the composite (curve C), as well as reducing the tendency for slow extension (creep) under long-term stress. While such composites are stronger than the unfilled polymer, they are not necessarily tougher, as cohesive failure of the polymer in the highly stressed interfacial regions can still result in brittle behavior. The brittleness may be reduced if the filler–matrix bonds are capable of breaking and reforming to redistribute this stress.

Treatments that can result in reduction of the stress gradients at the polymer–filler interfaces can improve the toughness of the composites. One method for achieving this is to surround each filler particle with a layer of modified matrix polymer that has a modulus intermediate to that of the filler and the bulk of the matrix (Figure 4.8; curve D). Ideally, such layers should have a graded modulus distribution; this has been claimed for the Ceraplast® fillers described in Section 4.7.2.

Criteria for the selection of fillers for particular applications are outside the scope of this book, but these have been adequately summarized elsewhere [37,38]. The reinforcement of polymers by finely dispersed fillers and the relationship between polymer morphology, molecular orientation, and tensile properties have also been reviewed [39].

The presence of small amounts of finely divided fillers can modify the development of spherulites in semicrystalline polymers such as the fiber-forming polyamides and the polyolefins and thereby alter the bulk tensile properties of these polymers. The fillers may also modify the properties of the polymer in zones adjacent to the filler surfaces, for example, by organization of the spherulites. The inherent modification of the viscoelastic properties of the matrix polymer in regions adjacent to a high-modulus surface is derived from steric factors and therefore should not be affected by chemical modification of the filler surfaces, provided that the modifying species or their byproducts are not desorbed and cannot themselves alter the bulk properties of the polymer. This is true for some systems, for example, calcite–polyolefin mixtures [40], but in others, acid–base interactions between the filler and the polymer can result in long–range effects in the matrix polymer (see Section 4.6).

4.3 MODIFICATION BY ADSORPTION

The modification of mineral or inorganic surfaces by the adsorption of organic molecules can occur via a variety of processes, introduced in Section 4.1. In this section we intend to describe modifications by the adsorption of acids, bases, their salts, and neutral compounds. The division of the adsorbates into acids, bases, and neutral compounds is based on a notional "litmus" test, although the chemisorption of neutral or even acidic compounds may largely result from their basicity, that is, their ability to act as proton acceptors or electron-pair donors.

Although organic molecules can be attracted to mineral surfaces by van der Waals forces, these forces are so weak that low-molecular-weight species may not be effectively adsorbed, unless the molecule contains functional groups that can react with surface species, resulting in chemisorption. The chemisorption processes discussed in this section include hydrogen bond formation, proton transfer, and coordination between the adsorbed species and the mineral surface. We shall concentrate on low-molecular-weight adsorbates; polymeric adsorbates are discussed in Section 4.6.

The surfaces of hypothetical minerals have been described in previous chapters by reference to "box" structures. One feature of all these structures is the presence of surface hydroxylic species which are capable of protonation, deprotonation, and of coordination with other metal ions; these hydroxylic species may include chemisorbed water. As the underlying metal cations are potential Lewis acids, their aquo or hydroxo ligands can be replaced by other ligands, which may include bases or the anions derived from weak acids. A semiquantitative estimate of the stability of the metal oxide–organic adsorbate complexes can be obtained from an examination of the stability of the complexes formed between the organic species and the hydroxo or aquo metal ions in solution [41]. Adsorbates having polarizable groups, particularly, carbonyl, carboxyl, amide, or nitrile derivatives, can form ion–dipole physisorption complexes with metal ions having high polarizability or coordination complexes with transition metal ions, as well as forming hydrogen bonds with surface hydroxylic species.

4.3.1 Adsorption of Acids and Their Salts

Carboxylic acid substituents comprise the mineral bonding groups in a large range of surface active agents used for modifying mineral surfaces; carboxylate–mineral interactions are also effective in the adsorption and bonding of polymeric materials to most of the commercial inorganic fillers and pigments. Most of these minerals have basic or amphoteric surface sites and are expected to form strongly bound surface carboxylates, as well as weaker hydrogen-bonded adducts, with the carboxylic acid-containing polymers. Chemisorbed anionic adducts are also formed by the analogous reactions of other organic acids, such as phosphonic or sulfonic acids, or sulfate hemiesters.

The treatment of calcium carbonate is one of the major applications for the modification by adsorption of acids. The surface of calcium carbonate is alkaline and can react readily with adsorbed fatty acids or their anions to form surface carboxylates. Reactions can also occur with solid fatty acids. Fatty acids are often added as modifying agents during grinding or milling of the carbonate rock. The most commonly used agents are C10 to C18 acids, such as lauric (C12) or stearic (C18) acid, or fatty acid mixtures derived from tallow or oils.

When fatty acids are adsorbed on calcite from solution, they initially form a monolayered array of alkyl chains oriented so that the carboxylate groups are adjacent to the mineral surface. Additional acid molecules may then be adsorbed in a second layer, with their chains oriented tail-to-tail to the first molecular layer. Subsequent layers may be adsorbed, with head-to-head and tail-to-tail orientation with respect to the adjacent layers. This alternating orientation is reflected in cyclic variations in the moisture absorption (hydrophobicity) and surface conductivity of treated calcite with increasing amounts of adsorbed acid [42]. In the case of the lauric acid treatment of calcite, a monolayer coverage of laurate molecules is insufficient to completely screen the polarizing effects of the underlying calcite surface. Maximum hydrophobicity requires the adsorption of the equivalent of three molecular layers of the acid [42].

Calcium carbonate is slightly soluble in water, but calcium laurate and stearate are less soluble. Treatment of aqueous suspensions of the mineral with sodium fatty-acylate solutions results in the diffusion of calcium ions from the lattice and the formation of a hydrophobic calcium acylate surface layer. The acylate layer may also comprise a series of molecular layers having alternating orientation with respect to adjacent layers. The first molecular layer, formed in the reaction with sodium laurate solutions, contains chemisorbed laurate ions and dissolved, unionized calcium laurate; subsequent layers consist of calcium laurate [43].

The adsorption of carboxylic acids on metal oxides such as alumina and titania can also result in the formation of surface carboxylate species. In contrast, the adsorption of carboxylic acids on acidic surfaces such as those of silica results in the formation of hydrogen bonds between the surface hydroxyl groups and the carbonyl groups of monomeric or associated dimeric forms of the acid [16]. The carboxylic acids can be readily desorbed from silica, whereas the chemisorbed carboxylates are tenaciously retained on the surfaces of basic minerals.

The adsorption of carboxylates on silica, silicates, and other weakly basic or negatively charged surfaces can be enhanced by the use of their oligomeric basic chromium or aluminum salts. These materials can undergo condensation and coordination polymerization with the surface hydroxyl groups to form hydrolysis-resistant carboxylate surface layers. These treatments are described in Section 4.5.2. Weakly basic or acidic fillers and pigments can also be modified by the addition of polyvalent cations to a carboxylate-containing dispersion of the minerals. This results in the precipitation coating of an insoluble acylate layer onto the particulate surfaces. This method of modification is frequently used for the preparation of oleophilic iron oxide or chromate pigments; titanyl or aluminum acylate coatings have also been used for the improvement of titanias [44]. Alternatively, aqueous suspensions of metal oxides may be treated with Al^{3+} ions, which are adsorbed and impart a positive charge to the particulate surfaces. The suspension is then treated with an anionic surfactant such as an alkylaryl sulfonate, forming a hydrophobic product that can be redispersed in nonaqueous media [45].

The interaction between adsorbed acids and the lamellar aluminosilicate clay minerals depends on the nature of the mineral and the exchangeable cations. Anion (ligand) exchange with hydroxo-aluminum ions exposed at the octahedral layer edges yields surface carboxylates. Similar reactions occur with adsorbed oxyaluminum species, which can be formed by the hydrolytic degradation of the mineral lattice or with other di- or polyvalent exchangeable ions on the mineral surfaces. Hydrogen bond formation can occur between the adsorbed acid and coadsorbed water associated with the charged surface species. Weak interactions may also occur between the silicate oxygen sheet and the alkyl or alkenyl chains of saturated and unsaturated aliphatic acids [46]. Ca-montmorillonite can intercalate aliphatic monocarboxylic acids to form two types of adducts. Acids with less than 10 carbons are held in the interlamellar spaces, with their alkyl chains lying nearly parallel to the silicate surfaces. Similar structures can exist in adducts containing C10–C18 acids, at temperatures below the melting points of the respective acids. At higher temperatures, these adducts have structures in which the interlamellar acid molecules are held in ordered, inclined arrays [47], analogous to those of some long-chain alkylammonium smectites.

The adsorption of carboxylic acids by montmorillonites in aqueous media is dependent on the pH of the medium and on the dissociation constant of the acid [148]. Negative adsorption, resulting from the repulsion of the carboxylate ions from the diffuse double layer, can occur at high pH. Positive adsorption of the acids occurs when the pH of the medium is 1 to 1.5 units above the pK_a value for the acid. Adsorption is greatest on the "hydronium" clays, presumably because of their content of highly charged exchangeable Al^{3+} ions or adsorbed oxyaluminum species.

Treatment of basic or amphoteric minerals by organic acids and salts can be used to impart hydrophobicity and to promote the wetting of their surfaces by organic solvents. The surface treatment of chalk by lauric or stearic acid results in greatly reduced rheological yield stress of the mineral suspensions in paraffin oil, a medium frequently used as a model for nonpolar polymers [49]. Such treatment also improves the compatibility between the filler and polymers such as PVC resins [50] and reduces the brittleness of most plastics formulations containing the treated mineral. Treatment with unsaturated acids can provide a hydrophobic, graft-forming mineral surface, which can promote reinforcement when the fillers are used in radical-cured plastics and elastomers. Similarly, dicarboxylic acids can provide a carboxylic acid functional surface, capable of forming grafts by condensation with polymer reactive centers during the curing of epoxy, polyurethane, or similar resins [51]. Treatment with diacids can also improve the compatibility between a basic mineral such as calcite or metal oxides and "basic" polymers such as the polyamides and amino resins.

When unsaturated acids such as acrylic or linoleic acids are adsorbed on basic oxide surfaces at low surface coverage, they form chemisorbed carboxylate layer structures in which the molecules lie prone on the oxide surface. This configuration allows weak adsorptive interactions to occur between the surface hydroxyls, or Lewis acidic centers, and the π electrons of the unsaturated groups of the adsorbed acids [51]. At higher surface coverages, the adsorbed acids may form an erect, close-packed array which, in the case of acrylic and methacrylic acid treated minerals, can undergo radical-induced homopolymerization or graft copolymerization with other adsorbed or contacted monomers. [51].

Monocarboxylic acids are usually too weakly adsorbed on lamellar silicates for effective conversion of the latter to reinforcing fillers, although these can be produced by treatment of a di- or polyvalent cation-exchanged mineral, for example, Ba- or Pb-kaolinite, with a soluble acrylate salt of the same cation, followed by drying at elevated temperatures [52]. The

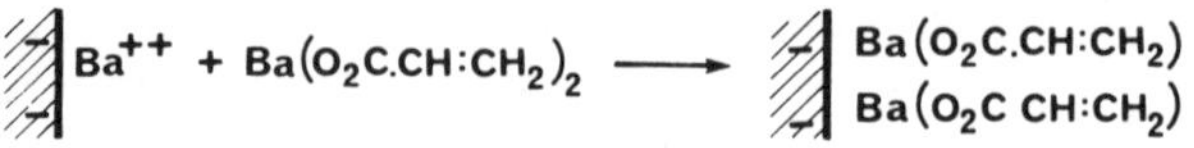

$$\text{—}\!|\;Ba^{++} \;+\; Ba(O_2C.CH\!:\!CH_2)_2 \;\longrightarrow\; \text{—}\!|\;\begin{array}{l} Ba(O_2C.CH\!:\!CH_2) \\ Ba(O_2C\;CH\!:\!CH_2) \end{array}$$

Figure 4.9

improved bonding of the acrylate or other acylate moieties probably arises from the formation of acylate ligands with the exchangeable cations. (See Figure 4.9.)

Chrysotile (asbestos) crystallites are formed with the octahedral layer as the external surface of the coiled structure. This basic layer can react with carboxylic acids to form chemisorbed carboxylates. Treatment of chrysotile dispersions with acrylic acid has been used for the preparation of reinforcing fillers [53]. The surface layer can also be attacked by surfactants, for example, sodium dioctyl sulfosuccinate. Treatment of aqueous chrysotile suspensions with excess sulfosuccinate is used to assist in the breakdown of massive aggregates in order to form dispersions of the primary colloidal tubular particles (Chapter 1). Treatment with lesser amounts of the dispersant results in the formation of fibrils having a hydrophobic monolayer of chemisorbed sulfosuccinate, which can assist the dispersion and improve the reinforcement of plastics and elastomers by the treated asbestos [54].

The reaction of moistened polymeric acids, for example, polyacrylic acid, with basic oxides such as magnesia or zinc oxide, or with acid-sensitive magnesium or zinc orthosilicates or pyrosilicates, can be used in the formulation of cements and tile grouting [e.g., 55]. Hardening of these compositions occurs by attack of the organic acid on the mineral, followed by diffusion of the released divalent cations into the acid phase. This results in the formation of a cured composite containing the mineral particles in an ionic cross-linked polymeric matrix.

4.3.2 Adsorption of Bases

Basic molecules like the primary, secondary, and tertiary amines are strongly adsorbed on a wide variety of mineral surfaces. In many respects, the adsorption of organic amines parallels that of ammonia on the various minerals discussed in the previous chapters. As well as forming protonated adducts with strongly acidic surfaces, the amines can form hydrogen bonds with weakly acidic or basic hydroxylic surfaces and with hydrated exchangeable cations. They can also act as ligands for coordination with other exchangeable cations or with the lattice cations of metal oxides [41]. Discussion in this subsection will be limited to amines as examples, as these are virtually the only types of organic base used for the commercial surface modification of fillers and pigments.

Pure hydrated silica gel is insufficiently acidic to protonate even the strongly basic aliphatic tertiary amines. Chemisorption of these is limited to the formation of hydrogen bonded adducts. The chemisorption of primary or secondary amines on dehydrated silica can result in the ammonolysis of reactive surface siloxane bonds and the formation of surface silylamine species [16]. The acidity of silica can be greatly increased by the presence of cationic impurities such as Fe^{3+} or Al^{3+} ions, as these can form strong Bronsted acidic centers.

Although pure anhydrous alumina can act as a strong Lewis acid, that is, forming coordination complexes with amines, hydrated alumina is not a Bronsted acid and only forms hydrogen-bonded surface complexes with these bases. However, silica–alumina gels, exposed aquo-aluminum ions at the lattice layer edges of aluminosilicates, and adsorbed, hydrated, cationic oxyaluminum oligomers are all Bronsted acids that can protonate organic amines [56]. More basic oxides such as magnesia adsorb amines by hydrogen bonding, while transition metal oxides may also adsorb amines as ligands. The adsorption of amines on pure titanias resembles that on alumina, but impurities resulting from the manufacturing process, particularly, adsorbed chloride, sulfate, or bisulfate ions, can generate strongly acidic adsorption sites. The silica–alumina gel coatings used for modification of the dispersion and photochemical properties of commercial titania pigments can be formulated to provide an acidic surface that can promote the adsorption of amine used for the production of oleophilic pigments [57].

The adsorption of amines on clay mineral surfaces is largely determined by the nature of the exchangeable cations and the water content of the particular mineral. The adsorption of amines on a variety of natural kaolins has been reported to occur predominantly on layer edge sites and to involve the formation of localized adducts, rather than ionized alkylammonium species [58]. The interaction between alkylamines and beneficiated Al-kaolins results in the formation of alkylammonium ions [59], which are adsorbed adjacent to the cation exchange sites, located largely on the basal surfaces of the mineral. Hydroxonium-bridged amine dimeric species are believed to be formed in aniline–anilinium montmorillonite adducts (Figure 4.10; 4) [60]. Water bridges are also important in the bonding of aniline and other weakly basic amines or amides to the exchangeable cations 5 [61,62].

The adsorption of amine vapors by hydrated Ca-montmorillonite has been reported to involve the formation of symmetrical, hydrogen-bonded, protonated, dimeric species that can revert to unprotonated amine on drying of the mineral [63]. Protonation of the adsorbed amines, like that of adsorbed ammonia, is dependent on the presence of adsorbed water [64], with the anionic silicate surface interacting with, and stabilizing, the protonated species. The stability of the protonated amines increases with increasing anionic charge on the smectite silicate surface [65], a trend which is the opposite of that observed for intercalated alkyl alcohol complexes [66].

Adsorption of amines by clay minerals from aqueous solutions is com-

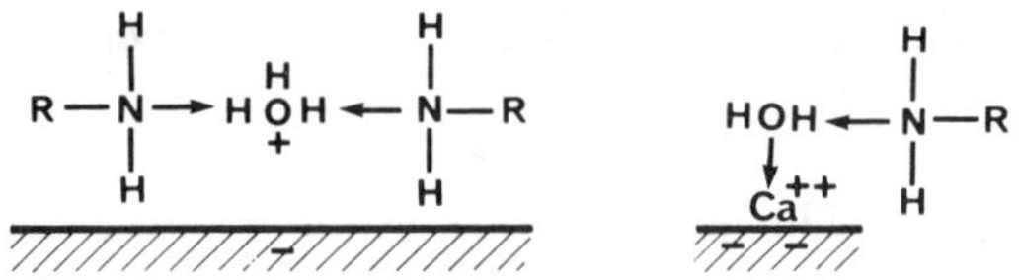

Figure 4.10

plicated by the occurrence of ion exchange. Protonation of bases in clay suspensions occurs to a much greater extent than that predicted from the pH of the aqueous phase. The adsorption of bases is dependent on the surface acidity of the mineral, rather than on the pH of the aqueous phase. Strong adsorption of amines can occur when the surface acidity is 1 to 2 units below the dissociation constant of the protonated base; the surface acidity of soil clays is typically 3 to 4 pH units below that of the aqueous phase [48].

Adsorption of amines on Cu^{2+}- and Co^{2+}-exchanged montmorillonites involves an exchange with one or two ligand water molecules. The stability of these adsorbed complexes increases with increasing chain length of the amine, which is the opposite of the trend in stability of these alkylammino Cu^{2+} complexes in solution [67]. The Cu^{2+}-exchanged minerals avidly adsorb and concentrate amines from dilute solution, whereas the homogeneous complexes are only stable in the presence of excess amine. The concentration, by adsorption on clays containing adsorbed copper ions, of amines and related aminoacids may have been an important factor in the prebiotic syntheses of peptides and other essential precursors (Section 6.3).

The principal aims of the amine treatment of the metal oxides and non-swelling clays are to produce hydrophobic, oleophilic surfaces that can result in improved wetting by, and dispersion in, nonpolar media; to reduce water adsorption by the oxide; to form graftable or other reactive centers on the mineral surface by use of multifunctional amines. Amine treatments are frequently used to neutralize the residual surface acidity of titanias [68] and kaolins [35]. Adsorbed aminoalcohols may also be used to assist dispersion of the pigments in aqueous media, particularly during post-treatments. Titanias do not disperse readily in neutral or acidic media, and amine surface modifiers such as triethanolamine can be used as additives for obtaining high-solids low-viscosity aqueous dispersions as well as improving the dispersibility of the pigments in polar organic media [69]. Aminoalcohols are also used as one of the components of the inorganic phosphate coatings applied to some varieties of titania pigment [70].

The amines most frequently used for imparting oleophilicity are primary or secondary C6 to C18 aliphatic amines. Improvement in adsorption can be obtained by use of polyamines; these are often derivatives of 1,2-diamines or 1,2-hydroxyamines which can form chelation complexes with metallic ions. The adsorption of the amine and the oleophilic properties of the treated minerals can often be improved by pretreatments which can increase the surface acidity. These means of activation include removal of the bulk of the adsorbed water [59]; adsorption of acidic anions, such as bisulfate ions, by the sulfuric acid treatment of titania, or of alumina-rich silica–alumina coated titanias, for example, [57,71,72]; adsorption of acidic, hydrated Al^{3+} ions, which can also occur during the alum flocculation of kaolins, or other anionic mineral dispersions.

Although amine treatments have been shown to be adequate for modification of acidic or dried mineral surfaces, a number of patent specifications claim that improved dispersion of minerals in organic media, and improvement in other properties of mineral organic composites, can be obtained by the adsorption of amine salts including those of relatively weak carboxylic acids. The adsorption processes in these cases can be complex and may include partial ion exchange, for example, with the compensating cations on the surfaces of kaolin. Unlike the ion-exchange treatments described in the next section, the procedures described in these patents do not allow for the removal of the displaced cations. Alkylammonium acylates are often more strongly adsorbed on inorganic surfaces, particularly those of amphoteric oxides, than either the alkylamines or the acids.

Specific examples of the modification by the adsorption of amine salts include the precipitation and adsorption of N-(long-chain alkyl)-alkylene diammonium dioleates on a variety of inorganic colored pigments, for prevention of agglomeration and improvement of the dispersion of the pigments in oil-based media [73]; the treatment of kaolin with the acetates or laurates of ethylene diamine, diethylene triamine, or other polyamines to improve the dispersion of the mineral in polar and nonpolar organic media, specifically, styrene and unsaturated polyesters [74]; and the treatment of kaolin with solutions of diethylaminoethyl methacrylate salts, followed by evaporation of the solvent, to produce an oleophilic reactive mineral filler, for use in styrene-unsaturated polyester resin systems [75]. General purpose organophilic kaolins can also be prepared by coating the particles with a mixture of aluminum and ethylene diammonium oleates [76].

4.3.3 Adsorption of Neutral Compounds

The amides, ketones, alcohols, and other neutral compounds discussed in this section could be more accurately described as bases other than amines because similar modes of adsorption are involved, namely, hydrogen bond formation, protonation, and coordination. Protonation may not be significant in the adsorption of these weak bases, except on strongly acidic surfaces such as those of the dehydrated aluminosilicates. The bonding of these bases to the surfaces of typical mineral fillers and pigments is relatively weak, unless multiple bonds can be formed, and the examples of useful modification of mineral surfaces by small neutral organic molecules

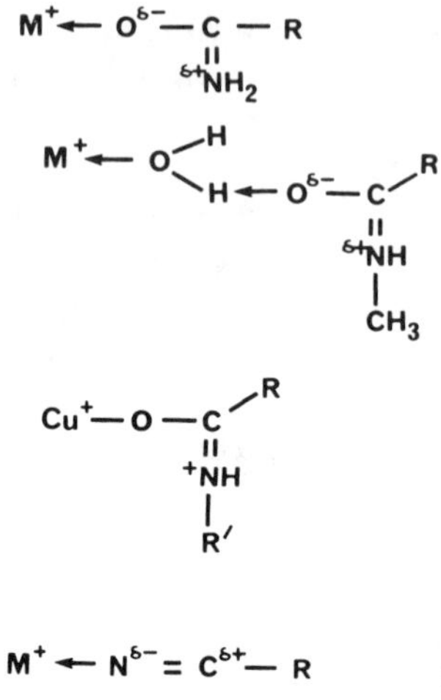

$$M^+ \leftarrow O^{\delta-} - C - R \qquad (6)$$

Figure 4.11

are limited in variety. These interactions are still important in mineral organic chemistry, and in the adhesion of polymers to mineral surfaces, where tenacious adsorption can be obtained by the formation of a large number of individually weak polymer–mineral bonds.

One feature of the complexes formed by the adsorption of compounds containing groups capable of mesomeric polarization, such as the various carbonyl derivatives (6) and nitriles (7), is the importance of ion–dipole interactions between these species and cations having high charge and small ionic radius, for example, Mg^{2+}, Ca^{2+}, or Cu^{2+} ions (see Figure 4.11).

Silica has been shown to adsorb nitriles, alkyl and aromatic ethers, phenols, aldehydes, ketones, amides, and esters by hydrogen bond formation between surface hydroxyl groups and electron pairs or π-electron clouds of the organic functional groups [16]. Hydrogen-bonded interactions also predominate in the adsorption of these species by hydrated alumina. Adsorption on activated, dehydrated alumina and other metal oxides often involves dissociation of the adsorbed molecule forming, for example, alkoxides from alcohols and π-allylic carbanions from alkenes; these reactions are discussed in Chapter 5. Adsorption on clay minerals is often dependent on coordination with, or hydrogen bonding via water bridges to, exchangeable cations.

The adsorption of some specific classes of neutral compounds is described below. The adsorption of silicone oils, used for the surface modification of coated titanias and other pigments, is described in Section 4.6.

i Alcohols and Ethers

Alcohols are strongly adsorbed by montmorillonite, and the use of glycol and glycerol in the determination of smectite structures has been described in Chapter 1. Low-molecular-weight alcohols, such as ethanol, can compete with water for adsorption as ligands on the interlamellar cations and can also act as effective dehydration agents for clays. The interactions between the intercalated alcohol molecules and the mineral appear to involve ion–dipole interactions, rather than hydrogen bond formation with the silicate surface; the nature of the cation has been reported to have a greater effect on the stability of the complex than does the charge density of the silicate surface [66].

Oligomeric ethers can form weak intercalation complexes with alkali or alkaline cation-exchanged montmorillonites and other smectites, but the organic species are readily displaced by adsorbed water. Macrocyclic crown ethers such as 15-crown-5 or dibenzo-18-crown-6, which can coordinate alkali or alkaline cations, can irreversibly displace water molecules associated with the exchangeable cations of montmorillonites, forming stable intercalation complexes [77]. The intercalated crown ethers are not displaced by solvents or salt solutions. The complexes are reported to maintain the cation exchange capacity of the original mineral in saline media.

Long-chain alcohols intercalated in smectites can form regular, inclined arrays of alkyl chains [78], although the organization and basal spacings of the intercalation complexes are more variable and temperature sensitive than those of the alkylammonium smectites. Studies of the structures of alkanol–alkylammonium smectite complexes have indicated that this variability arises from changes in conformation of the alkyl chains with temperature, with the formation of kink-blocks [79]. These changes in conformation may also be reflected in temperature-sensitive variations in the properties of surfaces modified with long-chain surfactants.

Polyols having low volatility, such as trimethylolpropane or pentaerythritol, or ethylene oxide oligomers, are often used for the surface treatment of titanias to improve dispersion properties of the pigments [e.g., 80]. They can also be used to control the surface acidity of minerals by displacement of ligand waters and other potential sources of protons adsorbed on Lewis acid surface sites [81]. The mode of action of the polyols in the "improvement" of titanias is often not clear. The polyols may be added during the final fluid energy milling of the pigment, when they can act as grinding aids in order to prevent the reformation of strong aggregates or agglomerates [82]; they can also limit dust formation during handling of the treated pigments. The polyols can act as wetting agents for dispersion of the pigments in polar media, but the neutral polyols are only weakly adsorbed and are displaced when the treated titanias are mixed with alkyd resins or with moderately polar polymer solutions [83].

ii   Aldehydes, Ketones, and Esters

These carbonyl derivatives, like the alcohols and ethers, form hydrogen-bonded adducts with the surfaces of silica, alumina, and titania, although enolate complexes may also be formed, particularly in the case of 1,3-

diones which can form chelates with metal ions [84]. The carbonyl compounds can be intercalated in smectites to form weak complexes with the interlamellar cations, either via water bridges to strongly hydrated cations such as Ca^{2+} or by ion–dipole interactions, for example, to Na^+ ions [84]. These carbonyl species are Lewis bases and have been used for the control of the surface acidity of clay minerals [81].

The adsorption of carbonyl compounds and of other bases by smectites is important in agriculture, as many herbicides and pesticides can be adsorbed by clay soil constituents, and this can modify the availability of these agents [48]. Degradative reactions catalyzed by the clay surfaces are also important in the detoxification of the residues (see Section 5.2.5).

iii   Nitriles

Nitrile can form a variety of adducts with mineral surfaces, the most important probably being those formed by hydrogen bonding with surface hydroxyls and by coordination of the nitrile group with metal cations. The interactions between the nitrile groups and metal ions is enhanced in the interlamellar environment of smectites [85]. The stability of the complexes can be correlated with polarizability of the cations, being strong in the case of Cu^{2+} and weak in the case of K^+-smectites [86].

The adsorption of nitriles on partially dehydrated basic oxide surfaces can also involve the formation of a mixture of adduct species having different thermal stabilities. The nitrile molecules may be linked to the oxide surface by hydrogen bonds between a nitrogen electron pair, by coordination of the nitrile group to a surface metal oxide ionic pair, and by bonding between an α-hydrogen and other surface oxide ions. One adduct species (Figure 4.12) containing all these bonds is believed to be a key intermediate in the surface catalyzed hydrolysis of aliphatic nitriles to amides [87].

iv   Amides

The interactions of amides with mineral surfaces have considerable interest because of the large range of amide–mineral surface interactions that occur in natural and artificial systems. These range from the absorption and reactions of natural aminoacids and peptides on clay minerals to the flocculation of mineral dispersions by synthetic polyamides and to the properties of the numerous synthetic polymer–mineral composites such as those formed from pigmented or filled polyamides or urea or urethane polymers.

Amides and polyamides are weak proton acceptors, but can be protonated by acidic sites having Hammett acidity, $H_0 < 1$. These acidic surface sites occur on dehydrated silica–alumina and in the interlamellar regions of montmorillonites. Adsorption may also involve hydrogen bond formation with surface hydroxyl groups and ion–dipole interactions with exchangeable cations. The interlamellar protonation of acetamide in Ca-montmorillonite involves water molecules associated with the Ca^{2+} ions.

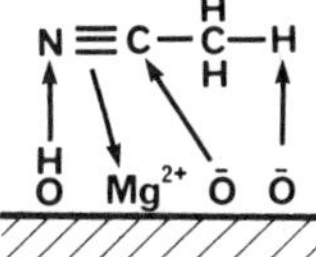

Figure 4.12 The modes of bonding involved in the chemisorption of acetonitrile on magnesia.

The protonated amide can revert to the unprotonated form on dehydration of the complex [88].

The adsorbed, protonated amide molecules can form hydrogen bonds with the carbonyl groups of other coadsorbed amides, aldehydes, or ketones [89,90]. Coordination can also occur between the amide carbonyl group and metal ions; this interaction is promoted by N-alkyl substitution of the amide. Ureas can form analogous protonated and coordinated adducts with interlamellar cations and associated water molecules [91].

The formation of interlamellar complexes between kaolin and amides, such as those of formamide, acetamide, and urea, has been described in Chapter 1. Dimethyl sulfoxide [92], and other polar, neutral organic molecules can be similarly intercalated. These intercalation adducts have low stability, and the intercalated molecules are readily removed by evaporation or by washing the adduct with water or other polar solvents.

v Amino Acids

Amino acids contain both acidic and basic centers and may exist as internal ammonium carboxylate salts or zwitterions. Their adsorption from aqueous media by aluminosilicates can involve ion exchange with adsorbed cations which, in the case of Al^{3+} and Ca^{2+} ions, may be readsorbed by association with the carboxylate groups [93,94]. Alternatively, chelation can occur between α- or β-amino acids and adsorbed transition metal ions [95]. These amino acids are strongly bound to mineral surfaces, and their long-chain alkyl derivatives, for example, can be used as amphipathic surfactants for the dispersion of minerals in aqueous media. 3-(Laurylamino)-propionic acid (8) can be used as an amphoteric dispersant for anticorrosive pigments, such as zinc chromate, in paint formulations [96] (see Figure 4.13).

Aminosulfonic acids and the zwitterionic sulfobetaines (9) have been used as the mineral bonding moieties of surfactants and dispersants. The polymeric sulfobetaines can be readily prepared by the reaction of β-propiosultone with tertiary amino groups, for example, those of a dimethylaminoethyl methacrylate copolymer [97]. The resultant sulfobetaine can

$$C_{12}H_{25} - \overset{+}{N}H - CH_2 - CH_2 - CO_2^{-} \qquad (8)$$

$$\wedge\!\!\wedge\!\!\wedge\!\!\wedge - \overset{+}{N}R_2 - CH_2 - CH_2 - CH_2 - SO_3^{-} \qquad (9)$$

Figure 4.13

form stronger adsorption bonds with oxide and other polar mineral surfaces than those formed by the amino polymer.

4.4   MODIFICATION BY ION EXCHANGE

A major application for the montmorillonites (bentonites) and hectorites and, to a lesser extent, some other smectites and vermiculites is the use of their alkylammonium-exchanged derivatives as thickeners or gellants in organic liquid formulations such as paints [98], lubricating greases [99], cosmetics and pharmaceutical preparations [100], and other compositions that are required to have thixotropic flow behavior. The alkylammonium smectites are also useful as sorbents [5,101,102], for example, for the removal of aromatic pollutants from water [103] or for chromatographic separation of molecules.

There are numerous minor applications for the modification of mineral surfaces in aqueous systems by ion exchange. One is the preparation of colored pigments by the adsorption of cationic dyes, such as Methyl Violet or the Rhodamines, on a Na^+-montmorillonite [104,105]. The products are used as colorants for powders and in paper manufacture. Ion exchange on the surfaces of clays and other minerals is also important in many natural processes.

4.4.1   Alkylammonium Smectites

Cation exchange by alkylammonium ions on the surfaces of clay minerals was introduced in Chapter 1, and detailed reviews can be found in the references 106 and 107. The alkylammonium smectites are prepared from aqueous dispersions of the beneficiated minerals, usually in their Na^+-exchanged or "sodium activated" forms, by the addition of a small excess of the alkylammonium salt solution. This results in rapid ion exchange and, in most cases, the formation and flocculation of a hydrophobic product. The flocculated material is then repulped, thoroughly washed to remove inorganic ions and excess ammonium salt, and then air-dried at temperatures below 100°C [108,109]. Ion exchange favors the formation of complexes of the more basic amine and those having the longest alkyl chains [106,110]. Quaternary ammonium ions are usually used for the commercial preparation of organophilic smectites. These are more strongly adsorbed than the primary, secondary, or tertiary ammonium ions and yield derivatives with greater hydrolytic stability. Most contain one or two long-chain n-alkyl groups (10) or a long-chain alkyl group and a benzyl group (11); in each case, the balance of the four substituents is provided by methyl groups. Organophilic smectites can also be produced by exchange using alkylpyridinium ions (12) (see Figure 4.14).

The quaternary ammonium salts are often prepared from mixtures of

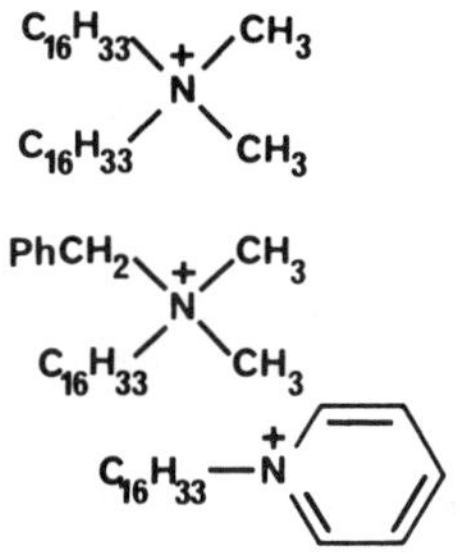

$$
\begin{array}{c}
C_{16}H_{33}\diagdown \overset{+}{N}\diagup CH_3 \\
C_{16}H_{33}\diagup \ \ \diagdown CH_3
\end{array}
\qquad (10)
$$

$$
\begin{array}{c}
PhCH_2\diagdown \overset{+}{N}\diagup CH_3 \\
C_{16}H_{33}\diagup \ \ \diagdown CH_3
\end{array}
\qquad (11)
$$

$$
C_{16}H_{33}-\overset{+}{N}\diagup\!\!\diagup \qquad (12)
$$

Figure 4.14

long-chain amines formed by reductive amination of fatty acids obtained from tallow (predominantly C18 acids) or from vegetable oils (C12 to C16 acids). The long-chain alkyl moieties of the frequently used dimethyldi "tallow"amine salts comprise about 70% octadecyl and 25% hexadecyl groups, plus smaller amounts of their lower (C12, C14) homologues. The well-known National Lead Co. Bentone® organoclays, Bentone-34, and Bentone-38 are the respective dimethyldioctadecylammonium derivatives of bentonite and hectorite. Bentone-24 is a benzyldimethyloctadecylammonium bentonite; similar products are now available from other manufacturers.

X-ray diffraction studies indicate that the interlamellar alkyl chains of the solvent-free alkylammonium derivatives of low-charge montmorillonites or hectorites, particularly those of the more bulky dialkyldimethylammionium ions or alkylbenzyldimethylammonium ions, lie prone on the silicate oxygen sheets [108,111]. This is in contrast to the alkylammonium or trimethylalkylammonium derivatives of the high-charge smectites such as the classical vermiculite, illite, and batavite derivatives described in Chapter 1. In these, the long-chain alkyl groups form well-ordered, compact, erect, interlamellar arrays.

The interlamellar spaces of the low-charge smectites may contain one, two, or three layers of adsorbed ammonium ions, depending on the amount adsorbed. Solvent-free single layer complexes may have typical basal spacings of 16 to 17 Å and an alkylammonium content equivalent to 50 to 70% of the cation exchange capacity of the mineral. Montmorillonites completely exchanged with alkylammonium ions have basal spacings of 22 to 24 Å, which corresponds to the formation of interlamellar structure of loosely packed double layers of alkyl chains; in many cases, these may incompletely cover the silicate surfaces. These organoclays readily adsorb alkylammonium ions, in excess of the mineral's exchange capacity, to complete the formation of a close-packed double layer. The excess ions are held by strong interaction with the oxygen sheets and are accompanied by counteranions which may be exchanged for hydroxyl ions during preparation of the complex; this excess salt can be removed by extraction with alcohol. Further quantities of ammonium salt can be intercalated by the double layer complexes, with the formation of a third pseudolayer; the basal spacings of these complexes are about 29 Å, but are often poorly defined [111,112].

Nitrogen adsorption isotherms (at 78°K) on Bentones 34 and 38 and on Bentone-18C (an octadecylammonium bentonite) indicate that adsorption only occurs on the exterior surfaces of the particles, whereas adsorption of methanol and water can also occur on the interlamellar silicate surfaces [101,113,114]. The sorption capacity of the organoclays for small molecules may be increased if all excess ammonium salts are extracted. Larger molecules are normally adsorbed on the external surfaces of the organoclay particles, and are only adsorbed in the interlamellar spaces if they are polar and can swell the organoclay, or if it has been activated or preswollen by the adsorption of some other, polar species. Smectites exchanged with the less bulky methylammonium or tetramethylammonium ions and having an incomplete coverage of the silicate surfaces by the alkyl chains may adsorb aromatic hydrocarbons that can interact with the silicate oxygen sheets [103,115]. Ion exchange between alkylammonium smectites can occur in weakly polar or nonpolar media if a small proportion of a swelling additive such as methanol is present to assist in the solvation of the alkylammonium ions [116].

Alkylammonium smectites, containing layers of well-ordered interlamellar alkyl chains, may intercalate long-chain alcohols with swelling of the organoclay and the formation of double layer complexes [117]; these have been briefly described in Chapter 1. The alkylammonium smectites can also swell by adsorption of other liquids, the degree of swelling often being measured by the sedimentation volume of the organoclay in the particular liquid or mixture of liquids. Values of gel volumes formed by dodecylammonium bentonite in various liquids have been reported [118]. These show that only slight swelling occurs in aromatic or aliphatic hydrocarbons, whereas moderate swelling can occur in ethanol, acetone, and ethyl acetate, with much greater swelling occurring in dodecanol or nitrobenzene.

The intercalation and swelling of the hydrophobic organosmectites requires solvation of both the cationic centers and the alkyl chains. This can be achieved by the addition of alcohols and the other polar swelling liquids; water or hydrocarbons only solvate part of the alkylammonium ions, and cannot, by themselves, cause appreciable swelling of the organoclays. X-ray diffraction studies of the adsorption of methanol by alkylammonium montmorillonites have shown that, at a critical partial pressure of the alcohol, there is a transition from the prone to an erect configuration of alkyl chains. This change is accompanied by a sharp increase in the organoclay's sorbency [119].

Very extensive swelling and disaggregation of the smectite platelets can occur in mixtures of alcohol and hydrocarbon. The degree of swelling increases with an increase in the chain-length of the alkylammonium ions, in the homologous series from ethylamine to dodecylamine, with decreasing chain-length of the activating alcohol. The swelling reaches a maximum value for some particular percentage content of alcohol in the solvent mixture. This is usually 5 to 10% in the case of methanol or ethanol used in an alkylammonium smectite/hydrocarbon/alcohol system [118].

The extensive swelling in polar liquids results from the dissociation of the alkylammonium–silicate ion pairs and the formation of a diffuse structural layer of solvated cations adjacent to the silicate surfaces [118]. The extent of swelling and the thickness of the diffuse structural layers is partly determined by the chain length of the ammonium ions, the dielectric constant (ϵ) of the solvent, and the electrostatic interactions and osmotic forces operating in the interlamellar region. The solvated layers formed by an octadecylammonium montmorillonite in nitrobenzene ($\epsilon = 35$) may be up to 100 Å thick [120], whereas swelling by diester plasticizers ($\epsilon = 4$ to 8) is limited by the molecular size of the ester, the increase in basal spacing being equivalent to twice the length of the plasticizer molecules [121].

The dimethyldialkylammonium smectites show qualitatively similar behavior to the alkylammonium smectites and will only swell in hydrocarbon mixtures if the mixture contains a small percentage of an activating solvent such as methanol [122,123]. The presence of some residual water is also apparently necessary for swelling in organic media. The organophilic Bentones®, for example, contain about 2% by weight of adsorbed water, approximately two solvating water molecules per ammonium ion. These waters can be removed by drying the organoclay under high vacuum, but the anhydrous materials swell less readily in alcohol–hydrocarbon mixtures or in a number of other solvents that can swell the hydrated organoclay [124].

The alkylammonium smectites are useful sorbents, but generally the material needs to be activated by a polar swelling agent for maximum sorbency. The organoclays can adsorb alcohols from aqueous media. The adsorption isotherms indicate that the interlamellar organic phase behaves as a solvent, with activity higher than that of carbon tetrachloride [125]. The organoclays have also been used as shape-selective sorbents for vapor phase chromatographic separation of organic molecules, adsorption occurring predominantly on the external particulate surfaces. The best chromatographic efficiency has been obtained using derivatives of smectites, such as an alkylammonium vermiculite [126], which have a high-charge density originating in the tetrahedral layers. These sorbents have been used in the separation of aliphatic and aromatic hydrocarbons and in the separation of isomeric aromatics [5,127,128].

α-ω-Alkylene diammonium smectites have been proposed as size selective sorbents, as they have defined limits to their swelling, which is governed by the chain length of the diammonium species [5,101]. Bulky diammonium derivatives, such as those of triethylenediamine, have an added advantage in that the layers cannot collapse on removal of solvent from the organoclay, but form a porous structure that can adsorb small molecules like ethylene or 2,4-dimethylpentane. Triethylenediammonium montmorillonite, for example, has an interlamellar spacing of 5.5 Å, with a lateral spacing between the "pit-prop" molecules of about 6 Å. These porous structures may have catalytic properties and can, for example, promote the hydrolysis, or esterification, of adsorbed molecules [129,130].

4.4.2   Organosmectite Gellants

The regular array of mineral layers in an alkylammonium smectite can be disrupted by milling the organoclay with a mixture of hydrocarbon oil and a small quantity of methanol to form a thixotropic gel. The gellation is caused by structure formation, similar to that of an inorganic smectite in aqueous media. This results from ionic interactions between the oppositely charged contacting edges and faces of the silicate platelets. The concepts of the structure formation in a nonpolar medium containing a polar activator are largely due to Granquist and McAtee [116]. They proposed that, at low concentrations, the activator (methanol) molecules are adsorbed to form a thin zone, approximately 3 Å thick, at the silicate surface. This zone has a sufficiently high dielectric constant for the solvation and dissociation of the alkylammonium smectite ionic pairs and the development of an ionized double layer. As the methanol is largely adsorbed, the bulk of the liquid phase remains nonpolar, and the alkylammonium ions are restrained to the immediate vicinity of the silicate surfaces. However, the alkyl chains are also solvated by the nonpolar phase and, as a result, the ions are mobilized and can diffuse across the silicate surfaces. The ionization of the alkylammonium groups in an alcohol-containing hydrocarbon suspension of the organosmectite can be demonstrated by their exchange with other ammonium ions [116,131].

The swelling and mobilization of the alkylammonium groups and ionic repulsion between the adsorbed layers allows the smectite platelets to be disagglomerated by mechanical shearing. These can subsequently flocculate to form three-dimensional thixotropic gel structures, probably by edge–face interactions. If too much activator is added, the excess may remain in the bulk liquid phase, thereby increasing its dielectric constant and allowing the compact solvated ionic layers to expand into more diffuse structures. The resulting development of the diffuse double-layer repulsion between contacting particles can destabilize the thixotropic structures. This results in a reduction of the rheological yield stress for the suspension, compared to the peak value obtained with the optimum amount of activator. In the absence of the activator, the ammonium ions remain tightly bound to the exchange sites on the silicate surface. The presence of adsorbed moisture in the organoclay can improve the structure formation, as the solvating water molecules can weaken the electrostatic bond between the ammonium ions and the silicate surface, thus facilitating the subsequent dissociation of the ionic complex in the activator-containing medium [124].

The maximum rheological yield stress of an alkylammonium bentonite in a toluene–methanol medium has been found to occur when the amount of adsorbed alkylammonium ions corresponds to the maximum structural double layer coverage of the silicate surfaces (as distinct to the amount corresponding to 100% ion exchange). Adsorption of further excess salt results in electrostatic repulsion between the platelets and a rapid reduction in the strength of the thixotropic structures [132]. The alkylammonium bentonites in methanol–petroleum oil media show a similar relationship between gel strength and degree of exchange [112].

The industrial applications of the organophilic alkylammonium smectites are numerous, some of the principal applications being: (1) the preparation of temperature-resistant greases from mineral oils; (2) the thixotropic thickening of paints to improve their ease of application and to control sedimentation of pigments; (3) the manufacture of high-temperature-resistant varnishes; and (4) the viscosity modifying agents in inks, plastigels, polishes, and liquid resin compositions. Organosmectites containing the most bulky long-chain ammonium ions are generally the most effective gellants. Dialkyldimethylammonium smectites are usually selected for dispersion in, or gellation of, organic media having low polarity, for example, aliphatic or aromatic hydrocarbon-containing compositions, whereas the alkylbenzyldimethylammonium smectites may be chosen for media having moderate polarity, for example, esters or vinylaromatic monomers [133]. Some applications of alkylammonium smectites, largely selected from recent patent abstract literature, are described below.

The use of finely divided inorganic gellants in the manufacture of lubricating greases is possible because particles less than 1 μm in width are effectively nonabrasive, being smaller than the normal irregularities between the sliding surfaces [99]. Although organophilic silicas may be used as gellants, organophilic smectites or attapulgites containing quaternary ammonium ions provide gel structures of greater stability in hydrocarbon greases. The mineral–gelled greases, unlike the soap-gelled products, do not melt and usually show good thermal stability and water resistance; however, they may accelerate rusting, unless the organoclay was rigorously freed from chlorides. The gel strength and stability of the greases can be further improved by the use of amino polymers, polyamines, and polyammonium salts, in addition to the monoquaternary salts, and by the preparation of the organophilic smectites from ammonium phosphates rather than ammonium chlorides. In each case, the organophilic smectite has to be activated by the addition of a small proportion of methanol, isopropanol, formamide, low-molecular-weight polyols, or of some other structure-promoting compound. Although the organoclay gelled greases have good rheological properties at elevated temperatures, some compositions have been reported to have inadequate high-temperature stability [132]. This instability possibly arises from clay-catalyzed degradation of the alkylammonium ions. The rate of this degradation depends on the source and particular variety of the smectite.

One disadvantage of the use of highly dispersed pigments in paints and other liquid media is that when the pigment settles, the sediment forms a hard cake that is often more difficult to redisperse than less efficiently dispersed, more flocculant pigments. The redispersion of sedimented pigments can be assisted by the inclusion of a small quantity of an organoclay gellant, which not only reduces the rate of sedimentation but also causes the sediment to have a loosely packed structure which can be readily redispersed [134]. Improved oleophilic pigment compositions have been claimed to result from the flocculation of an aqueous, dispersed pigment–smectite mixture by the addition of long-chain alkylammonium salts [135].

Alkylammonium smectites have been used in the stabilization of marginally immiscible liquid mixtures, for example, mixtures of high- and low-molecular-weight polyols [136], as thickeners in nail polishes [137], and in styrene–polyester [133,138] resins. The alkylammonium smectites have also been used to prevent the crazing of acrylic automotive lacquers [139] and to limit the penetration of surface-coating compositions into porous substrates [140]. Other novel uses include their application as an ink receptor in porous polystyrene sheets, used as paper substitutes [141], and as a reservoir for ink solvents such as ethylene glycol in plasticized-vinyl ink/pencil erasers [142]. Short-chain quaternary ammonium smectites have also been used as sorbents and as softening agents in nonionic detergent formulations [143].

Ion exchange has also been used for the preparation of mineral–polymer composites, examples of which are described in Sections 4.6 and 4.7. These examples include the use of adsorbed initiators for radical polymerization or adsorbed cationic monomers. Mineral–polymer composites have also been prepared by the ion-exchange reaction between amine-terminated polystyrenes and smectites, to give products which can be compression molded into plastics articles or used as thickeners in greases or fillers in plastics or elastomeric compositions [144,145]. Alkylammonium bentonites, dispersed in the methyl methacrylate monomer, have also been claimed to improve the hardness and to increase the glass transition and yield temperatures of cast acrylic sheet [146]. The alkylammonium bentonites can also provide good reinforcement for natural and synthetic elastomers [147].

Although most of the commercial organosmectite gellants employ alkylammonium ions as the solvatable cationic component, alkylphosphonium derivatives such as dodecyltriethylphosphonium montmorillonite may also be useful as organophilic gellants and thickeners [148]. Other "onium" smectites, for example, oxonium, thionium, or arsonium derivatives, are frequently mentioned in the patent literature relating to manufacture and utilization of alkylammonium smectites [e.g., 149], although the effectiveness and commercial value of these alternative organocationic smectites remains to be demonstrated.

4.5 MODIFICATION INVOLVING COVALENT BONDING

The hydroxylic species which occur on the surfaces of most mineral fillers and pigments can be used as anchor groups for the attachment of covalently bound surface modifying agents. These agents can be divided into three groups: silanes, metal alkoxides and coordination complexes, and organic condensation agents such as the epoxides and isocyanates. Each group will be discussed in turn.

Although these various reagents can condense with the surface hydroxyl groups of the hypothetical minerals introduced in the previous chapters, it is often questionable whether covalent bonds are, in fact, formed with the surfaces of real minerals that have not been rigorously outgassed. In many cases it is probable that the reactions of the modifying agent, or of some intermediate adduct, with adsorbed water molecules and other surface hydroxylic species results in the formation of tenaciously adsorbed, hydrogen-bonded hydrolyzates. The two modes of chemisorption often cannot be readily distinguished, and it is both convenient and practical to discuss them together.

4.5.1 Reaction with Silanes

Surface treatment with reactive silanes is one of the principal methods for converting mineral or inorganic particulates or fibers into materials bearing covalently bound functional groups, capable of graft formation when used as fillers or reinforcements in plastics compositions; over 3000 tons of silanes are used annually for these purposes [38]. The reactions of silanes with inorganic surfaces have been extensively studied, and their chemistry and technology have been frequently reviewed [e.g., 150–153].

Organosilanes containing halogen, alkoxide, acylate, and amino groups attached to silicon can react with surface hydroxylic species to form covalently bound surface siloxy derivatives (Figure 4.15). The chloro and alkoxy derivatives are most frequently used as surface-modifying agents for mineral pigments, fillers, and reinforcing fibers. Many of the modifying agents contain three hydrolyzable substituents on each silicon center which, in principle, would enable the formation of a chemisorbed cross-

$$\equiv\!\!-OH \quad + \quad XSiR_3 \quad \longrightarrow \quad \equiv\!\!-O\text{-}SiR_3 \quad + \quad HX$$

Figure 4.15

linked siloxane network by the condensation of adjacent adsorbed silan-triols with each other as well as with the surface hydroxyl groups. The chlorosilanes are considerably more reactive than the alkoxysilanes, but yield hydrogen chloride as a potentially detrimental byproduct from the surface silylation reactions.

The silane modifying agents also contain one or more nonhydrolyzable alkyl or aryl groups which are used to provide compatibility with organic matrices. These may also contain substituent groups that can provide reactive centers for graft formation between a polymeric matrix and the mineral surface. These reactive substituents can be tailored for specific applications and include vinyl, acrylate and methacrylate, chloro, amino, epoxy, and mercapto groups.

The minerals and other inorganic substances that can be usefully modi-fied by treatment with silanes include glass and silica fibers, titania and inorganic-coated titanias, silica and alumina, clays and calcined clays, the oxidized surfaces of some metals and a variety of minerals, the ideal struc-tures of which contain few surface hydroxyl groups. This latter group of minerals includes talc, calcium carbonate, and barium sulfate [152–154]. The stability of the mineral–silane condensation products towards hydrol-ysis varies, acidic silica surfaces providing the most stable adducts. Water-resistant covalent bonds are not usually formed with alkaline min-eral surfaces, but the bonding of silanes to calcium carbonate, for example, may be improved by acid pretreatment of the mineral [155].

The reactions between the silanes and inorganic surfaces are more com-plex than the scheme shown in Figure 4.15 indicates. Studies of the reac-tions of alkoxysilanes with silica have shown that the silane is first physically adsorbed and then may be slowly hydrolyzed if adsorbed water is also present. Reaction between the adsorbed alkoxysilane and surface silanol groups does not occur at room temperature, or readily below 100 to 150°C, unless catalyzed by coadsorbed acidic or basic species [156,157]. In contrast, chlorosilanes can react readily with surface hydroxylic species, the reaction being catalyzed by the byproduct hydrogen chloride. Although alkoxysilanes may be used as surface modifying additives in nonaqueous formulations, they are frequently hydrolyzed in dilute acid or alkali and the resultant silanol solution used for the treatment of the fillers or fibers.

In the sequence of reactions between a chloro or alkoxy silane and an inorganic surface, reaction with adsorbed water takes precedence, fol-lowed by condensation between adjacent adsorbed silanols, and then con-densation between adsorbed and surface silanols (Figure 4.16) [158]. Con-densation polymerization of silanol molecules may be more rapid when these molecules are adsorbed on oxide surfaces than when they are in solu-tion or in a precipitated form [159]. This is due to the catalytic effects of neighboring surface hydroxylic species that can act as proton acceptors. Siloxane network and surface siloxane bond formation is accelerated by

$$RSi(OR')_3 \xrightarrow{\text{H}_2\text{O}} RSi(OH)_3$$

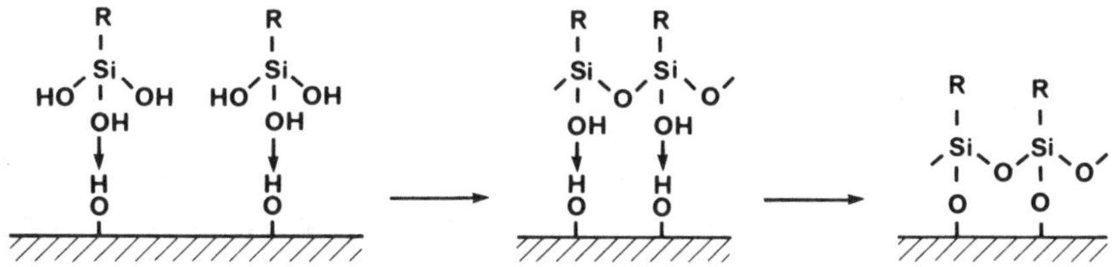

Figure 4.16 Hydrolysis and chemisorption reactions of silicate esters on hydroxylic surfaces.

heat and by dehydration of the hydrogen-bonded adsorption complex. However, the condensation is reversible and, in the presence of water, the grafted siloxanes tend to revert to hydrogen-bonded adsorbed silanols [150,160,161]. Reversible hydrolysis, or the rearrangement of hydrogen bonds between silanol groups and the surface, may be one means for the slow relaxation of applied stresses at reinforcing filler–matrix polymer interfaces [151].

The reactions between the silanes and alumina or other oxide surfaces appear to follow a similar route, although the adducts may be more sensitive to hydrolysis. The relative reactivities of the chloro and alkoxy silanes towards metal oxides differ from those towards silica. Titania, for example, reacts equally readily with the chloro and alkoxy silanes, but reaction with acetoxysilanes results in the formation of a surface acetate instead of a surface silyloxy adduct [162] (see Figure 4.17).

Silanes are expensive agents and tend to be used in specialized applications or for those that involve grafting of a polymer to a reinforcing filler having a low specific surface area, for example, in the priming of glass fibers, for use in fiber-reinforced styrene–polyester resins or for the partial surface treatment of micron-sized particulates used for elastomer pigmentation or reinforcement. The silanes are often most effective when used in amounts that correspond to an approximate monolayer coverage of the mineral surfaces. With the possible exception of the methacryloxypropylsilane derivatives, which can readily undergo vinylic homopolymerization in the adsorbed state, most silanes used for the surface modification

$$R-\underset{|}{\overset{|}{Si}}-O_2C.CH_3 \;+\; \equiv SiOH \;\longrightarrow\; R-\underset{|}{\overset{|}{Si}}-O-Si\equiv \;+\; CH_3.CO_2H$$

$$R-\underset{|}{\overset{|}{Si}}-O_2C.CH_3 \;+\; \equiv TiOH \;\longrightarrow\; R-\underset{|}{\overset{|}{Si}}OH \;+\; \equiv TiO_2C.CH_3$$

Figure 4.17

tend to form condensation homopolymers that lack cohesive strength; thick adsorbed layers of such silane condensates can degrade the mechanical properties of the mineral-containing composite materials.

Vinyl and methacryloxypropyl silane derivatives are commonly used to enable radical graft formation between mineral surfaces and polymers or polymerizing monomer systems. The reagents are principally used for the pretreatment, or priming, of silica fibers, glass fibers, and cloth for use as reinforcements in radical cured styrene/unsaturated polyester compositions. Good adhesion between the fibers and the matrix polymer is required in these systems, as the fibers carry most of the tensile stress in the composite. The vinyl and methacryloxypropyl silanes differ in reactivity towards radicals, although the adsorbed forms of both can copolymerize with vinylic monomers. The methacryloxypropyl derivatives are more reactive towards radical attack and are capable of both homopolymerization and copolymerization, or graft polymerization, with other monomers. Adsorbed vinylsilanes do not readily homopolymerize by radical processes and have been reported to act as virtual chain termination agents in the polymerization of some styrene-based compositions.

Homopolymerization of the adsorbed methacryloxyalkyl groups can result, in some circumstances, in the formation of a reinforcement-enhancing intermediate polymer that has a bilayer structure that lacks any extensive bonding to the surface via the siloxane groups. Methacryloxypropylsilantriol, formed by the prehydrolysis of the methoxide in dilute acetic acid, appears to be adsorbed on silica to form head-to-head (silyl-to-silyl) multilayers, the dominant interaction with the surface occurring by hydrogen bond formation between the ester carbonyl groups and surface silanols [159]. Despite their more effective bonding to silica surfaces, the vinylsilanes have been reported to be less effective as priming agents for some styrene-based formulations than the alternative methacryloxyalkyl derivatives and to produce weaker fiber-reinforced styrene–polyester composites [160].

Methacryloxypropylsilane-treated kaolins have been reported to give better composites with polyethylene than the vinylsilane-treated kaolins. This may be due to the formation of an organized polymer zone at the filler–polyethylene interface, resulting from the interaction between a homopolymerized methacrylate coating and the matrix polymer [163]. In contrast, the vinylsilane-treated filler does not undergo homopolymerization, but is grafted by matrix polymer radicals during compounding; this was reported to result in less effective reinforcement [163]. Solvent-resistant grafting of styrene monomer to methacryloxypropylsilane-treated kaolins [164] and calcium carbonate [155] has been demonstrated by the formation of cross-linked composites that merely swell when treated with benzene; the solution-polymerized styrene-silane copolymers are soluble in this solvent.

Other functional silanes such as primary or secondary aminoalkylsilanes, or epoxyalkylsilanes, can be used for treatment of reinforcing fillers

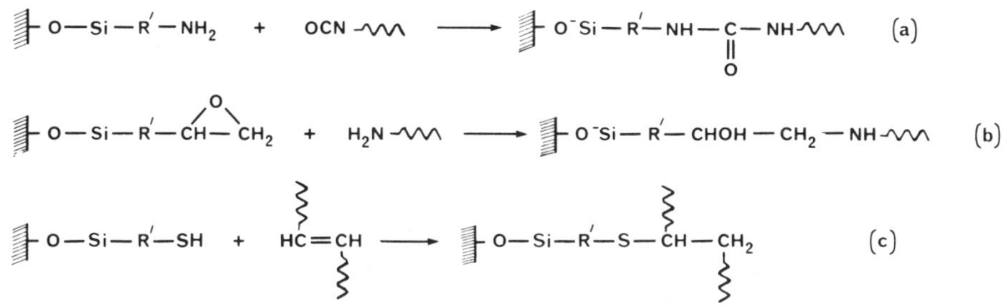

$$\equiv Si-O-Si-R'-NH_2 \;+\; OCN\!-\!\!\sim\!\!\sim \;\longrightarrow\; \equiv Si-O^-Si-R'-NH-\underset{\underset{O}{\|}}{C}-NH\!-\!\!\sim\!\!\sim \qquad (a)$$

$$\equiv Si-O-Si-R'-\overset{O}{\overset{\diagdown}{CH}\!-\!CH_2} \;+\; H_2N\!-\!\!\sim\!\!\sim \;\longrightarrow\; \equiv Si-O^-Si-R'-CHOH-CH_2-NH\!-\!\!\sim\!\!\sim \qquad (b)$$

$$\equiv Si-O-Si-R'-SH \;+\; HC\!=\!CH \;\longrightarrow\; \equiv Si-O-Si-R'-S-CH-CH_2 \qquad (c)$$

Figure 4.18   Some examples of bonding of polymers to reactive, silanized mineral surfaces.

for use in polymer or prepolymer systems containing suitable reactive centers. Aminoalkyl silanes, for example, can be used for treating minerals for graft formation with chlorinated polymers, carboxylate-containing polymers, and phenolic or aminoplast compositions, or in polyurethane (Figure 4.18a), or epoxy resin systems. Epoxyalkylsilyl surfaces can form grafts with epoxy resins, or with amine (Figure 4.18b), amide, acid, or other resin systems that contain reactive hydrogens. Silanes can be used for grafting ethylene oxide oligomers to silica, by reacting the isocyanate-terminated polyether with a dry, aminoalkylsilane-treated derivative of the mineral. Alternatively, the isocyanate-terminated polymer can be condensed with the aminoalkylsilane and the product used for treating an aqueous suspension of the mineral [165].

Commercially treated kaolin fillers containing adsorbed mercaptopropylsilanes [166] or aminoalkylsilanes [167] have been marketed as reinforcing fillers, for use in carbonless vulcanized elastomer formulations (Figure 4.18c), and as replacements for reinforcing channel blacks [168,169]. The kaolins contain approximately 0.5% of the silane and, according to the patents, may be prepared by treatment of aqueous slurries of the clay with the silane, followed by spray drying at 320°C.

Hexamethyldisiloxane, in the presence of catalytic amounts of water and hydrogen chloride, can be used to form trimethylsiloxane groups on the surface of silica [170]. The same reagent, or trimethylchlorosilane, in a hydrochloric acid–isopropanol mixture can be used to replace the octahedral layers of chrysotile, vermiculite [171], or halloysite [172], with the formation of two-dimensional sheets of trimethylsilylated polysilicic acid. The analogous reactions with allyldimethylchlorosilane [173] or vinylmethyldichlorosilane [174] can yield unsaturated oleophilic organosilicates. These can be used as reinforcing fillers in elastomers to produce composites having very high elastic moduli.

4.5.2   Reaction with Metal Alkoxides

While silanes are widely used for the surface modification of reinforcing fibers and siliceous fillers, they have been reported to be less effective for the modification of some other common fillers such as calcium carbonate [175]; they are also manufactured with only a limited, albeit versatile, range of functional groups. Because of this, treatment with metal acylate coordination complexes, or with the more recently developed alkoxymetal acylates, may provide a preferable alternative for the modification of many mineral fillers.

The earliest commercial organometallic surface modifying agents were basic chromic acylate complexes, which were used for the chemisorption of carboxylates on silica, silicates, and other weakly basic or negatively charged surfaces. The familiar adhesion promoting and waterproofing agents, Quilol® and Volan® (du Pont) [176] and Scotchgard® (3M) [177], are the respective long-chain fatty acid, methacrylic acid, and perfluorinated aliphatic acid derivatives, formed by reaction between the carboxylic acids and alcoholic solutions of basic chromic chloride, prepared by the oxidation of isopropanol with chromyl chloride. (The putative structure of Quilol® is shown in Figure 4.19.) Volan® has been used extensively for the sizing of glass fibers used in the manufacture of reinforced polyesters.

These chromium acylates are coordination complexes, in which the carboxylate groups are linked to pairs of solvated hydroxo-bridged chromic ions [178]. The monomers and their partially condensed oligomers are soluble in organic or aqueous–organic solvents, but they can form insoluble, adherent polymeric layers on hydroxylic surfaces. The chloride ions are weakly coordinated with the chromic ions and can be displaced by water molecules, forming an acidic aquo-complex which, on drying and elimination of hydrogen chloride, condenses with neighboring complex molecules and with surface hydroxyl groups. The chromium complex solutions are usually neutralized with aqueous ammonia or amines before application, and the chromate-treated materials dried at 100 to 150°C to complete the condensation reactions [178].

Metal alkoxides, notably those of titanium and aluminum, are also capable of condensation with adsorbed water or the surface hydroxylic species on a wide variety of inorganic materials to form firmly bonded adducts. In this respect, they are as versatile as the reactive silanes and have the added advantage that they can form strong bonds with the surfaces of nonsiliceous minerals. The principal difference between the silanes and the alkoxide modifying agents lies in the nature of the bond between the metal (or silicon) centers and the functional organic substituents that interact with the organic matrix. With the exception of some specialized reagents derived from silicate esters, the functional organic moieties of the adsorbed silanes are attached by hydrolysis-resistant Si–C bonds; in contrast, those of the common alkoxide-based reagents contain hydrolyzable Ti–O–C or Al–O–C bonds.

The syntheses of the alkoxide reagents are less complicated than those of the reactive silanes. A wide variety of agents for general or specialized use can be prepared by metathetical reactions between the readily avail-

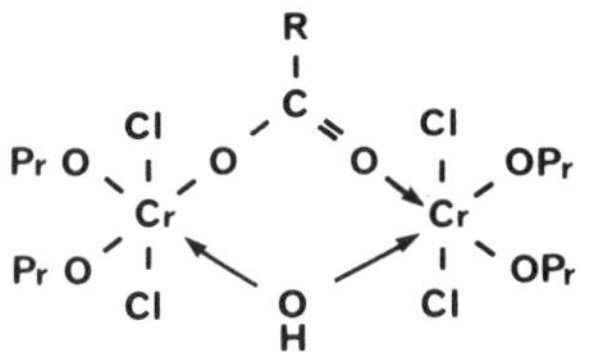

Figure 4.19 The putative structure of Quilol®, a complex chromium acylate surface modifying agent; R represents a long-chain alkyl group.

able metal alkoxides and the functional organic species that contain either hydroxylic or acidic substituent groups [179].

The commercial titanate modifying agents are usually prepared from isopropyl or *n*-butyl titanates, and the acidic functional species may include long-chain saturated and unsaturated carboxylic acids; sulfonic acids; acidic phosphate, pyrophosphate, and phosphite esters; and phenols, enols, and alcohols. The agents can contain up to three of these functional species, which may be of different types, with substituents capable of reacting with matrix polymers. The matrix-reactive substituents can include vinylic and acrylic groups, amines, aldehydes, and disulfides, but not free isocyanate or epoxy groups; the titanates can also be bound to carboxylic or hydroxylic matrix polymers by transesterification reactions.

The reaction between alkoxytitanium derivatives and adsorbed water or surface hydroxylic species is superficially similar to that of the alkoxysilanes (Figure 4.20a). However, the monomeric titanates are coordinately unsaturated and can form coordination complexes with Lewis bases such as alcohols, other hydroxylic species, esters, and acids. These coordination complexes are intermediates in titanate transesterification and hydrolysis reactions and in the condensation with surface hydroxylic species [179]. Adsorbed titanates can react rapidly with surface hydroxylic species at room temperature, forming covalent bonds with the mineral surfaces or with neighboring partially hydrolyzed titanate molecules. Despite claims in some technical reports that monoalkoxytitanium acylate treatments are insensitive to hydrolysis, chemisorption of these reagents on practical mineral samples usually results in some surface-catalyzed hydrolysis and resultant elimination of acylate groups; this presumably would also result in the formation of a cross-linked titanoxane network on the mineral surface (Figure 4.20b and c). The eliminated fatty acids or esters can act as inadvertant plasticizers in plastics systems that contain high loadings of alkoxytitanium acylate-treated fillers.

The titanium alkoxides are suitable for incorporation as general purpose modifying agents in organic media, but other derivatives have been developed for particular applications. These include chelated oxyacetates, for use in aqueous systems; ethylene glycolates, for use in admixture with ester plasticizers; and pyrophosphates, for adsorption on noncalcined clays and alumina [175]. A wide range of "mono-hydrolyzable" (monoalkoxy) and "dihydrolyzable" (dialkoxy) titanates has been developed [e.g., 180], and their applications have been described in numerous

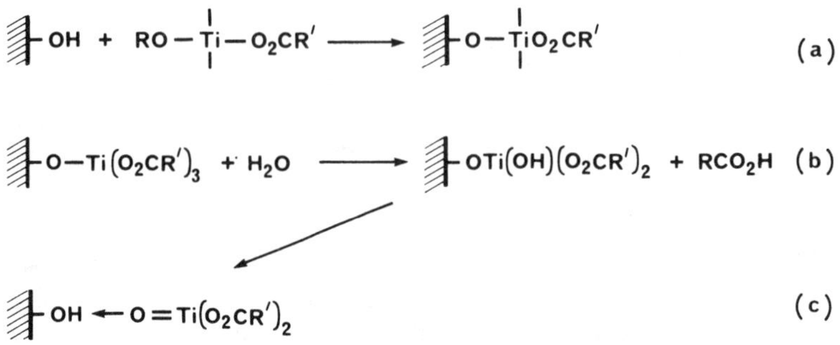

$$\equiv\!\!\!-OH \; + \; RO-\overset{|}{\underset{|}{Ti}}-O_2CR' \longrightarrow \equiv\!\!\!-O-\overset{|}{\underset{|}{Ti}}O_2CR' \qquad\qquad (a)$$

$$\equiv\!\!\!-O-Ti(O_2CR')_3 \; + \; H_2O \longrightarrow \equiv\!\!\!-OTi(OH)(O_2CR')_2 \; + \; RCO_2H \quad (b)$$

$$\equiv\!\!\!-OH \longleftarrow O=Ti(O_2CR')_2 \qquad\qquad (c)$$

Figure 4.20   Reactions of alkoxytitanium acylate modifying agents.

technical reports and reviews [e.g., 181–184]. Some of the functional
organic moieties used in these modifiers or "coagents" include stearate,
isostearate, dodecylbenzenesulphonate, dioctylphosphate, acrylate, methacrylate, ricinoleate, aminobenzenesulphonate, aminobenzoate, cumylphe-
nate, formylphenate, dimethylaminoethoxy, and ethylaminoethoxy
groups, or mixtures of these. Phosphite-coordinated titanates are also available for applications that require the inhibition of titanate-catalyzed reactions with the matrix polymers.

These titanate modifying agents can be used as surface modifying additives in liquid-based formulations or as pretreatments for fillers and pigments. Typical levels of titanate treatment range from 0.5 to 3.0% of the
weight of the mineral. Various pretreated kaolins, alumina, calcium carbonate, and other fillers are now commercially available [184].

The titanate treatment of fillers can result in significant improvements
in the various properties of a diverse range of systems. These include an
improvement in the wetting of the fillers by organic liquids and polymer
melts, which results in faster disagglomeration of the mineral particles
and reduced mixer energy requirements; reduction in the rheological yield
stress and viscosity of mineral organic liquid suspensions; reduction in the
melt viscosity of noncuring plastics systems; and reduced plasticizer
demand of filled plastisols. Long-chain acyl titanate-treated fillers and pigments are claimed to act as internal lubricants in plastics compositions,
and may be incorporated at high loadings without marked loss of tensile or
impact strength, elongation, or flexibility, when compared with the properties of the virgin polymers. Particularly favorable results have been
reported for high density polyethylene and polypropylene containing 30 to
60% of calcite, wollastonite, or barites, treated with 3% w/w mineral of
isopropoxytitanium triisostearate [175,185], and for polyethylene containing talc or calcite, also treated with this reagent or with ethylenedioxytita-
nium diisostearate [185].

The long-chain acyl titanate treatment of fillers may improve the flexibility of their polyolefin composites, but they can also result in loss of
interfacial adhesion. Oleophilic triisopropyltitanium oleate-treated delam-

inated kaolins have been manufactured [186,187] and used for the production of highly opaque, printable polypropylene films [188]. The opacity results from the formation of microvoids at the filler–matrix interfaces during biaxial orientation of the extruded filled polyolefin films. In contrast, treatment of the mineral fillers with alkoxytitanium acrylates, methacrylates, or other reactive acylates, which can form grafting bonds with the polymeric matrices, can result in improved interfacial adhesion and tensile strength for the filled polymeric composites.

The organotitanium treatments do have some disadvantages. Calcites and other white fillers which have been treated with titanates have been reported to discolor through oxidation when incorporated in some polymers. The titanium-containing coatings can form highly colored compounds with some of the antioxidants used in polyolefins. The titanium acylate modifying agents, unlike the alkylsilanes, are susceptible to hydrolysis or alcoholysis of their oleophilic, matrix-interactive moieties, either on storage of pretreated fillers or during melt-compounding of plastics compositions. This hydrolysis is potentially deleterious, particularly in the case of monoacyloxytitanate-treated minerals. A critical study of monohydrolyzable organotitanium treatments as adhesion promoters for paints on steel surfaces has shown that, as a result of this hydrolysis, any improvement or degradation in adhesion was due to interactions between the organic moieties of the reagents, the metal surface, and paint resin; the oxytitanium moiety served little apparent purpose [189].

The oxyaluminum monoacylates are more resistant to hydrolysis. Trimeric oxyaluminum acylates, prepared by hydrolysis of alkoxyaluminum acylates (Figure 4.21), have been patented for use in the preparation of oleophilic minerals [190]. These trimers, unlike their precursors, are probably only physically adsorbed on the mineral surfaces, although surface-catalyzed hydrolytic redistribution could result in the formation of chemisorbed polymeric oxyaluminum acylates. The dialkoxy-aluminum acylates, however, are particularly reactive towards adsorbed water and surface hydroxyl groups and can form strongly bound adducts with a variety of commercial fillers [191].

Dialkoxyaluminum methacrylate treatment of kaolins can be particularly advantageous. High density polyethylene containing the treated filler has been reported to have high elongation and flexibility, with tensile and impact strengths that can exceed those of the unfilled polymer. The enhanced mechanical performance of these filled polyethylenes may arise

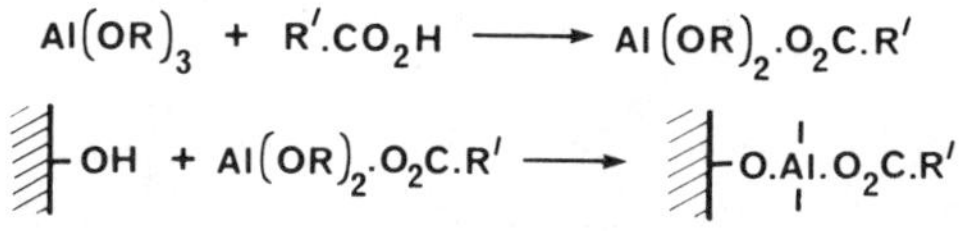

$$Al(OR)_3 \;+\; R'.CO_2H \;\longrightarrow\; Al(OR)_2.O_2C.R'$$

$$\text{—OH} \;+\; Al(OR)_2.O_2C.R' \;\longrightarrow\; \text{—O.Al.O}_2C.R'$$

Figure 4.21

from the formation of an intermediate ionomeric polymer layer at the filler-
-matrix interfaces which can serve as a zone of intermediate modulus and
which may allow some interfacial stress relaxation to occur by redistribution of the aluminum acylate bonds [192].

4.5.3   Reaction with Alcohols and Epoxides

The surfaces of minerals can be modified by means of a variety of condensation reactions between surface hydroxylic species and many of the
reagents that can form covalent bonds with organic hydroxylic compounds. Some of these reactions have been briefly outlined in the preced-
ing chapters and have been reviewed by Deuel [193], Kukharskaya and
Fedoseev [194] (silicate derivatives), and by Boehm [195] (other oxides).
Much of the early mineral organic chemistry is largely of academic interest, and we confine the present discussion to a selection of mineral organic
reactions that have either been patented for, or are relevant to, the commercial modification of mineral surfaces.

The surface silanol groups of a finely divided amorphous silica can be
esterified by reaction with primary or secondary alcohols at elevated temperatures, particularly if the silica has been activated by drying at 150 to
500°C (Figure 4.22a) [196–198]. The activation results from the removal of
adsorbed water and the condensation of adjacent silanols to form reactive
surface siloxane species. The esterification occurs rapidly at 150 to 200°C,
although the reaction with methanol requires higher reaction temperatures. Tertiary alcohols undergo dehydration, rather than esterification,
when heated with silica (Figure 4.22b).

The surfaces of the esterified silicas take on the characteristics of the
esterifying alcohols. Alkyl alcohols, for example, yield hydrophobic silicas
that can be used as reinforcing agents in plastics and elastomers and as
thickening agents for organic liquid compositions. Esterification with an
unsaturated alcohol, such as allyl alcohol, results in the formation of an
olefinic silica, suitable for grafting and cross-linking of monomers and
polymers [199]. Oleophilic silicas can also be produced by reaction between
the silica and unsaturated silanes, as described in Section 4.5.1, or by the
reaction between a "chlorinated" silica and an unsaturated alcohol. Esterification with aminoalcohols can result in the formation of an aminoalkyl-
grafted silica [200], and the reaction between activated silica and
epoxypropanol forms a glyceryl silica (Figure 4.22c) [201]. The reactions
between alcohols and porous silicas are usually incomplete, and the products contain some residual surface silanol groups.

The reaction between partially dehydrated crystalline aluminas and
alcohols can be used for the preparation of oleophilic aluminas. These can
be used as thickeners in greases, as reinforcing agents in elastomers, and in
a wide range of other applications [202]. Amorphous alumina and titania
can also be esterified by heating with alcohols. The esterification of

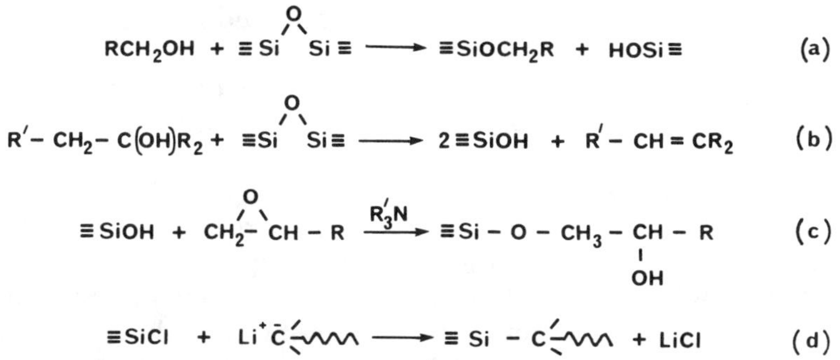

$$RCH_2OH \ + \ \equiv Si \overset{O}{\underset{}{\diagdown}} Si \equiv \ \longrightarrow \ \equiv SiOCH_2R \ + \ HOSi \equiv \qquad (a)$$

$$R'-CH_2-C(OH)R_2 \ + \ \equiv Si \overset{O}{\underset{}{\diagdown}} Si \equiv \ \longrightarrow \ 2 \equiv SiOH \ + \ R'-CH=CR_2 \qquad (b)$$

$$\equiv SiOH \ + \ CH_2 \overset{O}{\underset{}{\diagdown}} CH-R \ \xrightarrow{\ R_3'N\ } \ \equiv Si-O-CH_3-\underset{\underset{OH}{|}}{CH}-R \qquad (c)$$

$$\equiv SiCl \ + \ Li^+\bar{C}{\sim}\!\!\!\sim\!\!\!\wedge \ \longrightarrow \ \equiv Si-C{\sim}\!\!\!\sim\!\!\!\wedge \ + \ LiCl \qquad (d)$$

Figure 4.22   Surface modification of silica by esterification and alkylation reactions.

hydrated titania requires reaction temperatures of 230 to 280°C, but the surface alkoxides also undergo competitive decomposition at these temperatures [203].

Grafted silicas can be prepared by the reaction between "chlorinated" silica and organometallic reagents, although these reactions are of little commercial value. Chlorinated silica, for example, has been reacted with a living polymer, prepared by the butyllithium-initiated polymerization of butadiene, to form a polymer-grafted silica (Figure 4.22d) [204], although the grafting efficiency was reported to be low.

The condensation reactions of epoxides and isocyanates with hydroxylic and amino-compounds are widely used for the manufacture of thermoset or cured resins. Many of these compositions contain minerals as extenders, thickeners, pigments, or reinforcing fillers. Epoxides, isocyanates, and related functional species can also condense with the surface hydroxyl groups of minerals to form surface modified materials.

Epoxides, such as propylene oxide, adsorb readily on silica surfaces, but the adsorption is largely physical. Reaction with surface silanol groups does not occur except at elevated temperatures or unless a proton acceptor, such as a tertiary amine, is present as a catalyst [201]. As epoxy resin formulations usually contain amines as catalysts or coreagents and are often thickened with finely divided silica, graft formation by surface esterification is expected to accompany the resin curing reactions. The reaction between silica and allyl glycidyl ether has been used to prepare an oleophilic silica with surface allylic groups which can participate in graft forming reactions during vulcanization of elastomers containing the treated filler [205].

Studies of the adsorption and polymerization of glycidyl methacrylate on the surface of a porous glass have shown that epoxide groups of the polymer do not react readily with the surface hydroxylic species, whereas the residual monomer is hydrolyzed, forming dihydroxypropyl methacrylate and other products [206]. In contrast, amine-cured epoxy resins have been used for bonding of rock and the reinforcement of coal mine structures. The interaction between the epoxy resin and mine shale has been reported to occur by hydrogen bonding between the resin and the rock surface, and by formation of strong "primary" bonds through condensation between the hydroxy or epoxide groups of the prepolymer and shale silicates [207].

Epoxides can react with other hydroxylic surfaces to form grafted species. Titania, for example, reacts with propylene oxide under reflux to form free poly(propylene oxide), together with grafted polymer formed by reaction between the epoxide and surface hydroxyl groups [208]. The titania–epoxide adducts show improved dispersibility in benzene, but are susceptible to hydrolysis. The hydrolytic instability is common to all the mineral–epoxide adducts because the covalent bonding is similar to that of the surface alkoxides described earlier; the hydrolyzed products may still be tenaciously retained by hydrogen-bond formation or by dipolar interaction with the mineral surface. Surface modification of minerals with epoxides prior to incorporation in polymer systems has limited utility, and treatment of the mineral with a polyol may be the preferable alternative to surface modification with epoxides.

4.5.4   Reaction with Isocyanates

Alkyl and aryl isocyanates, such as those used in the production of polyurethanes, are generally more reactive towards surface hydroxylic species than the epoxides and can also yield a greater variety of products. These may include covalently bound surface urethanes, formed by reaction between surface hydroxyl groups and isocyanate molecules, and hydrogen bonded ureas, formed by reaction between the isocyanate and adsorbed water molecules. The relative proportions of these products and of the more complex condensation products depends on the nature of the mineral, its residual water content, and the reaction conditions.

Studies of the adsorption and reactions of isocyanates on silica have indicated that both alkyl and aryl isocyanates are initially adsorbed as hydrogen bonded complexes, but that only the alkyl isocyanates react to form surface urethanes [209]. Other studies [210] have shown that the silica surface urethanes are thermally and hydrolytically unstable, reacting with water to form ureas, and dissociating with the desorption of volatile isocyanates when heated in vacuum. The urea-forming reactions are concurrent with those that form urethanes from the "isolated" hydroxyl groups and are accelerated by the presence of adsorbed water or pairs of adjacent, hydrogen-bonded hydroxyl groups. The initial reaction with the isocyanate produces an unstable surface carbamate, which decarboxylates to form an amine; this then rapidly reacts with a second isocyanate molecule to form the urea. (See Figure 4.23.)

The reactions between magnesia and isocyanates are similar to those with silica, except that the formation of surface urethanes occurs more

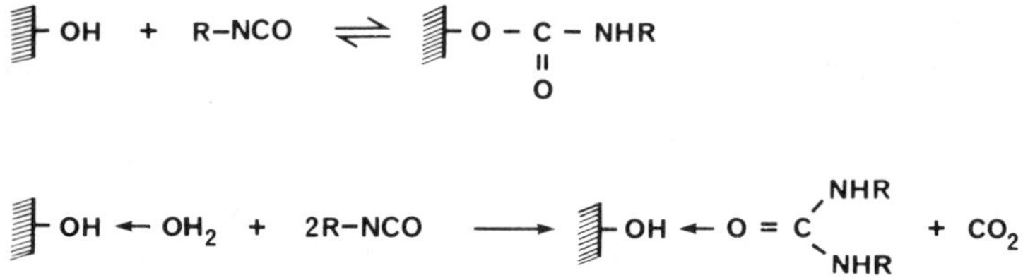

$$\equiv\!\!\!|\text{—OH} + \text{R—NCO} \rightleftharpoons \equiv\!\!\!|\text{—O} - \underset{\underset{\text{O}}{\|}}{\text{C}} - \text{NHR}$$

$$\equiv\!\!\!|\text{—OH} \leftarrow \text{OH}_2 + 2\text{R—NCO} \longrightarrow \equiv\!\!\!|\text{—OH} \leftarrow \text{O} = \text{C}\!\!\begin{array}{l}\nearrow\text{NHR}\\\searrow\text{NHR}\end{array} + \text{CO}_2$$

Figure 4.23

readily with the less acidic surface hydroxyls [211]. Ureas are also produced in these reactions, but the intermediate carbamates have greater stability on nonacidic surfaces. Whereas the ureas may be retained by hydrogen-bonding to an acidic silica surface, they are reported to be adsorbed as perturbed "carboxylate type" complexes on the basic surfaces of magnesia. The reactions described occur when isocyanates react with silica or magnesia at 20 to 100°C; more complex reactions occur at higher temperatures [210,211].

Isocyanates have been reported to form surface grafts with silicate minerals, although the formation of surface "urethane" bonded species is often only an assumption. Detailed studies of the reactions of hexamethylene diisocyanate with vermiculite have shown that the adduct contains approximately five molecules of the diisocyanate per 100 Å^2 of mineral surface, some 30 to 40% of which retain one free isocyanate group [212]. The isocyanates were reported to be bound by the formation of urethane or carbamate species with the hydroxyl groups associated with lattice edge aluminums and by the formation of hydrogen-bonded species. Other studies have shown that the terminal isocyanate groups of urethane prepolymers can react rapidly with adsorbed water on mineral surfaces to form ureas [213].

Examples of the modification of fillers, such as silica, titania, and other minerals with hydroxylic surfaces, by reaction with isocyanates include treatment of pigments with C6 to C18 alkyl monoisocyanates, followed by heating at 100 to 160°C, to form hydrophobic pigments [214]; and treatment with alkyl or aryl diisocyanates, or with vinylic or allylic isocyanates, to form oleophilic materials which can serve as reactive pigments, or as reinforcing fillers, for incorporation in plastics and elastomers [215,216].

Other reagents having mineral reactive groups with the general formula $-\text{X} = \text{C} = \text{Y}$ have been claimed to be effective in the surface modification of pigments. These analogues of the isocyanates include isothiocyanates, ketenes, thioketenes, ketenimines, and carbodimides [216], all of which can react readily with hydroxylic species and water. Some of these may actually form covalently bound adducts with mineral surface hydroxyl groups. Carbodiimides have also been patented for the modification of clays and silicas to make these materials oleophilic and hydrophobic [217].

Other methods for formation of covalently bound mineral–organic adducts include the formation of grafted polymers by the irradiation or mechanochemical processing of mineral–monomer mixtures; these reactions are outlined in Section 5.6. Other processes that involve radical transfer to mineral surfaces, such as the graft polymerization of monomers in the presence of unmodified minerals, have been frequently reported to yield covalently bound polymer, although in most cases the nature of the mineral–polymer bond is conjectural. Tenacious adsorption is only indicative of chemisorption, or alternatively, of "keying" of the polymer chains in surface irregularities; it is not proof of covalent graft formation.

4.6   MODIFICATION BY ADSORPTION OF POLYMERS

The surface modifications previously described involve the adsorption of, or the reaction with, low-molecular-weight materials by the minerals. Such treatments normally result in the formation of only thin layers of the organic modifying agents on the surfaces of the fibers or particles. In contrast, high-molecular-weight polymers are capable of forming thick, solvatable, adsorbed layers on the dispersed particles. Depending on their structures, the adsorbed polymers may act as dispersants, dispersion stabilizers (deflocculants), or as aids for the reinforcement of plastics and elastomers by the mineral fillers; other polymers may be used as flocculants for mineral suspensions.

The adsorption of polymers from solution and the reinforcement of filled composites may be improved by the selection of a filler or of a modifying treatment that can enhance acid–base interactions between the polymer and the mineral surface [218,219]. In filled composites, the adsorbed matrix polymer is considered to form a modified zone around the filler particles. The adsorption of a so-called "basic" polymer such as a polyester or ethylene–vinyl acetate copolymer (the ester groups of which can act as Lewis bases) on an acidic filler (silica), or of an "acidic" polymer, for example, poly(vinyl chloride), on a basic filler (calcite) is greater than the corresponding adsorptions of acidic polymer on acidic filler or of base on base. These differences are also reflected in the mechanical and rheological properties of their respective composites [218]. The acidity or the basicity of the fillers is usually modified by treatment with inorganic or low-molecular-weight organic species, but treatment with polymers has been used for enhancement of Bronsted acid–base interactions, for example, in pigmented paint films. The use of polycarboxylate-coated titania dispersions for the pigmentation of a basic polymer emulsion improves the wet-scrub resistance of the dried film. Premature coagulation on mixing the two oppositely charged dispersions is avoided by the addition of a volatile base (ammonia), which can evaporate during the drying of the emulsion film.

The adsorption of small amounts of long-chain polymers can result in the flocculation of mineral dispersions. Polymeric flocculants are widely used for water clarification and as industrial filtration aids; materials related to the polymeric flocculants are also useful as soil-conditioning agents. A wide variety of polymers and treatments has been described for the flocculation or dispersion of particulates or as reinforcing aids, but only a limited selection of examples can be discussed in this book. Mineral interactions with natural and synthetic polymers are of considerable importance in soil science, and these aspects have been reviewed by Theng [220].

4.6.1 Dispersion

The processes for stabilization of mineral dispersions by polymers have been outlined in Section 4.1.2, and have been shown to involve an initial wetting of the particulate surface by the liquid medium, followed by the mechanical disruption of the aggregates, and active stabilization of the resultant dispersion by the adsorbed polymer. There is no clear-cut distinction between polymeric dispersion aids and dispersants. Many polymeric dispersant or surfactant treatments are aimed largely at preventing the formation of agglomerates and for improving the wetting of pigments and fillers, for dispersion in plastics and elastomers, rather than for the stabilization of dispersions in media of low viscosity. Polymeric dispersion aids include adsorbed silicones, poly(alkylene oxides), and polymeric soaps, in addition to a variety of polymeric encapsulants, some of which are described in Section 4.7.

The relative merits of the various treatments are difficult to evaluate, particularly on the basis of examples and claims contained in the numerous patents relating to dispersion and dispersants. A number of patents relate to the coating of minerals with polymeric dispersants, to provide products which can be readily redispersed in organic solvents, but the comparative cost of these treatments limits their application to the more expensive pigments such as the titanias. However, low-cost polyester resin treatments have been used in the manufacture of some varieties of organophilic kaolins. Polymeric dispersion stabilizers are widely used in the formulation of mineral pigmented paints and other surface coatings, and in these applications the stabilizer is usually used as an ingredient of the paint formulation, rather than in the pretreatment of the pigment.

Treatment with silicone oils is used in the manufacture of some varieties of organophilic titania pigments. The adsorbed silicones reduce moisture adsorption and help to prevent agglomeration of the dry pigment during storage [69]. The silicones are usually poly(dimethylsiloxanes) or derivatives containing polar hydroxy or amino substituents that can bond to the mineral surface [221,222]. The chemisorption of the silicone on the titania or silica–alumina-coated titanias occurs largely by hydrogen bond formation, although acid-catalyzed condensation between surface hydroxyls and the cleaved siloxane polymers may also occur on heating the silicone-treated pigments. Maximum adsorption of the silicones from solution is obtained by predrying of the pigment and by use of a nonpolar solvent, such as hexane, as a suspending agent or diluent. However, the use of organic solvents may not be economically feasible in commercial practice. In most manufacturing processes, the silicones (or other organic additives) are introduced during the final fluid-energy milling treatment of the titania pigments, but some may be applied as water-soluble siliconates during the earlier wet-process treatments; the siliconates become insoluble and hydrophobic on acidification and condensation with the titania surface.

Silicone treatment can improve the dispersion of the titania pigments in media of low polarity, including polyolefins. The reduced moisture adsorption by the silicone-treated pigments can also result in improved durability of the pigmented paints and plastics. However, the silicones may hinder the dispersion of pigments in more polar media, such as the alkyd resin solutions used in paint formulations. These resins are polyesters, prepared by condensation between a triol (glycerol) and a dicarboxylic acid, such as phthalic acid, and are often modified by transesterfication with oxidizing vegetable oils [223]. They are good dispersants for titania pigments, particularly those having basic oxide coatings. Silicone treatment of the pigment reduces the interaction between the carboxyl groups of the alkyd resin and the pigment surface and prevents the resin from acting as an effective dispersion stabilizer [222].

Rutile pigments coated with a polyester resin, prepared from a 2-ethylhexanoic acid/maleic acid/phthalic acid mixture and 1,2,6–hexantriol, have been claimed to show improved dispersibility in poly(vinyl chloride) [224]. Other polyesters can be used as dispersion aids for minerals such as kaolin, talc [225], and titanias [153] in nonpolar and moderately polar media including polyolefins and poly(vinyl chloride) and can be used to improve the tensile and impact strengths, and the elongation-at-break of plastics containing the treated fillers. The polyesters used for the commercial pretreatment of minerals are often materials prepared for the formulation of styrene–polyester resin systems and consist of the condensation products of a mixture of dicarboxylic acids, polyols, and maleic anhydride. These polymers contain unsaturated centers that allow radical grafting to a matrix polymer via the adsorbed polyester.

A number of strongly bound polymers have been patented for improvement of the dispersibility of titanias, probably as dispersion aids rather than as dispersants. These include styrene–maleic anhydride copolymers, for dispersion in alkyd media [226]; polyethylene imine, or its reaction products with acrylonitrile or oxiranes, for dispersion in aqueous and polar organic media, particularly polyesters [227]; polymeric aluminum or zinc salts derived from partially hydrolyzed acrylic copolymers, for dispersion in hydrophobic media [228]; carboxylate-terminated polycaprolactone, for dispersion in polyethylene or polystyrene [229]; and poly(N-vinylpyrrolidone), for dispersion in polyurethane formulations [230]. Dispersants can be used to improve the mechanical properties of the mineral-containing compositions; poly(ethylene oxide) treatment of kaolin, silica, or calcium carbonate, for example, has been claimed to improve the adhesion between the filler and polyolefins [231], as well as improving the dispersibility of the minerals in the molten polymers.

For adsorbed polymers to be effective for the steric stabilization of dispersions, they need to be firmly attached to the mineral surfaces and to contain extended sequences of polymer segments that are only weakly adsorbed and that can form solvated extended loops or tails in the surrounding liquid phase. Although homopolymers can act as steric dispersants, most practical stabilizers are copolymers, such as the alkyd resins, or the widely used acidic acrylic copolymers.

The adsorption of a variety of linear (diol-diacid) polyester resins on titanias, alumina, ferric oxide, and silica has been shown [30] to result in the formation of a series of polymer segmental loops extending from the mineral surfaces. The remainder of the polymer segments are held on the surface by hydrogen bonding between the polymer carboxylate or hydroxyl groups and the surface hydroxylic species. The stability of the particulate dispersions in polyester solutions, measured by the sedimentation rates, can be correlated with the size of the segmental loops and, in turn, with both the amount of polymer adsorbed and the proportion of carbonyl groups specifically attached to the mineral surface; the apparent dimensions of the loops are not markedly dependent on the molecular weight of the polymer or the nature of its end-groups. In the examples cited [30], the estimated maximum lengths of the polymer loops were only of the order of 26 Å, which corresponds to a layer thickness of about 12 Å. Such thin adsorbed layers would be insufficient for the formation of "indefinitely" stable dispersions of micron-sized particles, but could be sufficient to prevent their rapid flocculation.

Commercial pigments have particulate dimensions similar in magnitude to the wavelengths of visible light. In the case of typical titanias, the equivalent particle diameters are about 300 nm. Particles with these dimensions may require coatings having a solvated thickness of at least 100 Å for effective steric stabilization [232]. Coatings of 30 to 50 Å thickness may suffice for commercial applications, as effective dispersion stabilization in many systems has been found to be more dependent on obtaining complete coverage of the mineral surface by the dispersant polymer than on further increase in the coating thickness. These thick coatings are usually obtained by the adsorption of copolymers which contain widely separated bonding groups on the polymer chains or of block copolymers which may consist of, for example, sequences of strongly bound polymer segments, to which are attached one or more extended sequences of solvatable, weakly adsorbed dispersion-stabilizing chains.

Dispersants containing widely separated bonding groups can be readily prepared by the random copolymerization of an acrylic monomer, for example, methyl methacrylate, with a small proportion of acrylic or methacrylic acid. Acidic methyl methacrylate copolymers, having degrees of polymerization of 500 to 1000 and containing 2 to 5 mol.% of acrylic or methacrylic acid, are the classic dispersants for titanias and other oxide pigments in paint formulations based on moderately polar aromatic hydrocarbon, ester, or ketone solvents. Although these dispersants are usually added to the pigment suspension during milling of the paint formulation, the use of pigments precoated with acrylic copolymer dispersants has been cited in some patents [e.g., 233]. The acrylic comonomers may be varied to optimize the polymer for particular film-forming resins and solvents. Dispersants for use in weakly polar hydrocarbon solvents, for example, may consist of a lauryl methacrylate copolymer, whereas hydroxyethyl methacrylate–methyl methacrylate copolymers can be used as dispersants in aqueous and polar organic media [234]. Other polymeric dispersants suitable for use in aqueous media include poly(acrylic acid), poly(vinyl alcohol), soluble cellulose derivatives, and poly(ethylene oxide) condensates.

Carboxylic acids are commonly used as the copolymerized bonding moieties and are particularly suitable for basic pigment surfaces, which include those of titania and most alumina and alumina–silica-coated titanias. Amines and alkylammonium species may be used for the dispersion of acidic minerals, while polymeric sulfobetaines, formed by the reaction between β-propiosultone and copolymers of dimethylaminoethyl methacrylate or vinylpyridine with acrylic monomers or styrene, can be used for the dispersion of titanias and a wide variety of other inorganic pigments [97]. Other polar comonomers, for example, amide, nitrile, or hydroxyl substituted species, may enhance the adsorption of an acrylic polymer, but do not bond sufficiently for the polymers to be effective dispersants in solvating media [235].

Acrylic copolymers containing alkoxysilanes as the mineral-bonding component, for example, a methyl methacrylate–trimethoxysilylpropyl methacrylate copolymer, have been found to be effective dispersants for titanias. Pigmented compositions containing silane-bonded dispersants have been claimed to be more resistant to photodegradation than similar compositions using carboxylic acid copolymer dispersants. The surface carboxylates, formed from the conventional dispersants, are readily oxidized by the titania, generating polymer radicals which can initiate the autoxidation and degradation of the contacted polymers. The surface silyloxy bonds cannot generate radicals in this manner and are more resistant to oxidation. Methoxysilane-terminated polystyrenes have also been used for the stabilization of oxide pigments in polystyrene media [236].

More effective coverage of the pigment surfaces can be obtained by the use of graft dispersants which consist of a polymer "backbone" to which are attached solvatable dispersant polymer chains containing statistically infrequent bonding groups in order to form a comb-like molecular structure. Comb dispersants have been extensively developed for the dispersion polymerization of polar monomers in nonpolar media [237]. A typical dispersant, for example, consists of a polar polymer backbone, which is to be adsorbed on the insoluble particulate surface; the polymer backbone is grafted with numerous nonpolar polymer chains that are capable of extending freely into the surrounding medium. These comb dispersants have been adapted for the dispersion of minerals by the incorporation of copolymerized mineral bonding groups, such as carboxylic acids, in the dispersant segments [228]. When adsorbed, these dispersant chains form a series of solvated loops on the mineral surface.

Other dispersants have been developed that contain a dual graft, or fishbone-like structure, which consists of an adsorbed polymeric backbone on which is grafted a series of dispersant polymer chains plus a separate series of oligomeric mineral-bonding chains. A dispersant for use in moderately polar solvents might consist of an acrylic copolymer backbone grafted with poly(methyl methacrylate) chains, while a dispersant for use in nonpolar solvents might contain grafted poly(lauryl methacrylate) or some other nonpolar polymer. The use of an oligo(alkoxysilane) or an alkyl oligosilicate as the second mineral bonding graft species allows the dispersants to be used with a wide variety of mineral pigments or fillers and improves the photostability of dispersed titanias. Rutile titanias coated with these disperants can have photostabilities equivalent to that of a premium grade alumina–silica-coated pigment [239].

4.6.2 Flocculation

The flocculation or coagulation of dispersed minerals is an important component of many beneficiation processes, particularly of the charged clay minerals. These colloidal dispersions are largely stabilized by electronic double layer repulsion between the particles and will flocculate when the double layers are attenuated. This can be achieved by several methods, the most common being by charge neutralization, according to the Schultz–Hardy rule, by the addition of an optimum amount of high-valence ions having charge opposite to that of the particle surfaces, as in the flocculation of negatively charged dispersed clay particles by addition of Al^{3+} ions. Other methods for flocculation of ionic stabilized dispersions include reducing the dielectric constant of the medium, for example, by addition of an alcohol or ketone to an aqueous dispersion, or increasing the ionic strength of the medium. Both these processes destabilize the dispersions by reducing the thickness of the double layer [240]. These processes can result in the coagulation of the mineral dispersions, that is, the formation of compact agglomerates.

Colloids in ionizing or nonionizing media can be flocculated by alternative processes which involve the bridging of numbers of dispersed particles by adsorbed high-molecular-weight polymer chains. This bridging can result in the formation of large aggregates which can be filtered or which can settle under gravity. Similar processes of particle-bridging by adsorbed polymers are involved in a number of soil conditioning and soil stabilization processes.

The flocculation of colloids by polymers occurs when the "tails" of long-chain molecules, adsorbed on one particle, are also adsorbed on neighboring particles. Repetition of this process builds up a three-dimensional network of bonded particles. As most flocculants rely on a multiplicity of relatively weak hydrogen bonds, or labile ionic bonds, for attachment to the mineral surfaces, redistribution of the polymer–particle bonds can result in the formation of denser aggregates on standing. Mechanical agitation of the flocs, however, can result in a further redistribution of the adsorbed polymers, causing individual molecules to preferentially coat single particles. This may ultimately result in the formation of a steric stabilized dispersion. Redispersion can also occur following the adsorption of excessive amounts of the flocculant.

Inadvertant flocculation can occur during dispersion of minerals if the techniques employed allow incompletely adsorbed dispersant molecules to bridge the suspended particles. Flocculation of mineral dispersions in polymer-free monomers may often occur during the critical early stages of polymerization, for example, in the manufacture of pigmented cast acrylic sheet from unthickened monomers. The flocculation is believed to result from the bridging of the pigment particles by the growing polymer chains and can be minimized by the use of strongly bound polymeric dispersants.

There have been numerous studies of the kinetics of flocculation of minerals by water-soluble polymers [e.g., 241–243]. The optimum amount of flocculant depends on a number of factors, including the chain length of the polymer and the size of the dispersed particles. Several small particles, each having dimensions of less than 400 Å, may be bridged by a single polymer chain, whereas larger particles require adsorption of multiple chains for flocculation [244].

Derivatives of polyacrylamide, or partially hydrolyzed polyacrylamide or polyacrylonitrile, are commonly used as commercial flocculants. Other polymeric flocculants include poly(vinyl alcohol)s, poly(ethylene oxide)s, modified celluloses, and a variety of their derivatives. The polymers usually have average molecular weights in excess of 10^6 daltons, equivalent to more than 30,000 segments per chain. This can enable the adsorbed polymer molecules to bridge widely separated particles and to form multiple points of attachment to the particle surfaces. Although the individual hydrogen bonds holding the polymer are weak, in excess of 10,000 of these bonds can be formed per polymer chain, which is equivalent to a large "energy of attachment" [245]. This results in the irreversible chemisorption of the polymer, although reversible redistribution of individual bonds may still occur on the mineral surface. The adsorption of polyacrylamides on clay dispersions is very rapid, and there is no apparent molecular-weight-dependent preferential adsorption of particular polymer fractions [243].

The polyacrylamide flocculants are often extensively hydrolyzed, and the resultant carboxylic acid groups assist in the flocculation process, either by nonspecific charge neutralization of cationic dispersions, like those of chrysotile, or by specific site bonding, for example, with layer edge Al^{3+} ions, or with exchangeable Ca^{2+} ions adsorbed on clay minerals. These anionic flocculants are frequently used at low pH, which allows maximum adsorption by the unionized carboxylic acid groups; the optimum level of anionic flocculant treatment is also pH-dependent. The adsorption of 20% hydrolyzed polyacrylamides on Al-montmorillonites is greatest at pH 6.8, which corresponds to the pK_a value of the polymer [246]. In this system, the adsorption largely involves interactions between the amide groups and the mineral surface, either by hydrogen bond formation or by strong ionic interactions resulting from the protonation of the amide groups by edge, or exchangeable, Al^{3+} ions. The carboxylate ions serve principally to ensure the maximum elongation of the nonadsorbed polymer segments, for the trapping of neighboring particles. High-molecular-weight hydrolyzed polyacrylamides are also used for the flocculation of kaolins, the optimum degree of hydrolysis being 30 to 40% [247].

Specific site bonding can also occur between many natural polymeric flocculants and soil constituents, for example, between "humic" acids, formed from decaying vegetable matter, and Al^{3+} or Fe^{3+} ions adsorbed on the clay mineral components. These strong interactions can assist in the formation of a porous, friable structure, in contrast to the heavy, impervious structure characteristic of many clayey soils [220].

Cationic flocculants, such as the polymers and copolymers of dimethyldiallylammonium salts [248–250], or quaternized alkylaminomethylacrylamide copolymers, are useful for the flocculation of anionic dispersions; in the case of dispersed clay minerals, they are probably adsorbed by ion exchange.

4.7 MODIFICATION BY GRAFT POLYMERIZATION AND ENCAPSULATION

Mineral particles can be encapsulated by the precipitation of polymers in the presence of suspended minerals or by spray-drying of polymer solutions containing the mineral. However, these methods tend to result in uneven coatings and the formation of polymer-bonded, coated agglomerates of mineral particles. In contrast, graft polymerization initiated by species adsorbed on the mineral surfaces, or involving their copolymerization, favors the formation of uniform polymeric coatings and the encapsulation of individual particles without flocculation of the dispersed mineral. Graft polymerization can also be used for the deposition of insoluble, cross-linked polymers.

An extensive review of the literature on the grafting of organic polymers (and monomers) on inorganic surfaces, covering the period 1946 to

1968, has been published by the United States Atomic Energy Commission [251]. Other useful reviews are contained in references 252,253.

Preformed polymers can be grafted to mineral surfaces by condensation reactions between surface species, for example, hydroxyl groups, and reactive substituents on the polymer. Polymers can also be grafted to minerals by use of ionizing radiation, or by mechanochemical processes, methods which are outlined in Chapter 5. Although graft polymerization can be obtained by use of processes involving ionic propagating species, discussion in this section will be largely confined to systems involving radical propagation. Some examples of ionic surface-initiated encapsulation are described in Chapter 5.

Radical graft-encapsulation may involve one or more of the following techniques: initiation of the polymerization of monomers by a surface-bonded radical species, copolymerization with a chemisorbed monomer, and grafting to polymer matrix via adsorbed monomers. These techniques will be described in turn.

4.7.1   Grafting by Surface-Initiated Polymerization

A variety of surface species can initiate the radical polymerization of adsorbed or contacted monomers. These include adsorbed derivatives of conventional initiators, such as organic peroxides or azo compounds, and species formed by hydrogen abstraction by, or electron transfer to, other radical species, as in the formation of surface oxy-radicals by abstraction of hydrogen from surface hydroxyl groups (Figure 4.24). In many published examples, the nature of the mineral–polymer bonds is unclear, and often the only relevant information available is the proportion of nonextractable polymer present in the product, this polymer being conveniently presumed to be grafted to the mineral surfaces.

The probable nature of the polymer–mineral bonds in these systems often can be predicted from examination of the initiator residues, which form the end-groups of the primary polymer chains, in related homogeneous polymerization systems; these have been studied for a wide range of initiators [254]. Persulfate initiators, for example, yield a mixture of sulfate and hydroxyl end-groups, the former forming an ionic bond with basic mineral surfaces. In other cases, such as an adsorbed Fe^{II}/hydroperoxide initiated polymerization, the formation of nonextractable polymer has to be ascribed to the intermediacy of surface radicals, such as those shown in Figure 4.24a, or else to "keying" of the propagating radicals into surface irregularities. A terminal hydroxyl group, the initiator residues in such a system, is unlikely to form a sufficiently tenacious bond with the mineral surface to prevent desorption of a long polymer chain. Polymerization in the presence of an adsorbed surfactant, such as lauryl sulfate, may also result in mineral graft formation. This may occur via surfactant-derived radicals resulting from hydrogen abstraction by the initiating or propagating radicals (Figure 4.24b).

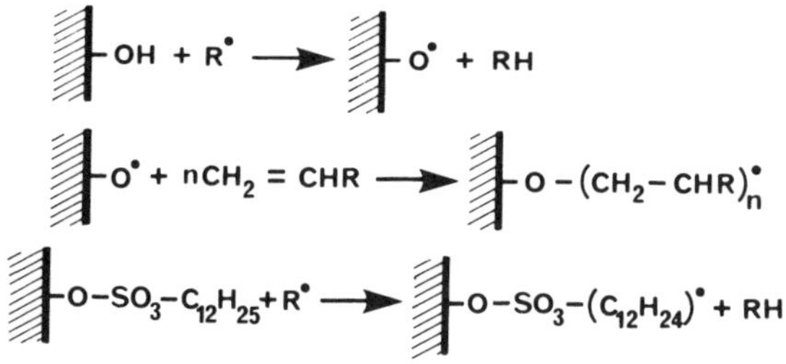

$$\text{[mineral]}-OH + R^\bullet \longrightarrow \text{[mineral]}-O^\bullet + RH$$

$$\text{[mineral]}-O^\bullet + nCH_2 = CHR \longrightarrow \text{[mineral]}-O-(CH_2-CHR)_n^\bullet$$

$$\text{[mineral]}-O-SO_3-C_{12}H_{25} + R^\bullet \longrightarrow \text{[mineral]}-O-SO_3-(C_{12}H_{24})^\bullet + RH$$

Figure 4.24   Redox-initiated graft polymerization on mineral surfaces.

Some of the earliest examples of intentional grafting via adsorbed radical initiators were reported by Dekking [255,256], who used ion-exchange processes to bond amino- or ammonium-substituted azo-initiators to the surfaces of clay minerals, such as kaolin, and other acidic or anionic mineral surfaces. These initiators included aminoazobenzene and azobisisobutyramidinium hydrochloride (ABIA) (Figure 4.25a). Heating the initiator-exchanged mineral results in the formation of chemisorbed radicals that can react with monomers to form grafted polymer chains. Some nongrafted polymer may also be formed through chain-transfer reactions. In the absence of monomer or other reactive species, neighboring radicals can undergo rapid mutual destruction, although a proportion of the bound radicals may be stabilized because of their isolation from adjacent radicals. These stabilized radicals are reactive, and polymerization of monomers can be initiated by the addition of the radical-bearing mineral. The reaction of methyl methacrylate with preheated ABIA–kaolin, for example, yields a product consisting largely of a polymer-grafted kaolin and containing only a small proportion of ungrafted polymer [256].

Polymers can be grafted to cationic surfaces by the use of adsorbed anionic initiators, such as azobiscyanoisovaleric acid (Figure 4.25b) on calcite [257], or adsorbed persulfate ions on chrysotile [258]. Persulfate ions can also be used for the grafting of neutral or anionic minerals, such as talc or kaolin [259]; a vinyl acetate-grafted kaolin, prepared by this method and used as a filler in an ethylene-vinyl acetate copolymer, formed a composite having 70% greater tensile strength than that of one containing untreated kaolin.

Grafted polymers can also be obtained by redox processes, for example, by treatment of a dispersion of the mineral containing an adsorbed peroxide, with a reductant in the presence of a monomer. When kaolin containing adsorbed t-butyl hydroperoxide, for example, is dispersed in an aqueous mixture containing phosphate, an alkylaryl sulfonate dispersant, a monomer (ethyl acrylate), and a dithionite (hydrosulfite) as a reductant, the resultant reaction yields an emulsion of polymer-encapsulated mineral particles, claimed to be useful as a surface coating composition [260]. In the cited example, half of the encapsulating polymer could not be extracted and was presumed to be grafted to the mineral surface.

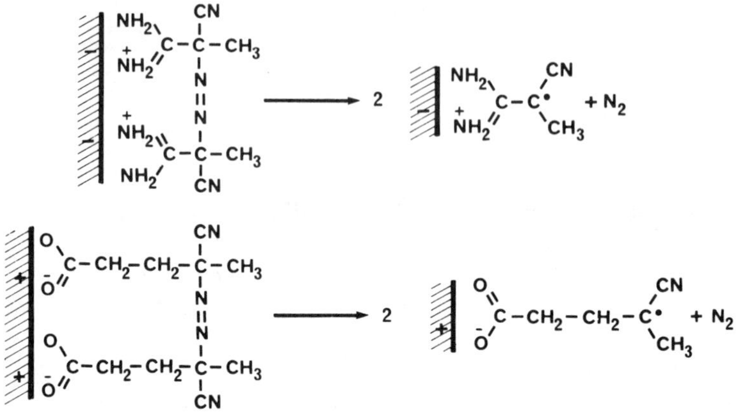

Figure 4.25   Cationic (a) and anionic (b) azo-initiators and their derived radicals.

Redox-initiated graft polymerization can occur on minerals containing adsorbed ionic reductants; these ions may include sulfite or dithionite or alternatively, adsorbed Fe^{2+} ions. Kaolin and silica [261] and asbestos and talc [262], for example, after treatment with ferrous sulfate solutions, can react with hydrogen peroxide or persulfates to initiate the graft polymerization of a range of acrylic and vinylic monomers (Figure 4.26). Acid pretreatment of chrysotile can markedly increase the yield of grafted polymer formed by this method, as removal of the outer brucite layer leaves an anionic surface for adsorption of the transition metal ions. A number of reports, principally from Japanese sources [e.g., 263,264], describe the use of sulfur dioxide-treated minerals, as initiating fillers, for the graft polymerization of styrene or acrylic monomers; in many cases the oxidizing component of the presumed redox initiating system could be trace peroxides present in the monomer, or residual atmospheric oxygen. Adsorbed sulfur dioxide has also been used for initiation of the polymerization of monomer–peroxide coinitiator mixtures adsorbed on particulates [265].

4.7.2   Grafting via Adsorbed Monomers

The graft polymerization may employ minerals pretreated with surface-bonding monomers in order to form reactive surfaces that are capable of copolymerization with other monomers; examples of these treated minerals include the unsaturated, silanized silica fibers (described in the previous chapter) and acrylic acid (or sorbic acid)-treated titanias [51,266].

Alternatively, the mineral-containing polymerization mixture may include a strongly adsorbed monomer that can modify the surfaces *in situ*. These monomers include acrylic acid for basic surfaces, vinylpyridine for acidic surfaces, and alkylamino acrylates, acrylamide or glycidyl acrylate for a variety of hydroxylic surfaces. Some specific examples of such polym-

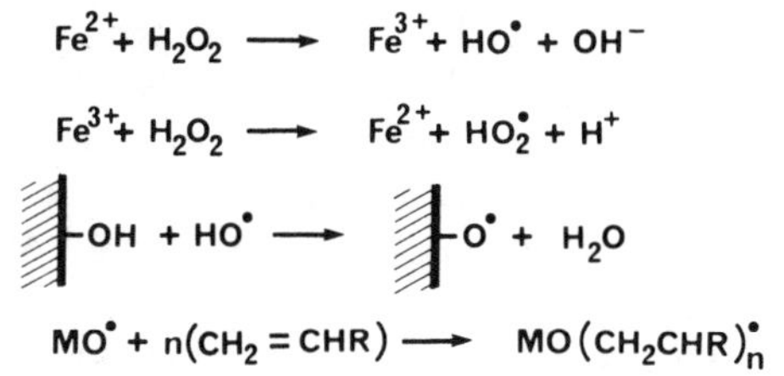

$$Fe^{2+} + H_2O_2 \longrightarrow Fe^{3+} + HO^{\bullet} + OH^-$$

$$Fe^{3+} + H_2O_2 \longrightarrow Fe^{2+} + HO_2^{\bullet} + H^+$$

$$\text{—OH} + HO^{\bullet} \longrightarrow \text{—O}^{\bullet} + H_2O$$

$$MO^{\bullet} + n(CH_2 = CHR) \longrightarrow MO(CH_2CHR)_n^{\bullet}$$

Figure 4.26 Some reactions for graft formation by chain transfer to surface species.

erization coating of minerals are calcite or titania with an acrylic acid-
–methyl methacrylate copolymer [267], calcium carbonate with diacetone
acrylamide copolymers [268], and chrysotile with a *t*-butylaminoethyl
methacrylate–1,2-ethylene dimethacrylate copolymer [269]. In all these
cases, grafting occurs by copolymerization between the chemisorbed
unsaturated species and the propagating radicals formed in the supernatant phase.

A considerable number of other patents describe the preparation of
polymer-coated particulates by polymerization of monomers in the presence of powdered minerals or mineral dispersions in weak solvents. Examples of this technique are the precipitation polymerization of short-chain
acrylic monomers on to pigments suspended in a nonswelling, weak solvent such as hexane; the solvent-free encapsulated pigments can be readily
dispersed in suitable solvents, as required, at a later date. As the monomers used in these examples often do not include species capable of strong
adsorption on the mineral or of reaction with matrix polymers, graft formation between the matrix and the mineral is minimal. The treatments are
primarily intended to improve the dispersibility and related properties of
the minerals in plastics compositions, or to provide prefilled or pre-
pigmented surface-coating emulsions or plastics molding compositions.
Any mechanical improvement of their composites with other polymers
results from improvement in dispersion or from the binding of the filler to
the matrix by entanglement between the chains of the encapsulating and
matrix polymers. Reinforcing fillers of this type can be prepared by polymerization coating of chrysotile dispersions, in sulfosuccinate [270] or polya-
nionic [271] surfactant-containing media, with polyacrylates or
polystyrene; the products are suitable for compression molding into reinforced plastics articles or for incorporation in other polymers.

A novel class of encapsulated reinforcing fillers have been patented by
the Amicon Corporation [272–274]; the postulated mechanism for reinforcement by these Ceraplast® fillers has been outlined in Section 4.2.3.
The fillers are based on a clay mineral, such as kaolin, and their use has
been claimed to convert low-cost polyolefins to composite materials having mechanical properties equal to those of the nylons and other costly,
engineering grade plastics. The fillers can be prepared by milling a mixture
of the acidic surfaced clay, a basic monomer, a divinylic monomer, and two
initiators, one of which decomposes at low temperatures and the other of

which decomposes at the higher temperature used for the subsequent melt-compounding of the treated filler in the polyolefin. The respective organic components, for example, might be 2-vinylpyridine, 1,3-butylene dimethacrylate, isopropyl percarbonate, and 2,5-dimethyl-2,5-bis(t-butylperoxy)hexane. The total weight of monomers used is equal to 1 to 5% of the weight of the kaolin. When this mixture is heated at moderate temperatures in an inert atmosphere, the percarbonate-initiated polymerization of the monomers forms a highly cross-linked, high-modulus coating around each kaolin particle. The coating contains residual unsaturation because a considerable proportion of the divinylic monomer molecules would not have partaken in cross-link formation during this first stage of copolymerization; the coating also retains the unreacted high-temperature initiator. When the encapsulated filler is compounded in molten polyolefin, the second initiator decomposes, forming radicals that can abstract hydrogen from the matrix polymer to form polymer radicals; these can react in turn with the residual vinylic groups of the coating polymer to form a grafted composite.

A second variety of Ceraplast® filler contains an adsorbed, unreacted mixture of a basic (or acidic) mineral-bonding monomer, a divinylic monomer, and a high-temperature initiator; formation of the encapsulating modified polymer zone occurs during the compounding of the filler and adsorbed monomers with the molten polyolefin [275]. In these systems, the thickness of the modified zone and the possible gradation of modulus will depend on the rate of diffusion of the initiator and divinylic monomer from the filler surfaces. This procedure can result in the formation of an extended modified zone around the filler particles. An independent examination of Ceraplast® compositions has shown that the second variety can provide the greatest improvement in tensile and impact properties of polyethylene, probably through the more extensive modification of the matrix polymer surrounding the filler particles [192]. A wide range of other monomer and initiator combinations is disclosed in the Amicon patents, but the relatively high cost of the monomers and initiators has so far limited the commercial viability of these modified fillers. Alternative reinforcing fillers which utilize the initiating capability of the kaolin surfaces have been patented [72,276]; these are described in Section 5.2.1.

Modification of polymers by peroxide-initiated cross-linking during compounding is frequently used to increase the softening point and tensile strengths of polyolefins. This technique can be used with formulations containing fillers bearing adsorbed unsaturated species to obtain reinforced, grafted composites. Grafted polymer layers can be formed *in situ* on kaolin particles, for example, by melt-compounding a mixture of the clay, polyolefin, maleic anhydride, and dicumyl peroxide [277]. Graft formation with adsorbed monomers can also occur during melt-compounding in the absence of added initiator, by the reactions of radicals formed by mechanochemical processes such as shearing degradation of the polymer melts. High grafting efficiency can be obtained in this manner if the mineral surface carries a highly reactive chemisorbed monomer such as an acrylic ester or salt. Other graft-promoting additives, especially useful for imparting toughness to polyolefins containing high filler loadings, include the cyclic acrylamide, 1,3,5-triacryl-hexahydrotriazine [278], or maleated (maleic anhydride-grafted) butadiene polymers [279].

Modifying coatings can be formed by polymerization of a termonomer composition onto the surface of clays dispersed in aqueous media [280]. The monomer mixture used in this method contains an inexpensive diluent such as styrene, a polar bonding monomer such as acrylonitrile or an aminoalkyl acrylate, and a cross-linking monomer such as divinyl benzene. The relative proportions of the three monomers can range from 98:2:0.1 to 1:1:0.2. These coatings contain little residual unsaturation that could be used for grafting; they are mainly intended for improvement of dispersion and compatibility between the mineral and organic media.

Whereas the use of high-modulus coatings can improve the mechanical properties of filled, low-modulus thermoplastics, improvement of filled, brittle, high-modulus thermoset resins is best obtained by the use of lower modulus coatings that can provide better stress distribution between the high-modulus matrix and filler and which can arrest the catastrophic propagation of cracks. This technique has been used, for example, for the improved reinforcement of epoxy resins by glass fibers or spheres, which had been pretreated with an aminoalkylsilane and then coated with a layer of a polyester-plasticized epoxy resin having a modulus, when cured, only 20 to 80% of that of the cured matrix resin [281]. Chrysotile fibers coated with a rubbery (low modulus) layer also provide better reinforcement for brittle polystyrene resins than fibers coated with higher modulus polymers [271].

The modified fillers discussed in this chapter were, for the most part, obtained by some form of pretreatment before incorporation into a composite product. However, modification of the mineral surfaces can occur during compounding. This may result from the reactions of formulation ingredients, for example, of monomers and initiators, from reactions with degradation products such as carbonyl derivatives or thermomechanical-generated radicals or from reactions with impurities in the matrix polymers. These reactions and interactions are beyond the scope of this book, but should not be overlooked, as they may have considerable consequences in the mechanical and chemical properties of the polymer–mineral interfaces and, in turn, on the performance of the composite material.

REFERENCES

1   A. W. Adamson, *Physical Chemistry of Surfaces,* 3rd ed., Wiley, New York, 1976.
2   D. H. Everet and R. T. Podoll, in *Colloid Science,* Vol. 3, Chemical Society, London, 1979, p. 63.

3   P. Connor and R. H. Ottewill, *J. Colloid Interface Sci.*, **37**, 642 (1971).

4   H. van Olphen, *Introduction to Clay Colloid Chemistry*, 2nd ed., Wiley, New York, 1977.

5   R. M. Barrer, *Zeolites and Clay Minerals as Sorbents and Molecular Sieves*, Academic, London, 1978.

6   B. Vincent, *Adv. Colloid Interface Sci.*, **4**, 193 (1974).

7   E. Dickinson and M. Lal, *Adv. Mole. Relaxation Interaction Processes*, Vol. 17, Elsevier, Amsterdam, 1980, p. 1.

8   M. Lal and R. F. T. Stepto, *J. Polym. Sci., Polym. Symp.* **61**, 401 (1977).

9   F. R. Eirich, *J. Colloid Interface Sci.*, **58**, 423 (1977).

10   V. T. Crowl, in D. Patterson, Ed., *Pigments*, Elsevier, Amsterdam, 1967, p. 176.

11   A. Doroszkowski and R. Lambourne, *J. Colloid Interface Sci.*, **26**, 214 (1968).

12   S. E. Rickert, C. M. Balik, and G. A. J. Hopfinger, *Adv. Colloid Interface Sci.*, **11**, 149 (1979).

13   A. Doroszkowski, R. Lambourne, and J. Walton, *Colloid Polym. Sci.*, **255**, 896 (1977).

14   Yu. A. Eltekov and A. V. Kiselev, *J. Polym. Sci., Polym. Symp.* **61**, 431 (1977).

15   C. H. Rochester, *Powder Technol.*, **13**, 157 (1976).

16   C. H. Rochester, *Adv. Colloid Interface Sci.*, **12**, 43 (1980).

17   P. Schuster, G. Zunde, and C. Sandorfy, Eds., *The Hydrogen Bond*, North-Holland, Amsterdam, 1976.

18   H. Knoezinger, Hydrogen Bonds in Systems of Adsorbed Molecules, in P. Schuster, G. Zunde, and C. Sandorfy, Eds., *The Hydrogen Bond*, North-Holland, Amsterdam, 1976, Chap. 27.

19   V. C. Farmer, *Soil Sci.*, **112**, 62 (1971).

20   G. D. Parfitt, *FATIPEC Cong., 14th.*, 107 (1978).

21   G. D. Parfitt, Ed., *Dispersion of Powders in Liquids*, Elsevier, London, 1969.

22   V. Buttignol and H. L. Gerhardt, *Ind. Eng. Chem.*, **60**, 68 (1968).

23   A. L. Smith, in G. D. Parfitt, Ed., *Dispersion of Powders in Liquids*, Elsevier, London, 1969.

24   J. N. Israelachvili and B. W. Ninham, *J. Colloid Interface Sci.*, **58**, 14 (1977) and following symposium papers.

25   P. Sennett and J. P. Olivier, *Ind. Eng. Chem.*, **57**, 33 (1965).

26   R. H. Ottewill, *J. Colloid Interface Sci.*, **58**, 357 (1977).

27   R. F. Conley, *J. Paint Technol.*, **46**, 51 (July 1974).

28   D. H. Napper, *J. Colloid Interface Sci.*, **58**, 390 (1977).

29   F. Th. Hesselink, *J. Polym. Sci., Polym. Symp.* **61**, 439 (1977).

30   G. R. Joppien and K. Hamann, *J. Oil Colour Chem.*, **60**, 412 (1977).

31   J. B. Smitham and D. H. Napper, *J. Colloid Interface Sci.*, **54**, 467 (1976).

32   F. Th. Hesselink, A. Vrij, and J. Th. Overbeek, *J. Phys. Chem.*, **75**, 2094 (1971).

33   B. C. Loft and D. H. Solomon, *J. Macromol. Sci. Chem.*, **A6**, 831 (1972).

34   J. W. Goodwin, in *Colloid Science*, Vol. 2, Chemical Society, London, 1975, p. 246.

35   D. H. Solomon and H. H. Murray, *Clays Clay Miner.*, **20**, 135 (1972).

36   R. Houwink and H. K. de Decker, Eds., *Elasticity, Plasticity, and Structure of Matter*, 3rd ed., Cambridge University Press, Cambridge, 1971.

37   T. H. Ferrigno, H. S. Katz and J. V. Milewski, Eds., *Handbook of Fillers and Reinforcements for Plastics*, Van Nostrand–Rheinhold, New York, 1978, p. 11.

38   R. B. Seymour, *Additives for Plastics*, Academic, New York, 1978.

39   Yu. S. Zuyev, *Polym. Sci. USSR*, **21**, 1315 (1980).

40   R. E. Felter, *J. Polym. Sci., Polym. Lett. Ed.*, **12**, 583 (1980).

41   R. Kummert and W. Stumm, *J. Colloid Interface Sci.*, **75**, 373 (1980).

42   O. I. Ivanishchenko and Yu. P. Gladkihk, *Colloid J. USSR*, **41**, 660 (1979).

43   J. Szczypa, S. Chibowski, and K. Kuspit, *Trans. Inst. Min. Metal., Sect. C*, **88**, 11 (Mar, 1979); *Chem. Abs.* **91**, 77217 (1979).

44   D. W. Pritchard and T. J. Wiseman (British Titan), Ger. Offen. 2,220,836 (November 9, 1972).

45   T. Mitsui, *Amer. Perfumer Cosmetics*, **80**, 23 (Aug. 1965).

46   L. K. Zgadzai, E. K. Varfolomeeva, G. L. Bolshedvorova, and V. V. Vlasov, *Colloid J. USSR*, **41**, 663 (1979).

47   G. W. Brindley and W. F. Moll, Jr., *Amer. Miner.*, **50**, 1355 (1965).

48   G. W. Bailey, J. L. White, and T. Rathberg, *Soil Sci. Soc. Amer. Proc.*, **32**, 222 (1968).

49   E. P. Plueddemann and G. L. Stark, in R. B. Seymour, *Additives for Plastics*, Vol. 2, Academic, New York, 1978, p. 49.

50   W. Janczak, *Angew. Makromol. Chem.*, **89**, 47 (1980).

51   K. Nollen, V. Kaden, and K. Hamann, *Angew. Makromol. Chem.*, **6**, 1 (1969).

52   E. A. Hauser and E. M. Dannenberg (Research Corp.), U.S. Patent 2,401,348 (March 17, 1942).

53   P. L. Smith (Union Carbide), British Patent 1,176,693 (January 7, 1980).

54   T. Otouma and S. Take (Nippon Asbestos) Jap. Kokai 74 34,485 (March 29, 1974); *Chem. Abs.*, **81**, 92493 (1974).

55   S. Crisp, S. Merton, A. D. Wilson, J. H. Elliot, and P. R. Hornsby, *J. Mater. Sci.*, **14**, 2941 (1979).

56   K. V. Raman and M. M. Mortland, *Soil Sci. Amer. Soc. Proc.*, **33**, 313 (1969).

57   J. H. Hodgkin, D. G. Hawthorne, J. D. Swift, and D. H. Solomon (C.S.I.R.O.), U.S. Patent 3,834,923 (September 10, 1974).

58   R. F. Conley and M. K. Lloyd, *Clays Clay Miner.*, **19**, 273 (1971).

59   D. H. Solomon, J. D. Swift, and A. J. Murphy, *J. Macromol. Sci. Chem.*, **A5**, 587 (1971).

60   L. Heller and S. Yariv, *Isr. J. Chem.*, **8**, 391 (1970).

61   S. Yariv, L. Heller, and N. Kaufherr, *Clays Clay Miner.*, **17**, 301 (1969).

62   S. Khan and V. Bansal, *J Colloid Interface Sci.*, **78**, 554 (1980).

63   V. C. Farmer and M. M. Mortland, *J. Phys. Chem.*, **69**, 683 (1965).

64   P. Cloos, R. D. Laura, and C. Badot, *Clays Clay Miner.*, **23**, 417 (1975).

65   J. R. Feldkamp and J. L. White, *Dev. Sedimentology*, **27**, 187 (1979); *Chem. Abs.*, **90**, 175345 (1979).

66   R. H. Dowdy and M. M. Mortland, *Clays Clay Miner.*, **15**, 259 (1967).

67   W. Bodenheimer, L. Heller, and Sh. Yariv, *Int. Clay Conf. Proc.*, **1**, 251 (1966).

68   C. F. Cole and S. Powell (British Titan), Ger. Offen. 2,112,560 (October 14, 1971); *Chem. Abs.*, **76**, 87305 (1972).

69   H. Rechmann, *Farbe Lack*, **72**, 1063 (1966).

70   American Cyanamid, Australian Patent Application 28644/71.

71   J. H. Hodgkin and D. H. Solomon, *J. Macromol. Sci. Chem.*, **A8**, 621 (1974).

72   J. D. Swift, D. G. Hawthorne, B. C. Loft, and D. H. Solomon (C.S.I.R.O.), U.S. Patents 3,849,149 (November 19, 1974) and 3,963,511 (June 15, 1976).

73   W. L. Riegler and R. L. Betcher (Armour Co.), U.S. Patent 2,852,406 (September 16, 1958).

74   J. R. Wilcox (Minerals and Chemicals Philipp Corp.), U.S. Patent 2,982,665 (May 2 1961).

75   T. H. Ferrigno (Minerals and Chemicals Philipp Corp.), U.S. Patent 3,236,802 (February 22, 1966).

76   Georgia Kaolin Co., British Patent 956,317 (April 22, 1964).

77   E. Ruiz–Hitzky and B. Casal, *Nature*, **276**, 596 (1978).

78   G. W. Brindley and S. Ray, *Amer. Miner.*, **49**, 106 (1964).

79   A. Weiss, *Chem. Ind.*, 382 (1980).

80   British Titan, Australian Patent Application 13,898/70.

81   B. O. Baum (Union Carbide), U.S. Patent 3,425,980 (February 4, 1969).

82   W. Twist, J. Peacock, and D. M. Porter (British Titan), Ger. Offen. 2,220,822 (November 9, 1972); *Chem. Abs.*, **78**, 45189 (1973).

83   T. J. Wiseman, in G. D. Parfitt and K. S. W. Sing, Eds., *Characterization of Pigment Surfaces*, Academic, London, 1976, p. 200.

84   R. L. Parfitt and M. M. Mortland, *Soil Sci. Soc. Amer. Proc.*, **32**, 355 (1968).

85   Yu. I. Tarasevich, V. P. Telichkun, and F. D. Ovcharenko, *Proc. Acad. Sci. USSR, Dokl. Phys. Chem.*, **182**, 659 (1968).

86   A. Tsunashima, M. Tachiki, K. Kodaira, and T. Matsushita, *Hokkaido Daigaku Kogakuba Kenkya Hokuku*, **91**, 87 (1978); *Chem. Abs.*, **90**, 196971 (1979).

87   F. Koubowetz, J. Latzel, and H. Noller, *J. Colloid Interface Sci.*, **74**, 322 (1980).

88   S. A. Tahoun and M. M. Mortland, *Soil Sci.*, **102**, 248 (1966); *Soil Sci.*, **102**, 314 (1966).

89   H. E. Doner and M. M. Mortland, *Clays Clay Miner.*, **17**, 265 (1969).

90   T. Stutzmann and B. Siffert, *Clays Clay Miner.*, **25**, 392 (1977).

91   M. M. Mortland, *Clay Minerals*, **6**, 143 (1966).

92   S. Olejnik, L. A. G. Aylmore, A. M. Posner, and J. P. Quirk, *J. Phys. Chem.*, **72**, 241 (1968).

93   J. J. Fripiat, P. Cloos, B. Calicis, and K. Makay, *Proceedings of the International Clay Conference, Jerusalem, 1966*, Vol. 1, Israel Program for Scientific Translations, Jerusalem, 1966, p. 223.

94   P. Cloos, B. Calicis, J. J. Fripiat, and K. Makay, *Proceedings of the International Clay Conference, Jerusalem, 1966*, Vol. 1, Israel Program for Scientific Translations, Jerusalem, 1966, p. 233.

95   W. Bodenheimer and L. Heller, *Clay Miner.*, **7**, 167 (1967).

96   Nippon Paint Co., Jpn. Kokai Tokkyo Koho 80,104,365 (August 9, 1980); *Chem. Abs.*, **94**, 32298 (1981).

97   A. E. Baker, D. J. Wluka, and H. W. Tankey (ICIANZ), Ger. Offen. 2,126,221 (January 5, 1972); Balm Paints Ltd., French Patent 1,578,934 (August 22, 1969).

98   P. Nylen and E. Sunderland, *Modern Surface Coatings*, Wiley–Interscience, New York, 1965.

99   B. W. Hotten, *Adv. Petroleum Chem. Refining*, **9**, 99 (1964).

100   C. Kahn and C. J. R. Eichhorn, *Cosmet. Perfum.*, **89**, 31 (Dec. 1974).

101   R. M. Barrer and A. D. Millington, *J. Colloid Interface Sci.*, **25**, 359 (1967).

102   R. M. Barrer and D. M. MacLeod, *Trans. Farad Soc.*, **51**, 1290 (1955).

103   M. B. McBride, T. J. Pinnavaia, and M. M. Mortland, *Adv. Environ. Sci. Technol.*, **8** (Part 1), 145 (1975).

104   W. S. Baldwin and J. L. Keen (General Mills), U.S. Patent 3,597,304 (August 3, 1971).

105   R. Kano (Matsuura Shoten), Japan. Kokai 75 72,926 (June 16, 1971); *Chem. Abs.*, **83**, 117186 (1975).

106   B. K. G. Theng, *Chemistry of Clay–Organic Reactions*, Hilger, London, 1974.

107   A. Weiss, *Angew. Chem. Int. Ed.*, **2**, 134 (1963).

108   J. W. Jordan, *J. Phys. Colloid Chem.*, **53**, 294 (1949).

109   J. W. Jordan (National Lead), U.S. Patent 2,531,440 (November 28, 1950); *Chem. Abs.*, **45**, 1340a (1951).

110   E. F. Vansant and G. Peeters, *Clays Clay Miner.*, **26**, 279 (1978).

111   G. Lagaly and A. Weiss, *Kolloid-Z., Z. Polym.*, **243**, 48 (1971).

112   Yu. L. Ishchuk, O. I. Umanskaya, A. A. Yaniv, N. V. Vdovenko, V. N. Moraru, and M. I. Balyta, *Colloid J. USSR*, **41**, 358 (1979).

113   W. H. Slabaugh and L. S. Carter, *J. Colloid Interface Sci.*, **25**, 359 (1967).

114   W. H. Slabaugh and G. H. Kennedy, *J. Colloid Sci.*, **18**, 337 (1963).

115   Yu. I. Tarasevich, *Adsorbtsiya Adsorbenty*, **5**, 58 (1977); *Chem. Abs.*, **89**, 153209 (1978).

116   W. T. Granquist and J. L. McAtee, Jr., *J. Colloid Sci.*, **18**, 409 (1963).

117   G. Lagaly and A. Weiss, *Kolloid-Z., Z. Polym.*, **248**, 968 (1971).

118   J. W. Jordan, B. J. Hook, and C. M. Finlayson, *J. Phys. Colloid Chem.*, **54**, 1196 (1950).

119   W. H. Slabaugh and P. A. Hiltner, *J. Phys. Chem.*, **72**, 4295 (1968).

120   G. Lagaly, H. Strange, and A. Weiss, *Kolloid-Z., Z. Polym.*, **250**, 675 (1972).

121   L. V. Berezov, N. V. Vdovenko, V. V. Guzeev, V. P. Lebedev, and V. M. Tseikinskii, *Colloid J. USSR*, **41**, 7 (1977).

122   I. Dekany, F. Szanto, L. G. Nagy, and G. Foti., *J. Colloid Interface Sci.*, **50**, 265 (1975).

123   I. Dekany, F. Szanto, and L. G. Nagy, *Progr. Colloid Polym. Sci.*, **65**, 125 (1978).

124   J. V. Kennedy and W. T. Granquist, *NLGI (Nat. Lubr. Grease Inst.) Spokesman*, **29**, 138 (1966).

125   M. S. Stul, A. Maes, and J. B. Uytterhoeven, *Clays Clay Miner.*, **26**, 309 (1978).

126   M. Taramasso, G. Lagaly, and A. Weiss, *Kolloid-Z., Z. Polym.*, **245**, 580 (1971).

127   A. V. Kiselev and Y. I. Yaskin, *Gas Adsorption Chromatography* (Transl. J. E. S. Bradley), Plenum Press, New York, 1969, p. 191.

128   L. R. Snyder, *Principles of Adsorption Chromatography*, Dekker, New York, 1968, p. 171.

129   M. M. Mortland and V. Berkheiser, *Clays Clay Miner.*, **24**, 60 (1976).

130   J. Shabtai, N. Frydman, and R. Lazar, *Proc. Int. Congr. Catal. 6th.* **2**, 660, (1976); *Chem. Abs.*, **87**, 157589 (1977).

131   J. L. McAtee, Jr., *Clays Clay Miner.*, **9**, 444 (1962); *Clays Clay Miner.*, **10**, 153 (1963).

132   F. D. Ovcharenko, V. N. Moraru, and S. A. Markova, *Proc. Acad. Sci. USSR, Dokl. Phys. Chem.*, **241**, 596 (1978).

133   C. M. Finlayson (National Lead), Ger. Offen., 2,912,070 (October 11, 1979); *Chem. Abs.*, **92**, 42873 (1980).

134   N. A. Ghanem, F. F. Abd El-Mohsen, and S. E. Zayyat, *J. Oil. Colour Chem. Assoc.*, **60**, 87 (1977).

135   D. J. Voskuil and Z. Konieczny (PPG Industries), U.S. Patent 3,793,044 (February 19, 1974); *Chem. Abs.*, **81**, 79475 (1974).

136   B. G. Barron (Dow Chemical), U.S. Patent 3,945,939 (May 20, 1974); *Chem. Abs.*, **84**, 181187 (1976).

137   F. W. Busch, Jr. (Chesebrough–Pond), U.S. Patent 3,864,294 (February 4, 1975); *Chem. Abs.*, **81**, 122961 (1974).

138   A. A. Ostwald and H. W. Barnum (Exxon), Ger. Offen., 2,541,557 (April 22, 1976); *Chem. Abs.*, **85**, 64037 (1976).

139   W. S. Zimmt, *Org. Coat. Plast. Chem.*, **36**, 131 (1976).

140   W. Frank, O. Bendszus, R. Kuechenmeister, and B. Peltzer (Bayer), Ger. Offen. 2.263.960 (December 8, 1977); *Chem. Abs.*, **88**, 52075 (1978).

141   K. Tani, N. Nihei, and C. Katoh (Japan Synthetic Paper), Jpn. Pat. 73 33,631 (October 16, 1973); *Chem. Abs.*, **80**, 146780 (1974).

142   W. Handl (Staedtler), Ger. Offen. 2,607,557 (September 1, 1977); *Chem. Abs.*, **87**, 168853 (1977).

143   P. R. H. Speakman (Procter and Gamble), Ger. Offen. 2,503,585 (August 14, 1975); *Chem. Abs.*, **83**, 149530 (1975).

144   H. G. G. Dekking (Union Oil), U.S. Patent 3,243,369 (March 29, 1966).

145   H. G. G. Dekking, *Nat. Conf. Clays Clay Minerals, 12th,* 1963, Pergamon, Oxford, 1964, p. 603.

146   I. A. Uskov, *Vysokomol. Soedin.,* **2**, 200 (1960); C.S.I.R.O. Transl. No. 6170.

147   L. W. Carter, J. G. Hendrichs, and D. S. Bolley (National Lead), U.S. Patent 2,531,396 (November 28, 1950); *Chem. Abs.*, **45**, 1804*d* (1951).

148   A. A. Ostwald (Exxon), U.S. Patent 4,136,103 (January 23, 1979); *Chem. Abs.*, **91**, 23376 (1979).

149   W. F. Fair Jr. (Koppers), U.S. Patent 2,805,954 (September 10, 1957); *Chem. Abs.*, **52**, 754a (1958).

150   M. R. Rosen, *J. Coatings Technol.,* **50**, 70 (Sep. 1978).

151   P. W. Erickson and E. P. Plueddemann, in *Composite Materials,* Vol. 8, Academic, New York, 1974, Chap. 2.

152   J. G. Marsden and L. P. Ziemionski, *Brit. Polym. J.,* 199 (Dec. 1979).

153   E. P. Plueddemann, in R. B. Seymour, *Additives for Plastics,* Vol. 1, Academic, New York, 1978, p. 123.

154   M. W. Ranney, S. E. Berger, and J. G. Marsden, *Soc. Plast. Ind., 27th Reinforced Plast. Conf. (1972), Paper 21-D.*

155   T. Nakatsuka, H. Kawasaki, K. Itadori, and S. Yamashita, *J. Appl. Polym. Sci.,* **24**, 1985 (1979).

156   Yu. V. Zhurdev, A. Ya. Korolev, Ya. I. Mindlen, and T. V. Menkova, *Colloid J. USSR,* **31**, 160 (1969).

157   J. V. Duffy, *J. Appl. Chem.,* **17**, 35 (1967).

158   H. Ishida and J. L. Koenig, *J. Colloid Interface Sci.,* **64**, 555 (1978).

159   H. Ishida and J. L. Koeng, *J. Colloid Interface Sci.,* **64**, 565 (1978).

160   H. Ishida and J. L. Koenig, *J. Polym. Sci., Polym. Phys. Ed.,* **17**, 615 (1979).

161   H. Ishida and J. L. Koenig, *J. Polym. Sci., Polym. Phys. Ed.,* **18**, 1931 (1980).

162   E. V. Broun, A. Ya. Korolev, L. M. Vinogradova, R. V. Artamonova, and T. V. Menkova, *Proc. Acad. Sci. USSR, Dokl. Chem.,* **187**, 626 (1969).

163   J. Gede and V. Mueller, *Polym. Sci. USSR,* **A18**, 836 (1976).

164   T. Nakatsuka, H. Kawasaki and K. Itadori, *Bull. Chem. Soc. Jap.,* **50**, 2829 (1977).

165   K. Bridger and B. Vincent, *Euro. Polym. J.,* **16**, 1017 (1980).

166   T. A. Grillo, *Rubber Age,* 37 (Aug. 1971).

167   J. Iaanicelli (J. M. Huber), U.S. Patent 3,567,680 (March 2, 1971); U.S. Patent 2,007,761 (January 9, 1970).

168   J. Iaanicelli (J. M. Huber), U.S. Patent 3,290,165 (December 6, 1966).

169   T. A. Grillo (J. M. Huber), U.S. Patent 3,834,924 (September 10, 1974).

170   A. G. Aristova, I. M. Zimmer, A. I. Gorbunov, and K. A. Andianov, *Dokl. Akad. Nauk SSSR,* **211**, 130 (1973).

171   L. Zapata, E. Mendelovici, J. Castelein, and J. J. Fripiat, *Bull. Soc. Chim. Fr.,* 54 (1972). (1972).

172 K. Kuroda and C. Kata, *Clays Clay Miner.*, **27**, 53 (1979).

173 L. Zapata, J. J. Fripiat, and J. P. Mercier, *J. Polym. Sci., Polym. Lett. Ed.*, **11**, 689 (1973).

174 L. Zapata, A. Van Meerbeek, J. J. Fripiat, M. della Faille, M. Van Russelt, and J. P. Mercier, *J. Polym. Sci., Polym. Symp.*, **42**, 257 (1973).

175 S. J. Monte, S. Sugerman, and P. F. Bruins, *J. Elast. Plast.*, **8**, 30 (Jan. 1976).

176 R. K. Iler (du Pont), U.S. Patent 2,683,156 (July 6, 1954); *Chem. Abs.*, **48**, 13182h (1954).

177 T. S. Reid (Minnesota Mining and Manufacturing), U.S. Patent 2,662,835 (December 16, 1953); *Chem Abs.*, **48**, 12791f (1954).

178 F. B. Hauserman, in *Metal-Organic Compounds*, Adv. Chem. Ser., No. 23, American Chemical Society, Washington, 1959, p. 338.

179 R. Feld and P. L. Cowe, *Organic Chemistry of Titanium*, Butterworth, London, 1965, p. 81.

180 S. J. Monte and P. F. Bruins (Kenrich Petrochemical), Ger. Ofen. 2,515,863 (October 23, 1975).

181 S. J. Monte and G. Sugerman, *Amer. Chem. Soc. Rubber Div. Meeting*, San Francisco, October 8, 1976, Paper No. 85; reprinted in Kenrich bulletin KR-1076-5.

182 S. J. Monte and G. Sugerman, *Org. Plast. Coatings Chem.*, **38**, 49 (1978); *Org. Plast. Coatings Chem.*, **38**, 202 (1978).

183 S. J. Monte, Coupling Agents—Titanates, in *Modern Plastics Encyclopedia, 1976–1977*, McGraw-Hill, New York, 1977.

184 S. J. Monte and G. Sugerman, *Soc. Plast. Ind. 35th Reinforced Plast. Conf.* (1980), Preprint, Paper 23-F.

185 C. D. Han, C. Sandford, and H. J. Yoo, *Polym. Eng. Sci.*, **18**, 849 (1978).

186 H. H. Morris, J. P. Oliver, and P. I. Prescott (Freeport Sulphur), Ger. Offen. 2,051,924 and 2,053,179 (May 6, 1971).

187 H. H. Morris and J. P. Oliver, U.S. Patent 3,660,134 (May 2, 1972).

188 H. H. Morris, P. I. Prescott, and R. J. Drexel, Jr., *Tappi*, **56**, 57 (1973).

189 Cleveland Society for Coating Technology, *J. Coatings Technol.*, **51** (655), 38 (Aug. 1979).

190 J. Rinse (J. W. Ayers and Co.), U.S. Patent, 2,979,497 (April 11, 1961).

191 D. G. Hawthorne (C.S.I.R.O.), U.S. Patent 3,905,936 (September 16, 1975).

192 D. G. Hawthorne and D. H. Solomon, *J. Macromol. Sci. Chem.*, **A8**, 659 (1974).

193 H. Deuel, *Makromol. Chem.*, **34**, 206 (1959).

194 E. V. Kukharskaya and A. D. Fedoseev, *Russ. Chem. Rev.*, **32**, 490 (1963).

195 H. P. Boehm, *Angew. Chem., Int. Ed.*, **5**, 533 (1966).

196 C. C. Ballard, E. C. Broge, R. K. Iler, D. S. St.John, and J. R. Mc Whorter, *J. Phys. Chem.*, **65**, 20 (1961).

197 R. K. Iler (du Pont), U.S. Patent 2,657,149 (October 23, 1953); *Chem. Abs.*, **48**, 2336*b* (1954).

198 R. K. Iler (du Pont), U.S. Patents 2,739,074—2,739,078 (March 20, 1956); *Chem. Abs.*, **50**, 11564 (1956).

199 I. E. Neimark and A. A. Chuiko, *S. P. E. Trans.*, 135 (April 1962).

200 A. A. Chuiko, V. A. Tertykh, N. V. Khaber, and E. A. Chuiko, British Patent 1,519,657 (August 2, 1978); *Chem. Abs.*, **90**, 154266 (1979).

201 V. V. Pavlov, V. A. Tertykh, V. N. Apanovich, and A. A. Chuiko, *Adsorbtsiya Adsorbenty*, **5**, 18 (1977); *Chem. Abs.*, **89**, 136228 (1978).

202 J. Bugosh (du Pont), U.S. Patent 2,944,914 (July 12, 1960); *Chem. Abs.*, **90**, 154266 (1979).

203 S. Okazaki and T. Kanno, *Nippon Kagaku Kaishi*, 1285 (1973); *Chem. Abs.*, **80**, 4967 (1974).

204 S. Yamashita and S. Kohjiya, *J. Appl. Polym. Sci.*, **17**, 2935 (1973).

205 L. Gonzalez Hernandez, L. Ibarra Rueda, J. Royo Martinez, and C. Charmorro Anton., *Rev. Plast. Mod.*, **33**, 507 (1977); *Chem. Abs.*, **87**, 54279 (1977).

206 B. Schneider, D. Doskovilova, J. Stok, M. Tulustakova, and J. Kalal, *Acta Polym.*, **30**, 283 (1979); *Chem. Abs.*, **91**, 92038 (1979).

207 R. A. V. Raff, J. A. Haugen, and M. F. Mostafa, *J. Appl. Polym. Sci.*, **17**, 1315 (1973).

208 S. Okazaki and T. Kijiya, *Nippon Kagaku Kaishi*, 2084 (1975).

209 A. A. Kachan and L. A. Negievich, *Usp. Khim. Poliuretanov*, 245 (1972); *Chem. Abs.*, **78**, 16503 (1973).

210 D. D. Eley, G. M. Kiwanuka, and C. H. Rochester, *J. Chem. Soc., Farad. I*, **69**, 2062 (1973).

211 D. D. Eley, G. M. Kiwanuka, and C. H. Rochester, *J. Chem. Soc., Farad. I*, **70**, 1099 (1974).

212 B. Siffert and H. Biava, *Clays Clay Miner.*, **24**, 303 (1976).

213 L. K. Zgadzai and E. K. Varfolomeeva, *Colloid J. USSR*, **40**, 102 (1978).

214 J. W. Eastes and T. F. Cooke (American Cyanamid), U.S. Patent 2,789,919 (April 23, 1957); *Chem. Abs.*, **51**, 10925c (1957).

215 T. A. Te Grotenhuis (General Tire and Rubber), U.S. Patent 2,780,612 (February 5, 1957); *Chem. Abs.*, **51**, 9183a (1957).

216 T. A. Te Grotenhuis, U.S. Patent 3,156,576 (November 10, 1964).

217 S. E. Gebura (Interpace Corp.) U.S. Patent 3,556,829 (January 19, 1971).

218 J. A. Manson and J. T. Williams, *Org. Coat. Plast. Chem.*, **42**, 175 (1980).

219 M. J. Marmo, M. Mostafa, H. Jinnai, F. W. Fowkes, and J. A. Manson, *Ind. Eng. Chem., Prod. Res. Dev.*, **15**, 206 (1976).

220 B. K. G. Theng, *Formation and Properties of Clay-Polymer Complexes*, Elsevier, New York, 1979.

221 K. G. Cooper, H. T. Cooper, and C. M. Rowland (Midland Silicones), Ger. Offen. 2,035,976 (April 22, 1971).

222 B. V. Ashmead, M. Bowrey, P. M. Burrill, T. C. Kendrick, and M. J. Owen, *J. Oil Colour Chem. Assoc.*, **54**, 403 (1971).

223 D. H. Solomon, *Chemistry of Organic Film Formers*, 2nd ed., Krieger, New York, 1967.

224 H. Rudolph, A. Boeckmann, K. Prater, and W. Gutsche (Farbenfabriken Bayer), Ger. Offen., 2,001,381 (July 22, 1971); *Chem. Abs.*, **76**, 128947 (1972).

225 H. Innami, R. Maeda, and K. Kasuya (Asohi Chem. Ind.), Japanese Patent 72 08,931 (March 15, 1972).

226 J. E. Nelson (Titangesellschaft), Ger. Offen. 1,814,569 (October 16, 1969).

227 J. Barksdale and W. P. Coker (Dow Chemical), U.S. Patent 2,425,855 (February 4, 1969).

228 I. Sugiyama and H. Tomozuka (Matsumoto Chem.), U.S. Patent 3,642,510 (February 15, 1972); *Chem. Abs.*, **76**, 142482 (1972).

229 La Porte Ind., Fr. Demande 2,156,867 (July 6, 1973); *Chem. Abs.*, **80**, 27936 (1974).

230 K. Wolf, M. Preuss, W. Kullick, A. Haus, R. Hoernle, and T. Mager (Bayer), Ger. Offen. 2,247,188 (April 25, 1974); *Chem. Abs.* **81**, 122980 (1974).

231 A. Galeski, R. Kalinski, and M. Kryszewski, Ger. Offen. 3,004,738 (August 28, 1980); *Chem. Abs.* **93**, 151229 (1980).

232 V. T. Crowl and M. A. Malati, *Discuss. Farad. Soc.*, No. 42, 301 (1966).

233 LaPorte Titanium, British Patent 1,155,567 (June 18, 1962).

234 P. R. Sperry, R. J. Wiersema, and K. Nyi (Rohm and Haas), Ger. Offen. 2,758,626 (July 20, 1978); *Chem. Abs.*, **89**, 131155 (1978).

235 G. J. Howard and C. C. Ma, *J. Paint Technol.*, **51**, (651), 47 (Apr. 1977).

236 R. Laible, *Farbe Lack,* **84**, 946 (1978).

237 D. W. J. Osmond and I. Heath (ICI Ltd.), U.S. Patent 3,317,635 (May 2, 1967).

238 G. H. Cox, D. W. J. Osmond, I. Heath, M. W. Skinner, and C. H. Young (ICI Ltd.), U.S. Patent 3,393,162 (July 16, 1968).

239 Commonwealth Scientific and Industrial Research Organization, British Patent, 1,328,136 (August 30, 1973).

240 N. DeRooy, P. L. DeBruyen, and J. H. G. Overbeek, *J. Colloid Interface Sci.*, **75**, 542 (1980).

241 W. E. Walles, *J. Colloid Interface Sci.*, **27**, 797 (1968).

242 N. Schamp and J. Huylebroeck, *J. Polym. Sci., Polym. Symp.* No. 42, 553 (1973).

243 A. P. Black, F. B. Birksen, and J. J. Morgan, *J. Colloid Interface Sci.*, **21**, 626 (1966).

244 R. K. Iler, *J. Colloid Interface Sci.*, **37**, 364 (1971).

245 W. F. Linke and R. B. Booth, *AIME Trans.*, **217**, 364 (1960).

246 B. Siffert and P. Espinasse, *Clays Clay Miner.*, **28**, 381 (1980).

247 M. Chmelir, A. Kuenschner, and E. Barthell, *Angew. Makromol. Chem.*, **89**, 137 (1980).

248 American Cyanamid, Netherlands Application 6,505,750 (December 30, 1965); *Chem. Abs.*, **64**, 17739*g* (1966).

249 Peninsular ChemResearch, British Patent 905,831 (March 10, 1959).

250 T. Ueda and S. Harada, *J. Appl. Polym. Sci.*, **12**, 2383 (1968).

251 R. A. V. Raff, *U.S.A.E.C. Report*, RLO-2043-2 (January 30, 1968), United States Atomic Energy Commission, Washington, 1968.

252 R. Kroker, M. Schneider, and K. Hamann, *Prog. Org. Coatings*, Vol. 1, Elsevier Sequioa, Lausanne, 1972, p. 23.

253 R. Laible and K. Hamann, *Adv. Colloid Interface Sci.*, **13**, 65 (1980).

254 S. R. Palit and B. J. Mandel, *Rev. Macromol. Chem.*, **3**, 226 (1969).

255 H. G. G. Dekking, *J. Appl. Polym. Sci.*, **9**, 1641 (1965); *J. Appl. Polym. Sci.*, **11**, 23 (1967).

256 H. G. G. Dekking (Union Oil), U.S. Patent 3,208,984 (September 28, 1965).

257 Yu. A. Zvereva, V. A. Popov, V. V. Guzeev, E. P. Shnarev, S. S. Ivanchev, and G. P. Gladyshev, *Proc. Acad. Sci. USSR, Dokl. Phys. Chem.*, **252**, 464 (1979).

258 A. S. Michaels (Amicon Corp.), U.S. Patent 3,519,594 (July 7, 1970).

259 M. M. Revyako, A. Ya. Markina, N. L. Tutaeva, V. S. Komarov, N. I. Pechkovskaya, and M. D. Belyakova, *Vesti Akad. Navuk BSSR, Ser. Khim. Navuk*, 97 (1979); *Chem. Abs.*, **91**, 5849 (1979).

260 P. Stamberger (Minerals Chemicals Philipp Corp.), U.S. Patent 3,068,185 (December 11, 1962).

261 N. L. Tutaeva, V. S. Komarov, and M. D. Belyakova, *Vesti Akad. Navuk BSSR, Ser. Khim. Navuk*, 105 (1977); *Chem. Abs.*, **87**, 152849 (1977).

262 N. L. Tutaeva, V. S. Komarov, and M. D. Belyakova, *J. Appl. Chem. (USSR)*, **51**, 2216 (1978).

263 T. Yamaguchi, T. Ono, Y. Saito, and S. Ohara, *Angew. Makromol. Chem.*, **53**, 65 (1976).

264 Lion Fat and Oil Co. Ltd., Ger. Offen. 2,309,150 (February 24, 1972); *Chem. Abs.*, **80**, 27665 (1974).

265 H. G. G. Dekking and E. M. Maxey (General Tire and Rubber), Ger. Offen. 2,058,265 (May 27, 1971); *Chem. Abs.*, **75**, 78272 (1971).

266 T. Nakatsuka, H. Kawasaki, and K. Itadan, *J. Appl. Polym. Sci.*, **23**, 3139 (1979).

267 K. Thinius and B. Hoesselbarth, *Plaste Kaut.*, **17**, 474 (1970).

268 G. M. Grundus, Jr., and S. E. Gebura (Interpace Corp.), U.S. Patent 3,763,084 (January 14, 1972).

269   J. C. Bolger (Amicon Corp.), U.S. Patent 3,519,593 (July 7, 1970).

270   D. G. H. Ballard, J. C. Padget, and R. C. Roberts (I.C.I. Ltd), British Patent 1,320,688 (June 20, 1973).

271   M. Xanthos and R. T. Woodhams, *J. Appl. Polym. Sci.*, **16**, 381 (1972).

272   G. J. Fallick, H. J. Bixler, R. A. Marsella, F. R. Garner, and E. M. Fettes, *Soc. Plast. Eng., Tech. Papers*, **14**, 540 (1968); *Modern Plastics*, **45** (5), 143 (Jan. 1968).

273   R. W. Hausslein and G. J. Fallick, *J. Appl. Polym. Sci., Appl. Polym. Symp.*, No, 11, 119 (1969).

274   H. J. Bixler and G. J. Fallick (Amicon Corp.), U.S. Patent 3,471,439 (October 7, 1969).

275   Amicon Corp., British Patent 1,223,286 (February 24, 1971).

276   D. G. Hawthorne, J. H. Hodgkin, B. C. Loft, and D. H. Solomon, *J. Macromol. Sci. Chem.*, **A8**, 649 (1974).

277   Champion Paper Co., British Patent 1,255,310 (December 1, 1971).

278   Nitto Funka Kogyo, Jpn. Kokai Tokkyo Koho, 80,110,138 (August 25, 1980); *Chem. Abs.*, **94**, 31598 (1981).

279   Furukawa Electric, Jpn. Kokai Tokkyo Koho, 80,112,248 (August 29, 1980); *Chem. Abs.*, **94**, 31584 (1981).

280   S. E. Gebura (Interpace Corp.), U.S. Patent 3,590,018 (June 29, 1971).

281   A. S. Kenyon and R. J. Slocombe (Monsanto), U.S. Patent 3,518,221 (June 30, 1970).

5

ORGANIC REACTIONS CATALYZED BY MINERAL SURFACES

In previous chapters we have described the variety of active groups which may be found on the surfaces of mineral pigments and fillers. We have also indicated their potential for initiating or catalyzing reactions with adsorbed or contacted organic molecules. In this chapter, we consider these and related reactions in more detail as well as demonstrating their practical importance in the utilization of mineral–organic systems. We shall be largely concerned with reactions that can occur at significant rates at temperatures below 300°C, a nominal maximum temperature to which most practical mineral–polymer systems might be subjected during processing or in service.

The mineral-catalyzed reactions described in this chapter are used primarily to illustrate reactions which are known to occur on surfaces of mineral fillers and pigments or which are similar to those that could occur on these materials, but which may not have been recognized. In many cases, we have had to concentrate on examples of an academic nature, as these are the only systems for which sufficient relevant details have been published. Whereas the colloidal and related physical chemical properties of many mineral organic systems have been extensively investigated, studies of organic reactions which can occur at filler or pigment surfaces, particularly in polymer matrices, have been comparatively limited in number and in scope. Hopefully, these aspects of mineral–polymer chemistry will receive more attention in the future.

5.1   CATALYSIS AND MINERAL SURFACES

The mineral-catalyzed reactions discussed in this chapter can be broadly divided into four classes, according to the type of catalytic species involved.

(i)   Reactions catalyzed by acidic species, such as those of the acidic aluminosilicate clay minerals, usually with the formation of cationic or radical-cationic intermediates.

(ii)   Reactions catalyzed by basic species, such as those of metallic oxides and carbonates, usually with the formation of anionic intermediates.

(iii)   Photolytic or electron-transfer reactions catalyzed by variable-valence species, such as transition metal ions, or on semiconducting surfaces, such as those of titania, with the formation of radical or radical-cationic intermediates.

(iv)   Reactions catalyzed by the polarization or other physical activation processes of the adsorbed molecules.

This classification is useful only as a guide because more than one class of chemical intermediates can be generated on a particular mineral surface. The Lewis acidic surface of alumina, for example, can adsorb a variety of alkenes as reactive, coordinated allylic anions, whereas the dehydration of alcohols and the adsorption of other alkenes can involve cationic carbenium ion formation. Aluminosilicate clay minerals, with their diversity of reactive centers, can also generate a variety of reactive intermediates, which may be cations, radical-cations, radicals, radical-anions, or anions, depending on the nature of the adsorbed organic species and the reaction conditions.

We have chosen to classify the mineral–organic reactions principally according to the nature of the reactive intermediates formed, rather than discussing the reactions of particular classes of minerals or organic compounds. We believe this approach gives a better appreciation of the range of reactions that can occur on pigment or filler surfaces.

5.2   REACTIONS VIA CATIONIC INTERMEDIATES

Heterogeneous reactions catalyzed by acidic minerals are widely used in the chemical and petrochemical industries and have been extensively studied. Most of these studies have been concerned with cracking catalysts and their models, but the reactive sites of these materials are often similar to those occurring on the clay minerals and other pigments and fillers. Many of the reactions occurring on cracking catalysts or related materials can therefore serve as useful models for understanding the reactions occurring on other mineral surfaces.

The examples discussed in this section mainly relate to the aluminosilicate minerals, kaolinite and montmorillonite, but we could expect analogous reactions to occur on the surfaces of other aluminosilicate minerals, aluminosilicate-coated pigments, and other materials bearing surface Bronsted or Lewis species with sufficient acidity.

We have already shown that the acidic mineral particles can be conveniently represented by box-like structures, the surface of which may carry Lewis acid centers, usually associated with exposed, coordination-deficient lattice Al^{3+}, or other metal cations, plus a variety of Bronsted acid sites. The latter include hydronium ions at exchange sites, ligand waters bound to polyvalent exchangeable cations, and acidic hydroxyl groups. We have also shown that the surface acidity may increase dramatically on drying of the mineral at moderate temperatures and that this may be due to the reduced hydration of the proton-donating species, which increases polarization of residual water molecules, or to the unmasking of Lewis acid sites. The most active of the fillers or pigments are the acidic aluminosilicate minerals, or materials that have been encapsulated with acidic alumina–silica gel coatings. Some weakly acidic oxides, such as silica, alumina, or titania, may be activated by surface treatment with their respective chlorides or with small amounts of sulfuric or phosphoric acid to form adsorbed, strongly acidic hydrogen sulfate or dihydrogen phosphate anions that are capable of polymerizing styrene [1,2].

The relative importance of Bronsted and Lewis acidic centers in mineral catalysis has been a controversial issue for many years, particularly in regard to the activity of cracking catalysts which, after activation at high temperatures, may contain predominantly Lewis acid sites. However, it is now generally recognized that most of the practical organic transformations may, in fact, actually involve residual Bronsted acid species [3]. This conclusion is of particular relevance, as strong Lewis acid species (as distinct from hydrated Lewis or potential Bronsted acid species) usually constitute only a minor proportion of the acidic centers on the surfaces of aluminosilicate minerals dried below 300°C. In contrast, while the active species on calcined aluminosilicates are often reported to be Lewis acidic centers, studies on calcined kaolins have shown that strongly acidic hydroxyl groups are tenaciously retained after calcination at tempera-

$$SH + RCH = CH_2 \longrightarrow R\overset{+}{C}H - CH_3 + S^- \tag{1}$$

$$SH + RCH(OH) - CH_3 \longrightarrow R\overset{+}{C}H - CH_3 + S^- + H_2O \tag{2}$$

$$S_x^x + R_2C = O \longrightarrow S^{\delta-} - O - \overset{\displaystyle R}{\underset{\displaystyle R}{\overset{|}{\underset{|}{C^{\delta+}}}}} \tag{3}$$

Figure 5.1

tures up to 900°C [4]; these hydroxyls are surface Bronsted acid species. The presence of Lewis acid species can be shown, for example, by infrared examination of a dehydrated mineral after treatment with pyridine, although selective poisoning of these sites may result in a corresponding decrease in the catalytic activity of the mineral. The products of the acid clay-mineral-catalyzed reactions usually correspond to those which would be initiated by protonation of one or more of the reagents.

The cationic intermediates can be formed by several processes, which include (see Figure 5.1):

(i)   Transfer of a proton from a Bronsted acid species to an unsaturated center in the adsorbed organic molecule (Eq. 1).

(ii) Protonation, followed by the elimination of a substituent group such as a hydroxyl (Eq. 2).

(iii) Coordination of a polar group, such as a carbonyl, to a Lewis acidic species (Eq. 3).

The yield of products obtained from a reaction catalyzed by an acidic clay mineral at moderate temperatures may be less than that obtained from industrial processes using zeolites or silica–alumina catalysts at elevated temperatures. This is partly due to the lower concentration of highly acidic centers on the surfaces of the clay, but also to the lower probability of complete regeneration of the catalyst, which results from the slow or incomplete desorption of the reaction products at the lower temperatures. This is most noticeable in condensation or elimination reactions when the products include species such as water or amines that are more basic than the reactants. While such reactions may be of little synthetic use, they are nevertheless important because many of the properties of filled or pigmented polymeric media can still be significantly affected by self-terminated reactions, particularly if these involve the destruction or neutralization of additives, such as initiators or inhibitors, or the cross-linking or scission of polymer chains.

In contrast, chain reactions, such as the polymerization of vinylic monomers, can yield large amounts of product from a comparatively limited number of catalytic centers. Once the active intermediates (e.g., carbenium ions) are formed, propagation by reaction with a succession of monomer molecules proceeds without further intervention by the catalyst. In this way, the cationic polymerization of susceptible vinylic monomers, such as styrene or vinyl ethers, can result in high conversions to polymer. These reactions have been extensively studied, both because of their commercial potential and because of their value as a tool in the investigation of mineral surface acidity.

5.2.1 Polymerization of Vinylic Compounds

A wide range of vinylic monomers can undergo polymerization when contacted with acidic minerals. These monomers are among those recognized as being susceptible to cationic polymerization and include vinylaromatics, vinyl ethers, conjugated dienes, and some alkenes [5–8]. Styrene has been the most extensively studied of these monomers. It is possible to compare its mineral-catalyzed polymerization with the analogous cationic polymerization in homogeneous media [5,6] or on acidic resins [9], for which detailed kinetic and product investigations have been reported.

Styrene polymerizes readily at room temperature when contacted with a variety of acidic minerals, particularly those having sites with Hammett acidity numbers, $H_0 < -3.0$. Such minerals include dry attapulgite, alum-flocced kaolins, and bentonites [10], as well as decationized zeolites [8], calcined clay aluminosilicates, and high-silica silica–aluminas [11]. Styrene contacted with these minerals undergoes a violent reaction in the absence of a diluent. The average molecular weight of the polystyrenes is typically on the order of 1000 to 5000 daltons, corresponding to average degrees of polymerization of 10 to 50. Cationic polymerization of styrene can occur on other minerals with acidic surfaces, although for weakly acidic minerals, the rate of cationic polymerization may be low and almost indistinguishable from those of other concurrent processes such as thermal and autoxidative polymerization.

The propagating species in the polymerization can be shown to be a carbenium ion by a number of diagnostic observations. These include the following:

(i) The polystyrene produced at room temperatures usually contains a large proportion of oligomeric material (dimers through pentamers); this is characteristic of cationic polymerization.

(ii) The correlation between surface acidity and polymerization activity of the catalysts.

(iii) The inhibition of the polymerization by the addition of small amounts of bases such as water or ammonia.

(iv) The absence of an electron spin resonance (esr) signal that might be indicative of a radical propagating species, although this test is not necessarily conclusive.

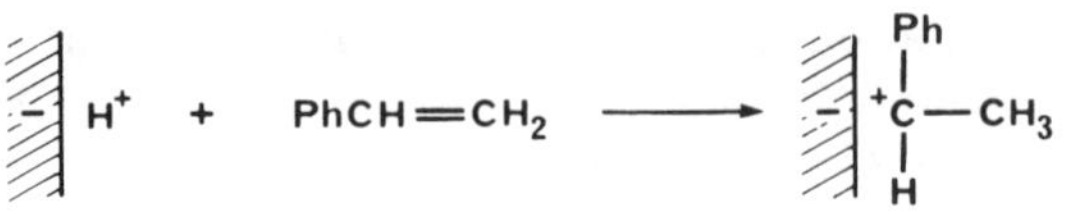

$$\text{—}\!\!\big|\, H^+ \;+\; PhCH{=}CH_2 \longrightarrow \text{—}\!\!\big|\,{}^+\!\!\underset{H}{\overset{Ph}{C}}{-}CH_3 \qquad (4)$$

Figure 5.2

(v)   The lack of inhibition of the polymerization on addition of small amounts of radical-trapping agents, such as quinones, and the lack of telomerization, shown by the virtual absence of halogen in the polymer, when the reactants include chlorinated solvents, such as chloroform. The polymerization of styrene in the presence of atmospheric oxygen is also contraindicative of a radical propagating species.

(vi)   The appearance of a yellowish-brown coloration on the surface of the mineral during the polymerization. This coloration is discharged on addition of base and may be indicative of the presence of an adsorbed, styrene-derived carbenium ion.

The nature of the initiation process has been the subject of controversy, but circumstantial evidence now indicates that initiation occurs at a Bronsted acid site by proton transfer to adsorbed monomer (Eq. 4 in Figure 5.2) [12]. The polymerization of styrene by montmorillonites is not inhibited by triphenylmethyl chloride, which can poison Lewis acid centers [13]. However, replacement of exchangeable Al^{3+} or H_3O^+ ions by Na^+ can greatly reduce the polymerization activity of aluminosilicates [14]. Studies of the vapor phase polymerization of styrene on an Al-montmorillonite have shown the importance of bound surface water molecules. The polymerization-initiating species appear to be ligand waters associated with exchangeable Al^{3+} [15], the maximum rate of polymerization occurring on minerals containing approximately 9% of adsorbed water; larger amounts of water act as an inhibitor of the reaction. Analogous correlations between water content and catalytic activity have been observed for 1-butene on silica–aluminas [16] and for 2-butene on an Al-kaolinite [17]. Another indication of Bronsted acid initiation is the rapid exchange that can occur between the β-hydrogens of styrene and a deuterated mineral surface [18].

The locations of the initiating sites on the surface of lamellar aluminosilicates, such as kaolinite or montmorillonite, still have not been conclusively established. Pretreatment of these minerals with reagents such as polyphosphates, which can mask exposed octahedral lattice aluminums, can also drastically reduce the polymerization activity, indicating that the active sites may occur at the lattice edges [7]. This result is still ambiguous, as these reagents also mask Al^{3+} ions or other complex-forming polyvalent ions that may be occupying exchange sites or present in amorphous aluminosilicate contaminants. These species are present on most practical samples of aluminosilicate minerals [14]. The presence of at least two vari-

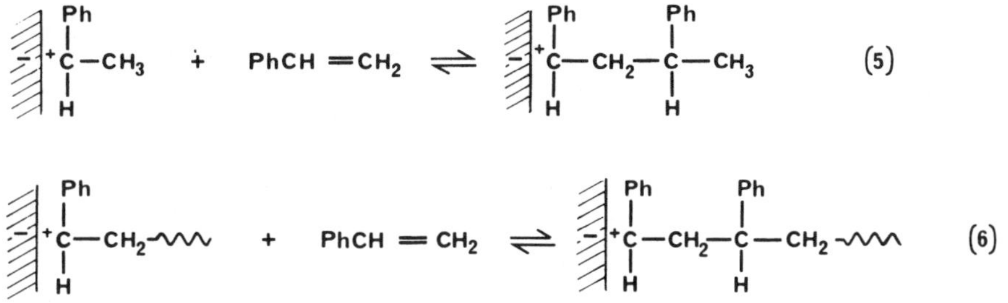

$$\text{(surface)}-\overset{Ph}{\underset{H}{\overset{|}{\underset{|}{C}}}}{}^{+}-CH_3 \;+\; PhCH=CH_2 \;\rightleftharpoons\; \text{(surface)}-\overset{Ph}{\underset{H}{\overset{|}{\underset{|}{C}}}}{}^{+}-CH_2-\overset{Ph}{\underset{H}{\overset{|}{\underset{|}{C}}}}-CH_3 \qquad (5)$$

$$\text{(surface)}-\overset{Ph}{\underset{H}{\overset{|}{\underset{|}{C}}}}{}^{+}-CH_2\text{-}\!\sim\!\sim\!\sim \;+\; PhCH=CH_2 \;\rightleftharpoons\; \text{(surface)}-\overset{Ph}{\underset{H}{\overset{|}{\underset{|}{C}}}}{}^{+}-CH_2-\overset{Ph}{\underset{H}{\overset{|}{\underset{|}{C}}}}-CH_2\text{-}\!\sim\!\sim\!\sim \qquad (6)$$

Figure 5.3

eties of Bronsted site on the surface of one commercial alum-flocced kaolin is shown by the bimodal molecular weight distribution of the polystyrenes [19]. Propagation of the polymerization proceeds by the addition of styrene molecules to the adsorbed initiating or propagating carbenium ions (Eqs. 5 and 6 in Figure 5.3). The polymer chains will continue to grow until terminated by transfer of a proton to another monomer molecule or by other processes described later in this section.

In a cationic polymerization, the propagating carbenium ions must be associated with counter-anions, the degree of interaction between the two being a function of the dielectric constant of the medium. If the polymerization is initiated by proton transfer from an acidic mineral, the propagating cation must remain close to the macro-anionic surface, unless some soluble anionic species is present to act as a counterion, and thus enable desorption of the growing chain. In systems where the monomer is absorbed and oriented in some regular conformation with respect to the propagating carbenium ion, the probability of a stereospecific polymerization is enhanced [20]. In an ideal case, this would result in formation of an isotactic polymer (1), in which all the monomer units in the chain are similarly oriented, or of a syndiotactic polymer (2), in which the units have an alternating orientation (see Figure 5.4). Random orientation of the monomer with respect to the propagating ion results in the formation of an atactic polymer.

Although styrene has been reported to form a syndiotactic polymer when contacted with an acid-activated, calcined bentonite [21], recent ^{13}C nuclear magnetic resonance (nmr) studies have shown that polystyrenes,

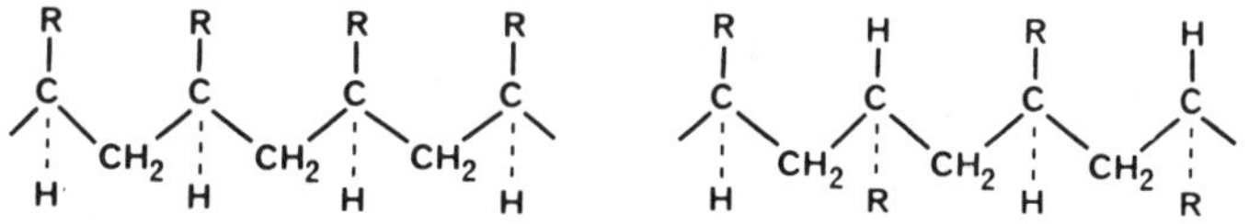

$$\underline{1} \qquad\qquad\qquad \underline{2}$$

Figure 5.4

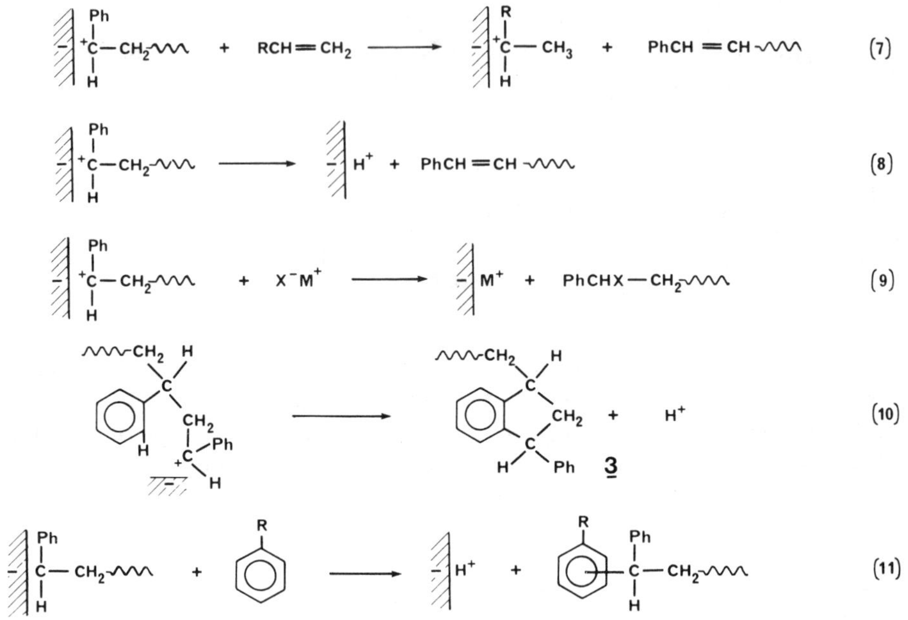

Figure 5.5

prepared using various crystalline and amorphous aluminosilicate catalysts, are essentially atactic [22]. Despite the absence of tacticity, kinetic studies of the polymerization process indicate that the propagation does occur largely or entirely at the mineral surface [23].

Chain termination usually occurs by more than one process, and the resultant polymer contains a variety of end-groups. The principal termination processes are (see Figure 5.5):

(i) Transfer of a proton from the propagating carbenium ion to another monomer molecule, thereby initiating the polymerization of another polymer chain (Eq. 7). This process introduces a terminal unsaturated group.

(ii) Proton transfer to the mineral surface (Eq. 8), thereby regenerating the catalytic center, or to a more basic species such as water or amines added to quench the reaction. This also results in terminal vinylic unsaturation.

(iii) Combination with an anion (Eq. 9). This is frequently an HO^- ion derived from a surface hydroxylic species, and results in a terminal hydroxyl group.

(iv) Intramolecular cyclization (Eq. 10), followed by proton transfer, forming a terminal indanyl group (**3**).

(v) Alkylation of an aromatic diluent or the phenyl rings of the polymer (Eq. 11). This results, respectively, in aryl end-group or branched chain formation.

Termination on aluminosilicate fillers occurs predominantly by processes (i) or (ii), resulting in polymers containing terminal unsaturation; process (ii) is the more likely, in view of the rapid hydrogen exchange that can occur between adsorbed styrene and the mineral surface. The polymers also contain indanyl end-groups, resulting from process (iv), although these may also be formed by the reprotonating and cyclization of "dead" polymer chains. Polymers prepared using the more acidic catalysts contain greater proportions of indanyl end-groups [22]. Alkylation of aromatic diluents, such as benzene or toluene, occurs to a minor extent if these are present. Hydroxyl end-groups, resulting from process (iii), are usually minor constituents or are absent in polymers formed using dried clay aluminosilicate catalysts [22], although they may be present in polymers prepared with other mineral catalysts [12].

The kinetics of the polymerization of styrene by acidic mineral surfaces are complex and show rate dependences that vary with monomer concentration [11,23]. Kinetic analyses are complicated by the variety of acidic initiating species that can occur on the mineral surfaces and their differing affinities for monomer and inhibitor molecules. Bittles, Chaudhuri, and Benson have provided a comprehensive kinetic analysis of the polymerization of styrene by an acid-modified montmorillonite [23].

Substituted styrenes may also undergo polymerization when contacted with acidic minerals. The rates of polymerization are qualitatively in accord with the relative reactivities measured in homogeneous acidic systems or predicted from the effects of the substituents on the electron density and basicity of the vinylic groups [24]. The order of reactivity for *para*-substituted styrenes is *p*-methoxystyrene > *p*-methylstyrene > styrene > *p*-chlorostyrene. Minerals having insufficient free strongly acidic sites for the polymerization of styrene at appreciable rates may readily polymerize the more basic methoxy or methyl-substituted styrenes. The polymerization of these more reactive monomers is also less susceptible to inhibition by basic impurities. α-Methylstyrene (Figure 5.6) can oligomerize when contacted with zeolite 5A and other acidic crystalline and amorphous aluminosilicates. Because of the steric effects of the geminal α-substituents, the homopolymer has a low ceiling temperature (approximately 50°C), above which the polymer is thermodynamically unstable and may depolymerize. Polymerization on zeolite 5A at 25°C can yield a solid oligomer with an average degree of polymerization of 4, whereas reaction at 150°C yields the dimers, 2,4-diphenyl-4-methylpent-1(and 2)-ene (4 and 5) [25,26]. The polymerization on an acid clay at 30°C yields the cyclic dimer (6) as the major product [27].

Other vinylaromatics that are known to polymerize on acidic minerals include vinylnaphthalene, divinylbenzene, and *N*-vinylcarbazole; vinylpyridines are not polymerized as protonation occurs at the nitrogen center, thus inhibiting the initiation process. 1-vinylnaphthalene is polymerized by acidic aluminosilicates in the liquid phase, but the frequent chain transfer results in a low degree of polymerization [28]. 1,1-diphenylethylene

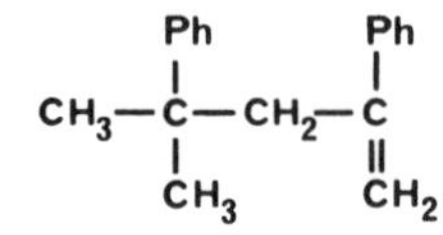

Figure 5.6

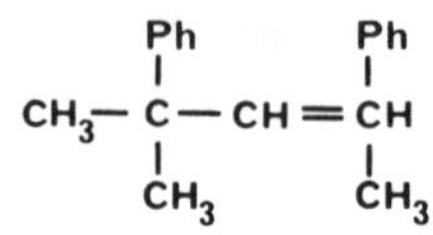

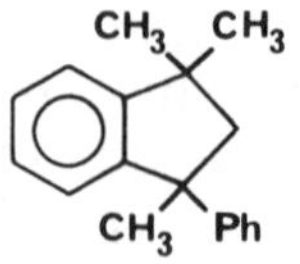

forms a dimeric indane when contacted with aluminosilicates. Stilbene and indene form cyclic dimers and trimers when adsorbed on a copper montmorillonite [29]; however, these reactions involve an oxidative process, and are described in Section 5.3.

Alkyl vinyl ethers and divinyl ethers are readily susceptible to cationic polymerization, for example, by zeolites [30], and can be polymerized to high conversion even by weakly acidic materials like nickel cyanide [31]. The polymers often have low-molecular weights, and the reaction catalyzed by aluminosilicate fillers is usually self-limiting. This results from hydrolysis of the monomer or from chain termination by HO$^-$ addition; these reactions yield acetaldehyde and alcohols, which are basic inhibitors and which can compete with the monomer for the acidic sites (see Figure 5.7). The polymerization processes for the vinyl ethers are probably similar to those of styrene, apart from the alternative termination mechanisms.

Other vinylic monomers that can be polymerized by aluminosilicates or other strongly acidic minerals include alkenes, such as the butenes, and conjugated dienes, such as butadiene, isoprene, and piperylene [32,33]. In each case a cationic propagation mechanism has been proposed, and the initiating mechanisms are probably similar to those of the polymerization of styrene. The propagation reactions may be more complex than those of styrene because of the greater probability of rearrangement of the intermediate carbenium ions [34].

Isobutene can be readily polymerized when contacted with zeolites [35]. Simple alkenes other than isobutene are not readily polymerized by cationic processes, unless the reactions are conducted at very low temperatures in the presence of Friedel–Crafts catalysts which can complex the monomer molecules; the products from acid-catalyzed reactions are usually extensively isomerized oligomers. Polymerization of these monomers on mineral surfaces occurs only slowly at temperatures below 100°C and appears to be limited to the formation of the equivalent of a few monolayers of polymer. However, extensive formation of oligomers may occur at higher temperatures [36], and the treatment of aromatic fractions with dried clays has been proposed for the polymerization and removal by adsorption of unsaturated contaminants [37].

The mechanisms for the polymerization of the simple 1-alkenes are not as unambiguous as those of the more reactive vinylic compounds. It is possible that Lewis acidic species are required as cocatalysts for the

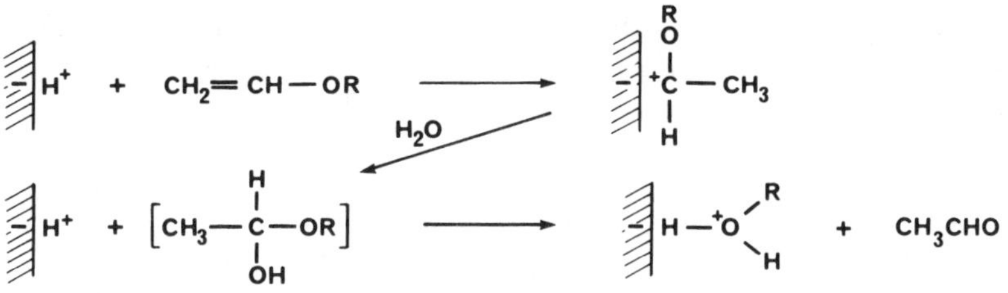

Figure 5.7

adsorption and activation of the alkenes, or that other activation processes such as polarization of the monomer in the surface field of the mineral
are necessary for polymerization. Most of the published examples of the
application of simple mineral catalysts relate to the use of zeolites or similar catalysts in the reforming of petroleum fractions [e.g., 34,38]. However,
some mineral fillers and pigments have sufficient surface acidity to polymerize adsorbed ethylene or propylene and thereby produce useful
improvements in the oleophilicity of the minerals [39].

Conjugated dienes are more reactive than the simple alkenes and can
polymerize rapidly, when contacted with acidic minerals, to yield useful
products. The propagation reactions are more complex than those of the
alkenes because the intermediate carbenium ions have allylic structures
(see Figure 5.8) and can undergo 1,2-, 1,4-, or 3,4-addition; 1,4-addition usually represents a minor pathway in the cationic polymerization.

The polymer chains have unsaturated groups and can undergo alkylation by propagating cations, either to form cyclic structures or to form
intermolecular cross-links. The mineral-catalyzed polymerization probably only incorporates five to ten monomer molecules before termination by
proton transfer occurs. However, cross-link formation between a series of
oligomeric chains can result in a polymer of high-molecular-weight that
can encapsulate the mineral particles. If the cross-linking is sufficiently
extensive, the polymer may be insoluble in organic solvents.

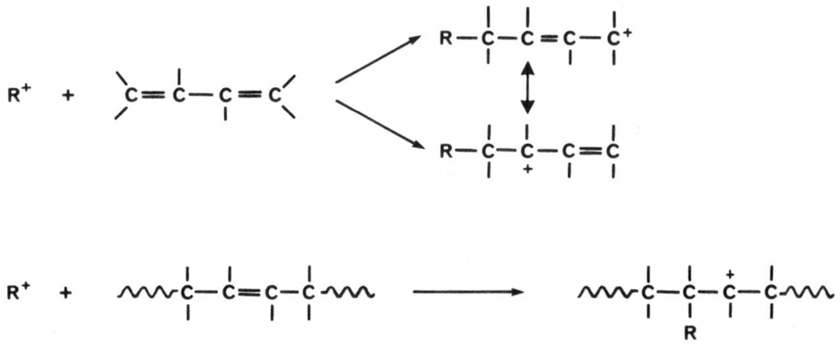

Figure 5.8

Kaolins and acidic bentonites can show a drastic reduction in catalytic activity following neutralization, by addition of amines, of only 10% of their acidic sites. Water can have a similar deactivating effect. A relationship between water content, surface acidity, and catalytic activity has been demonstrated for a number of clay mineral fillers. A kaolin containing 0.2% of adsorbed water, for example, and having surface acidity, $H_0 < -3.0$ to -5.6 [7], is not an effective catalyst for the polymerization of styrene, although it can still cause the rapid polymerization of more basic monomers such as p-methoxy-styrene [24]. Other species that can inhibit the polymerization of styrene include low-molecular-weight esters, alcohols, amides, aldehydes, and ketones [7]. Surprisingly, polyesters are reported to be ineffective inhibitors for the polymerization of styrene by clay mineral fillers. While some of the carboxyl groups are adsorbed on the strongest of the acidic sites, steric restrictions on the polymer chains may prevent adsorption on other active sites, which still remain accessible to the smaller styrene molecules.

α-Methylstyrene and other more basic vinylic monomers can also inhibit the polymerization of styrene by mineral surfaces. These inhibitors are preferentially adsorbed at the initiating centers, forming carbenium ion–macroanion pairs that do not propagate to the less basic styrene. This failure to propagate, with copolymer formation, is in contrast to the copolymerization which may occur in homogeneous systems containing catalysts and solvents which enable the formation of free carbenium ions.

Copolymerization of monomers on acidic surfaces has been alluded to in the patent literature, but there appear to be no reports of formal studies of the copolymerization of styrene or other monomers in this manner. Copolymerization of monomers of similar basicity, such as styrene and divinylbenzene, on acidic surfaces is feasible, but the inhibition of the polymerization of styrene by more basic monomers such as p-methoxystyrene, vinyl ethers, or N-vinylcarbazole, which can undergo homopolymerization in the presence of styrene, indicates that the copolymerization of monomers of differing basicity is unlikely at low-to-moderate reaction temperatures. Polymerization of the less reactive monomer is inhibited by the stability and preferential adsorption of carbenium ions derived from the more basic monomer.

When styrene is polymerized on a nonswelling acidic mineral such as kaolinite, virtually all the polymer is desorbed during the polymerization process, or can be removed by extraction with solvents. A small amount of polymer is usually tenaciously retained, but it is unlikely that this residue is grafted to the mineral surface, rather it is entangled or trapped in surface porosities. Styrene can be intercalated in montmorillonites by a process of displacement of some more-polar swelling liquid. The styrene can undergo a slow, acid-initiated polymerization during the intercalation process. This can propagate to the interlamellar absorbed styrene molecules, forming a nonextractable mineral–polymer intercalation complex containing stabilized interlamellar carbenium ions [40,41]. Treatment of this intercalation complex with amines can yield products having anion-exchange properties, indicating that an alkylation reaction between the polystyryl cations and the amine had occurred [42].

Apart from its scientific interest, the mineral-catalyzed polymerization of styrene has commercial value for the preparation of oligomers as additives, for example, in the processing of plastics and elastomers; the polymerization processes have been the subject of numerous patents [e.g., 43,44]. Although the use of natural minerals is usually cited, most of the patent examples employ synthetic silica–aluminas or zeolites, or extensively modified minerals that have large numbers of strongly acidic sites.

The formation of oleophilic coatings by adsorption and polymerization of monomers on acidic mineral particles can be used to modify the surfaces of acidic fillers and pigments, thereby improving their dispersibility in organic media and the mechanical performance of filler–polymer composites. These processes can provide alternatives to the surface modification treatments described in Chapter 4. Polydiene-coated minerals, in particular, have potential as reinforcing fillers.

The encapsulating polymers formed from the surface-catalyzed polymerization of dienes contain considerable residual unsaturation that can be used for the grafting of other polymers, for example, by radical-initiated reactions with a matrix polymer or with vinylic or acrylic monomers [33,39,45]. The unsaturated poly(1,4-alkadiene) coatings, for example, can undergo autoxidation on storage in air, a process which introduces polar groups into the polymer as well as increases the cross-link density. Autoxidation also introduces hydroperoxide groups, which can initiate grafting with neutral monomers or matrix polymers, as well as with basic monomers. The latter can be used to both neutralize the surface acidity and promote bonding between the polymer and the mineral.

Cross-link formation can also occur during the polymerization of non-conjugated alkadienes and of reactive divinylic compounds like the divinylbenzenes and *bis*-vinyl ethers. These monomers may be copolymerized with their monovinylic counterparts to increase the effective molecular weight or to insolubilize the resultant polymers. This has a particular advantage in the preparation of modified fillers. As the propagating carbonium ions are constrained to the vicinity of the mineral surface, the polymerization of a divinylic monomer or of a vinylic–divinylic monomer mixture by an acidic filler, suspended in an organic solvent, results in the encapsulation of the filler particles with polymer; little if any free polymer is formed in solution [46,47].

The catalytic activity of acidic minerals towards monomers can be disadvantageous in some circumstances, for example, in the drying of styrene with zeolites [48] or the use of kaolins as fillers in styrene–polyester molding compositions [49]. However, the surface acidity can usually be neutralized, or reduced sufficiently by the pretreatment of the mineral with a

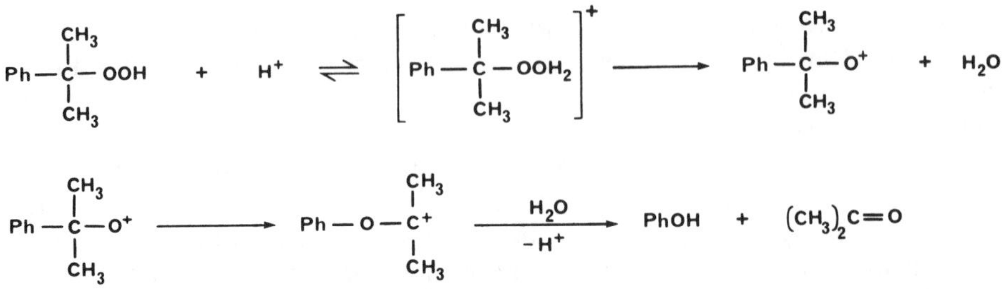

Figure 5.9 Reaction mechanism for the acid-induced cationic decomposition of cumene hydroperoxide.

small quantity of a base such as ammonia [e.g., 50] or a polyether [51], without affecting other properties of the mineral. As well as initiating the polymerization of vinylic monomers, acidic minerals may also catalyze the depolymerization of polystyrene and many other polymers [e.g., 52]. These reactions are the reverse of the cationic polymerization process and are discussed in Section 5.2.4.

Acidic minerals can also catalyze the polymerization of monomers such as ethylene oxide and various other condensation polymerizations; these will be discussed in Section 5.2.4. Acidic minerals may catalyze the radical polymerization of monomers, either by redox processes not directly related to the surface acidity or by the acid-catalyzed radical decomposition of peroxidic additives or impurities; these reactions are discussed in Section 5.3.

5.2.2 Peroxide Decomposition

Alkyl and alkylaryl peroxides and hydroperoxides have long been used for the curing or polymerization of mineral-filled plastics and elastomerics compositions; examples include "ketone peroxide" cured styrene–polyester compositions and dicumyl peroxide vulcanized elastomers. Technologists quickly found that compositions containing certain aluminosilicate clay minerals, or some silicas or acidic carbon blacks, required more peroxide for adequate curing than similar compositions containing neutral or basic fillers. These observations were among the earliest indicating that clay minerals might not be as inert as earlier experience, largely gained from aqueous clay systems, suggested. The active fillers were found to induce an acid-catalyzed heterolytic (nonradical) decomposition of the peroxides, and byproducts typical of such acidic cleavage could be recovered from the vulcanizates or from model formulations [53].

The acid-catalyzed cleavage of cumyl peroxide and hydroperoxide was well known at the time of these observations, and the decomposition of the hydroperoxide was in use for the large-scale commercial synthesis of acetone and phenol [e.g., 54]. This reaction is believed to occur by the mechanism shown in Figure 5.9.

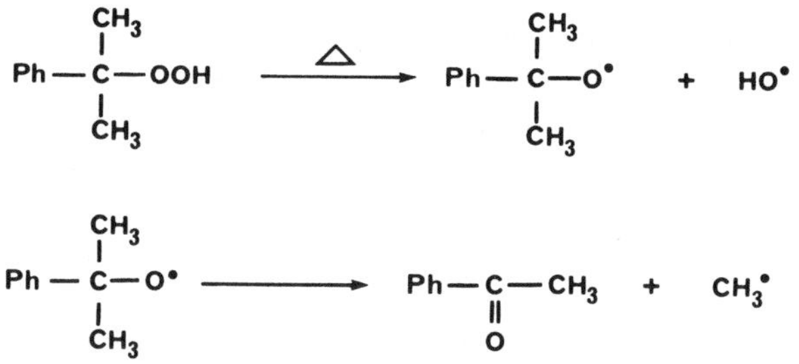

$$Ph-\underset{\underset{CH_3}{|}}{\overset{\overset{CH_3}{|}}{C}}-OOH \xrightarrow{\;\triangle\;} Ph-\underset{\underset{CH_3}{|}}{\overset{\overset{CH_3}{|}}{C}}-O^\bullet \;+\; HO^\bullet$$

$$Ph-\underset{\underset{CH_3}{|}}{\overset{\overset{CH_3}{|}}{C}}-O^\bullet \longrightarrow Ph-\underset{\overset{\|}{O}}{C}-CH_3 \;+\; CH_3^\bullet$$

Figure 5.10 The thermal, homolytic decomposition of cumene hydroperoxide.

A wide variety of acidic catalysts can be used in this reaction, although solid heterogeneous catalysts are favored for their ease of separation from the products. Various minerals, such as natural or activated kaolinites and montmorillonites, attapulgite, and zeolites, as well as synthetic silica–aluminas, are among the examples cited in the patent literature. The decomposition reaction results in high conversion of the hydroperoxide to phenol and is usually not accompanied by alternative radical-producing processes. In the case of cumyl hydroperoxide, the homolytic (radical) decomposition yields acetophenone as one product (Figure 5.10). The formation of acetophenone, or of phenol plus acetone, has been used for the quantitative differentiation between the two modes of decomposition of hydroperoxide adsorbed on mineral fillers and pigments [55].

The products from the industrial processes usually contain, as impurities, acetophenone, α-methylstyrene monomer and dimer, dimethylphenylcarbinol, p-cumylphenol, mesityl oxide, and hydroxyacetone. The acetophenone is probably a residue from the autoxidative preparation of the hydroperoxide from cumene, while the α-methylstyrene and the carbinol are decomposition products of dicumyl peroxide (see Figure 5.11), formed by condensation of the hydroperoxide [53,56]. This latter reaction can be used, with choice of a suitable aluminosilicate catalyst, for the production of α-methylstyrene [57,58].

The mineral-catalyzed decomposition of cumyl hydroperoxide can occur

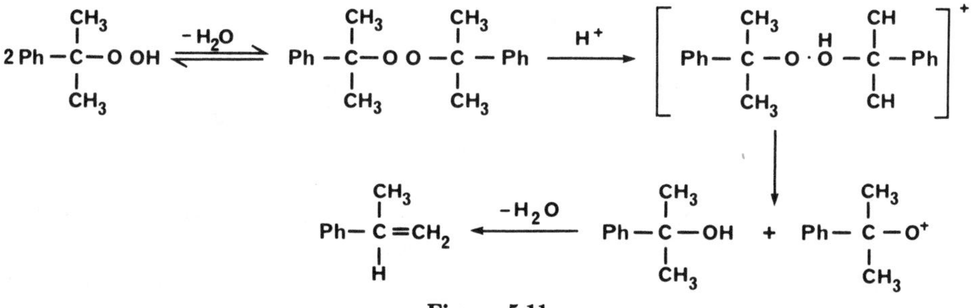

Figure 5.11

at centers considerably less acidic than those required for the polymerization of vinylic monomers [24], and the reaction is less readily inhibited by basic impurities or reaction products. The catalytic activity of the minerals generally correlates with the Hammett acidity of the surfaces. However, minerals appear to be particularly effective as catalysts, in comparison to conventional acids of similar strength. Kaolin, for example, has been reported to have 120-fold greater activity than a similar weight of p-toluene sulfonic acid [55]. This increased activity is due, in part, to the stabilization of the intermediate carbenium ions through adsorption on the mineral surface.

The facile cleavage and rearrangement is not confined to the cumyl peroxide and hydroperoxide, but can occur with other alkylaryl peroxides or hydroperoxides that contain at least one geminal aromatic group. p-(Hydroxypropyl)phenol, resorcinol, or hydroquinone can be produced from the corresponding hydroperoxypropyl-substituted precursors by reactions catalyzed by clays and other acidic minerals [59]. ar-Substituted phenols may be similarly prepared from ar-substituted benzyl hydroperoxides [60], but in these reactions and in the acid-catalyzed cleavage of other hydroperoxides having low thermal stability, the products may be accompanied by appreciable amounts of aldehydes or ketones that are formed by the competitive radical decomposition mechanisms.

Alkyl peroxides and hydroperoxides are more resistant to acid-induced cleavage, but still decompose when adsorbed on the surfaces of acidic minerals [55] (See Figure 5.12). The half-life of t-butyl hydroperoxide on an acidic kaolin is approximately 6 min at 30°C, compared to 30 sec for cumyl hydroperoxide. However, the cleavage of the alkyl peroxides may occur by simultaneous, competitive homolytic and heterolytic pathways. In the case of the acid-induced decomposition of t-butyl hydroperoxide, 20 to 30% of the cleavage may occur by a homolytic mechanism [61].

Diaroyl peroxides do not undergo appreciable acid-induced cleavage when adsorbed on dry aluminosilicate surfaces. A dry alum-flocced kaolin can adsorb benzoyl peroxide from a solution in toluene [55], but the peroxide can be quantitatively desorbed on the addition of water, even after contact times of 100 hr at 20°C [19]. This is in contrast with the action of conventional Lewis acids, such as $AlCl_3$ or $BF_3 \cdot Et_2O$, which promote the rapid decomposition of benzoyl peroxide to phenyl benzoate and CO_2, at room temperature, via an apparently nonhomolytic process [62]. Zeolites may promote the decomposition of aroyl peroxides, but in these instances the reactions appear to be homolytic, with the formation of adsorption-stabilized benzoyloxy radicals; these reactions are described in Section 5.4.

The acidic mineral fillers can interfere with the curing of styrene–polyester compositions, as well as affecting the vulcanization of elastomers. Retardation or inhibition of curing may occur by adsorption of the accelerators or other additives or by heterolytic cleavage of the peroxide or hydroperoxide catalysts. The acidic minerals can also adversely affect the

$$(CH_3)_3 COOH \xrightarrow{+H^+} (CH_3)_3 CO^+ + H_2O \xrightarrow{-H^+} (CH_3)_2 C=O + CH_3OH$$

Figure 5.12

storage of the styrene-diluted prepolymers by promoting premature polymerization of the styrene. Potentially acidic minerals used in these formulations need to be neutralized by one or more of the methods described in Chapter 4, for example, by treatment with polyphosphates, by neutralization with bases, or by encapsulation with a neutral inorganic gel coating.

The heterolytic decomposition of peroxides by acidic minerals can be used to advantage in improving the weather resistance of filled or pigmented plastics composites. Much of the polymer embrittlement and degradation, which limits the outdoor service life of plastic articles, is caused by ultraviolet-light-induced autoxidation, with the formation and homolytic cleavage of hydroperoxide intermediates. The inclusion of acidic fillers can result in nonradical cleavage of the intermediate hydroperoxides and disruption of the autoxidation chain reaction. The addition of 1% of an acidic kaolin, for example, into polyethylene pigmented with 10% of titania, results in a significant reduction in embrittlement on accelerated weathering, compared to that of a similarly pigmented polyethylene containing 1% of the neutral filler, talc [55].

5.2.3 Addition and Elimination

Acidic mineral surfaces can catalyze the dehydration of alcohols, the cleavage of amines, and the elimination of the corresponding acids from esters, halides, and mercaptans. The minerals may also catalyze the reverse reactions such as the addition of water or amines to alkenes. The rates of these reactions are governed by the acidity of the mineral surface and the stability of the intermediate carbenium ions formed in Bronsted acid-catalyzed eliminations, but may proceed at significant rates under a variety of reaction conditions. The dehydration of primary alcohols by kaolinites, zeolites, or activated clays, for example, occurs rapidly at 150 to 250°C, but may also occur to an appreciable extent on storage, at ambient temperatures, of minerals treated with long-chain alcohols. Alkylammonium montmorillonites may rapidly decompose on heating at 200 to 250°C, with elimination of the alkyl groups as alkenes [63]. In contrast, hexadecyltrimethylammonium derivatives of some grades of montmorillonite undergo noticeable decomposition after several months' storage at 20°C, with the evolution of trimethylamine, hexadecanol, and hexadecene (Figure 5.13) [64]. Slow elimination reactions are also probably involved in the geochemical transformation of lipid and other organic sediments into petroleum deposits.

The dehydration of alcohols may be accompanied by the formation of

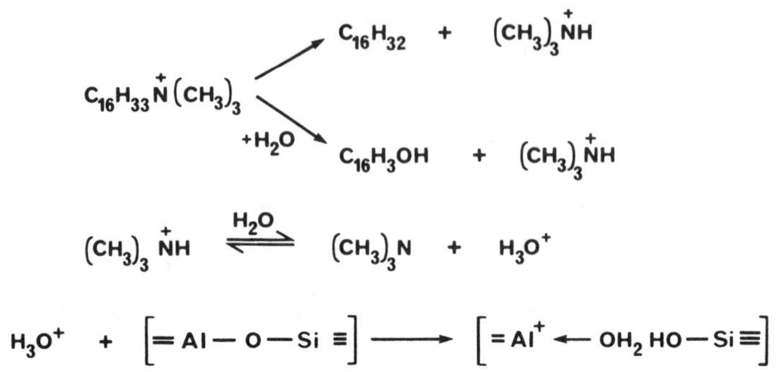

Figure 5.13 · Some modes for the decomposition of hexadecyltrimethylammonium ions on an acidic aluminosilicate surface.

ethers and by the rearrangement of the intermediate carbocations to form a mixture of isomeric alkenes; in the case of 1,2-diols, the products may also include aldehydes, dioxanes, and dioxolanes. The dehydration of butanol or higher alcohols by aluminosilicates or other Bronsted acid species usually produces isoalkenes, for example, *n*-butanol and *t*-butanol both yield isobutene (Figure 5.14). Dehydration of alcohols at elevated temperatures, typically above 200 to 250°C, may also result in polymerization, cracking, and hydride-transfer reactions and the formation of alkanes, carbon monoxide, and carbonaceous residues [65]. Heating of "single layer"

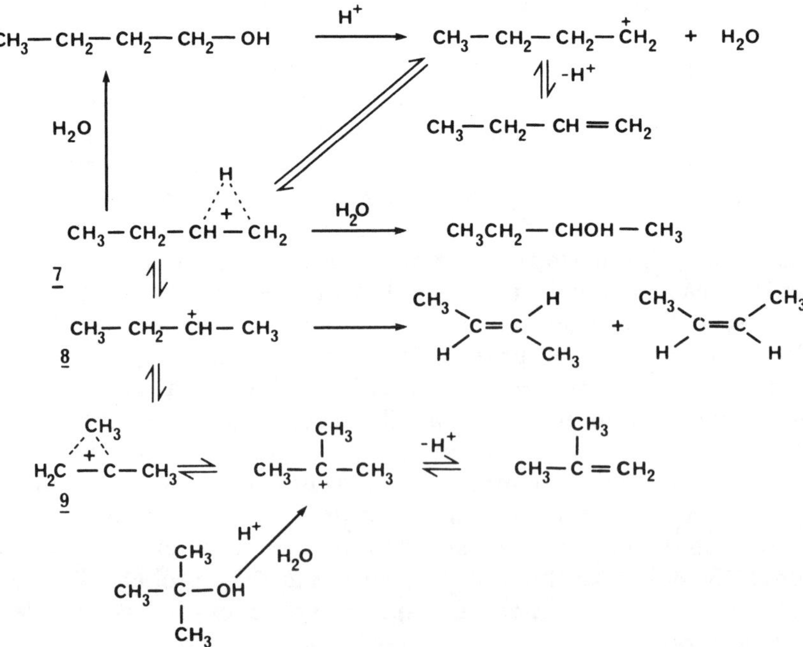

Figure 5.14   Acid-catalyzed dehydration and isomerization of butanols.

glycol–smectite or glycerol–smectite intercalation complexes, for example, above the boiling points of the respective alcohols, results in the formation of black solids which may be useful as pigments [66].

The mineral-catalyzed dehydration of alcohols is not confined to the aluminosilicates or acidic silica–aluminas, but can also occur in the cavities of the porous magnesium silicate mineral, sepiolite, at temperatures near 300°C. Ethylene is the main product from the dehydration of ethanol by this mineral, while Mn-exchanged sepiolites can catalyze oxidative dehydration of the alcohol to butadiene [67].

Pure Lewis acid-catalyzed dehydrations occur via chemisorbed anionic intermediates; these may undergo double-bond migration without skeletal isomerization. However, the carbonaceous impurities, which accumulate on practical Lewis acid catalysts, can act as proton donors, or Bronsted acids.

The rehydration of alkenes can yield mixtures of isomeric alcohols and ethers [e.g., 68], although ethers are often the dominant hydration products. The mineral surface may introduce steric effects that can result in selective hydration, for example, of 1-alkenes to *bis*-(*sec*-alkyl) ethers in the interlamellar regions of an acidic montmorillonite [69]. The formation of *sec*-alkyl ethers is in contrast to the interlamellar dehydration of the 1-alkanols, which form di-(alk-1-yl) ethers by the nucleophilic displacement of water from a protonated alcohol molecule by an adjacent unprotonated molecule [70].

Cleavage of adsorbed or intercalated amines by acidic minerals may be accompanied by various transalkylation reactions; if water is present, alcohols and ethers may also be formed [63]. These reactions are usually accompanied by the formation of polymeric materials that can limit the conversion of the amines. The transformation of alkylammonium ions or amines, on heating their smectite complexes in an inert atmosphere, results from a series of equilibrated reactions that are catalyzed by protons derived from highly polarized adsorbed, or ligand, water molecules [71] (see Figure 5.15). These reactions are favored by the low-diffusion-coefficients of the adsorbed amines.

The decomposition of an ethylammonium montmorillonite, for example, can yield ethylene, ethanol, diethyl ether, ethylamine, and diethylamine, while heating of an ammonium montmorillonite with ethylene may yield small amounts of various ethylamine derivatives. Decomposition of the amine complexes in the presence of air can also result in oxidation of the intermediate species and the formation of carbonyl compounds as byproducts. The degradation of alkylammonium smectites is relevant to the performance of their commercial applications as gellants. Decomposition of alkyltrimethylammonium bentonites on storage can result in the formation of unpleasant odors and in a marginal degradation in the rheological properties of the organoclay. However, the degradation is more important at elevated temperatures and is one cause for the poor high-

$$RNH_3^+ + H_2O \rightleftharpoons ROH + NH_4^+$$

$$R_2NH_2^+ + H_2O \rightleftharpoons ROH + RNH_3^+$$

$$R_3NH^+ + H_2O \rightleftharpoons ROH + R_2NH_2^+$$

$$R_4N^+ + H_2O \rightleftharpoons ROH + R_3NH^+$$

$$ROH_2^+ + H_2O \rightleftharpoons ROH + H_3O^+$$

$$R_xNH_{(4-x)}^+ + H_2O \rightleftharpoons R_xNH_{(3-x)} + H_3O$$

$$ROH_2^+ + ROH \rightleftharpoons R_2O + H_3O^+$$

$$ROH_2^+ \rightleftharpoons R'CH = CH_2 + H_3O^+$$

Figure 5.15 Hypothetical reactions involved in the transformation of intercalated alkylammonium ions by acidic smectites.

temperature performance of greases using gellants prepared from some bentonites.

The mechanisms originally proposed for the transformation of the alkylammonium smectites involved reactions between pairs of protonated species [71]. We suggest that alternative mechanisms deserve consideration. The reactions may involve the Bronsted acid-catalyzed formation of transient carbenium ions which could react with other, nonprotonated species which could exist in equilibrium with their protonated forms, but probably involve the displacement of water or amine by the reaction between protonated and unprotonated amine or alcohol molecules (Figure 5.16). Amines and alcohols can undergo a similar series of reactions when adsorbed on zeolites [72], probably via similar mechanisms. Acidic minerals can catalyze the conversion of an alcohol into an amine, for example, methylcyclopentanol into methylcyclopentylamine [73], or of aromatic amines into phenols [62].

Two types of mechanisms have been proposed for the dehydration of alcohols, respectively involving Bronsted and Lewis acid catalysis [74]. Bronsted acid catalysis occurs on the surfaces of aluminosilicates, silica, and possibly also on the surfaces of metal oxides "activated" by adsorbed, hydrated polybasic anions. Lewis acid catalysis occurs on the surfaces of metal oxides, but may also occur on aluminosilicate surfaces that have been dehydrated by calcination and used, for example, in low-conversion high-vacuum pyrolysis or desorption studies, so that the reaction products cannot accumulate on the catalyst surface.

In the Bronsted acid-catalyzed dehydration, the alcohol is first protonated by the surface acidic species and then eliminates water to form a carbenium ion (Figure 5.14). The rates of the elimination reactions generally reflect the relative stabilities of the carbenium ions, namely, (1) tertiary > secondary > primary and (2) long chains > short chains. The carbenium ion may eliminate a β-proton, with or without isomerization, it may alkyl-

$$R_x NH_{(4-x)}^+ \rightleftharpoons R_{(x-1)}NH_{(4-x)} + R^+$$

$$ROH_2^+ \rightleftharpoons R^+ + H_2O$$

Figure 5.16

ate a coadsorbed species, or it may reform the alcohol or an isomeric alcohol by reaction with water. Studies using amorphous silica–aluminas have shown that apparently all the surface acidic sites are capable of catalyzing the dehydration of alcohols, but that only sites having Hammett acidities, $H_0 < -6.3$ are involved in skeletal isomerization [75]. The elimination of water from adsorbed alcohols usually follows zero-order kinetics, indicating that the rate-determining step is the fission of the alkyl–oxonium bond [74]. The rate of elimination is often greater than that catalyzed by protonic acids of equivalent strength, as the mineral surface can stabilize the adsorbed carbenium ion and can also increase the subsequent rate of deprotonation through the availability of adjacent basic sites.

The dehydration of an alcohol on a Lewis acid involves the adsorption and coordination of the organic hydroxyl group with the Lewis centers which, in the case of alumina, are coordination-deficient Al^{3+} ions (see Figure 5.17). This is accompanied by hydrogen bonding with adjacent surface hydroxy groups. This complex differs from the Bronsted acid complex because the alcohol is chemisorbed as an "alkoxide," with protons being donated to the mineral surface. This has been demonstrated in nmr studies, which show the localization of the alkoxide, and the delocalization of the proton via a charge-relay process involving the Al–OH groups and O^{2-} ions [76]. The donated protons are acidic, and the mineral–alcohol adduct can act as a Bronsted catalyst [77]; purely Lewis reactions may not occur if reaction products are allowed to accumulate on the mineral surface. The actual structures of the chemisorption complexes probably differ for the various types and environments of the Lewis sites, but they are similar to the structure shown below. A similar process could occur at low coverage on aluminosilicates, with coordination of the alcohol to exposed edge or tetrahedral-layer Al^{3+} ions and transfer of protons to the basal oxygen sheets [78].

The actual mechanism for dehydration by Lewis acids is more complex than that indicated above. Apart from side-reactions that can form surface proton sources, the mode of dehydration of alcohols on alumina can vary

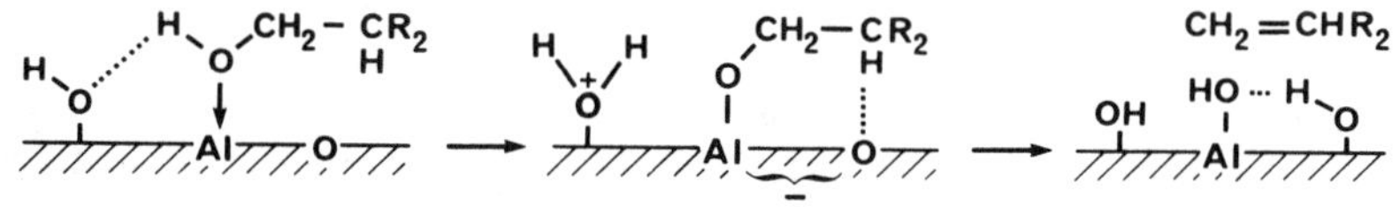

Figure 5.17 The dehydration of alcohols on alumina via a chemisorbed alkoxide intermediate.

according to the type of alcohol. Primary and secondary alcohols have been reported to undergo dehydration on alumina by the Lewis acid-catalyzed mechanism, whereas tertiary alcohols undergo dehydration by a pseudo-cationic process. In these dehydration reactions, some water must be present in the form of surface hydroxyl groups for the reaction to proceed. The presence of water can influence the course of the dehydration reactions. For example, in the dehydration of 2- and 3-methylbutan-2-ols [79], the addition of water increases the yield of β-olefin from the 2-methyl isomer, but decreases that from the 3-methyl isomer; water decreases the aprotic (Lewis) acidity of the alumina. This appears to favor the elimination of hydrogen from secondary, rather than from tertiary, carbons.

The dehydration of alcohols adsorbed on rutile may also occur via carbenium ion or by π-allylic intermediates [80]. In the presence of water or high alcohol concentrations, Bronsted acidity is developed, resulting in the formation of carbenium ionic intermediates and the formation of isomeric alkenes. In the absence of water or excess alcohol, π-allylic intermediates with partial carbenium ion character are formed rather than the π-allylic carbanions.

Galwey [78,81], in studies related to the geological genesis of petroleum, found that the desorption of various intercalated medium- and long-chain alcohols (from their dehydrated montmorillonite complexes at 180°C, under high-vacuum, Lewis acid-catalyzed reaction conditions) yielded similar mixtures of C5 to C8 n-alkenes as the volatile products. Desorption of the alcohols under a lower vacuum, which would facilitate the readsorption of the reaction products and favor the formation of Bronsted acid catalytic species, yielded mixtures of C4 to C8 isoalkanes. These contained small proportions of n-alkanes but negligible amounts of n-alkenes. Mixtures of C6 to C8 isoalkanes are formed in analogous reactions of dodecanol with kaolinite. These products are similar to those obtained from dehydration on illite and montmorillonite at higher temperatures [82], indicating that the same carbenium ionic intermediates were involved and that the product distribution is probably determined by the rates of desorption. These products differ from those normally obtained using nonswelling aluminosilicates as catalysts. Such drastic redistribution reactions usually only predominate at temperatures above 300°C. This illustrates the powerful catalytic environment that can exist on the basal surfaces and in the interlamellar regions of the aluminosilicate clay minerals.

The elimination of hydrogen sulfide from mercaptans is also believed to occur by mechanisms similar to those for the dehydration of alcohols; the reactions are less facile and, except under vigorous conditions, usually yield thioethers as the major products [83]. Esters can undergo Bronsted acid catalyzed cleavage, either via the formation of a carbenium ion or by a concerted mechanism, to yield alkenes plus acid, while amides can yield nitriles under anhydrous conditions [77]. In contrast, alkenes can condense with carboxylic acids intercalated in montmorillonites to form esters via a Bronsted acid-catalyzed reaction [84]. Carboxylic acids may be decarboxylated via protonated intermediates when heated with acidic minerals at high temperatures. Stearic acid has been variously reported to yield a mixture of linear and branched C15 and C16 alkanes [77], or a mixture of 2- and 3-methylated C5 to C7 alkanes when heated with aluminosilicates [82]. The difference in product compositions probably reflected the different experimental conditions.

Alkyl halides may eliminate hydrogen halides when heated with aluminosilicates, for example, zeolites [85], with the formation of carbenium ion intermediates which can alkylate coadsorbed species or result in polymer formation. The hydrogen halide produced may promote the reaction, although the decomposition may be promoted by the polarization of the halide in the surface field of the mineral rather than by protonation by some Bronsted acidic species. Dehydrohalogenations can occur on other oxide surfaces, and the addition of hydrogen halides to alkenes [86] or the formation of alkyl halides from alcohols [87] is also catalyzed by a wide range of mineral surfaces. The nitration of aromatic hydrocarbons by 65% nitric acid has been claimed to be catalyzed by montmorillonite [88], presumably by promotion of the formation of adsorbed NO_2^+.

Compounds lacking β-hydrogens or migrating groups, for example, benzyl or methyl compounds, may undergo α-elimination on mineral surfaces. Benzyl mercaptan, for example, yields *trans*-stilbene when heated with zeolites, possibly via carbene or carbenoid intermediates, while the dehydration of methanol adsorbed on calcined aluminosilicates or zeolites, above 250°C, results in the formation of dimethyl ether, which dehydrates to form ethylene and higher hydrocarbons. The hydrocarbons may result from the formation and decomposition of surface methoxides via carbene [89]. However, studies on the conversion of methanol or dimethyl ether on zeolites have shown that the methyl C–H bonds are initially nonlabile, thus ruling out carbene or oxymethylene intermediates. Instead, an oxonium species has been proposed as the intermediate in the formation of propylene and higher hydrocarbons and in the methylation of aromatic hydrocarbons by methanol over zeolite catalysts [90] (see Figure 5.18).

5.2.4   Alkylation, Isomerization, and Depolymerization

The carbenium ions, generated as intermediates in the polymerization, addition, and elimination reactions described in the previous sections, have been shown to be capable of undergoing extensive isomerization involving intramolecular hydride shifts and skeletal rearrangements. These ions may also be intermediates in acidic mineral-catalyzed intermolecular redistribution of hydrogen and skeletal fragments. Although zeolites and silica–aluminas are usually used as catalysts for these reactions [90,91], they may also be catalyzed by natural or acid-activated clay minerals [92]. Reactions on alumina or other metallic oxides may involve either,

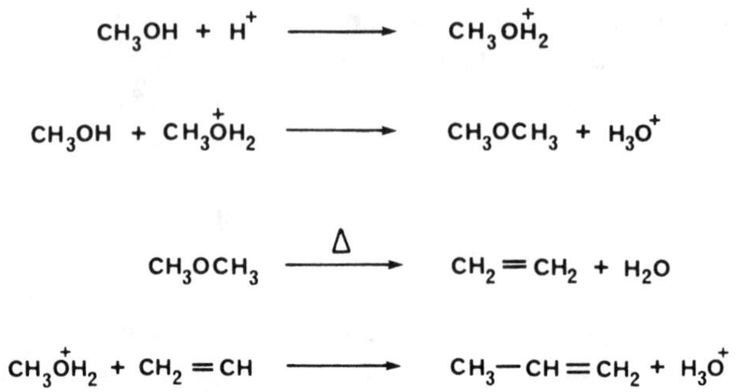

$$CH_3OH + H^+ \longrightarrow CH_3\overset{+}{O}H_2$$

$$CH_3OH + CH_3\overset{+}{O}H_2 \longrightarrow CH_3OCH_3 + H_3O^+$$

$$CH_3OCH_3 \xrightarrow{\Delta} CH_2 = CH_2 + H_2O$$

$$CH_3\overset{+}{O}H_2 + CH_2 = CH \longrightarrow CH_3 - CH = CH_2 + H_3O^+$$

Figure 5.18 One reaction sequence for the zeolite-catalyzed transformation of adsorbed methanol to higher hydrocarbons.

or both, carbenium ion (Bronsted acid-catalyzed) or π-allylic (Lewis acid-catalyzed) intermediates.

The acid-catalyzed isomerization of adsorbed organic molecules may take several different forms. Olefins adsorbed may undergo geometrical isomerization, double-bond migration, skeletal isomerization, condensation, and cracking reactions. Saturated aliphatics may undergo skeletal isomerization or cracking, while alkylaromatics can undergo substitutional isomerization, alkylation, and dealkylation. The variety of products formed in the bentonite-catalyzed isomerization of the terpene, limonene, illustrates the extent of rearrangements that can occur on clay minerals at moderate temperatures [93] (see Figure 5.19).

The course of the dehydration and isomerization of 4-methylpentan-2-ol by silica–aluminas can be correlated with the acidic strength of the effective catalytic sites (Table 5.1) [75]. These correlations do not necessarily apply to other systems, for example, the isomerization of butene on silica–alumina has been reported to occur only on sites having a Hammett acidity, $H_0 < -12$ [94]. Correlations between the acidity of titania, montmorillonite, or other silicates, and the isomerization of pinene, and the energy of desorption of ammonia, have also been reported [95]. Titania containing adsorbed surface bisulfate ions can behave as a superacid catalyst with active sites having Hammett acidity $H_0 < -14.5$. This material is capable of catalyzing the skeletal isomerization and redistribution of butane or isobutane at 50°C [94].

The extent of these various reactions is determined not only by the catalyst and reagents, but also by the reaction temperature and residence time. These factors determine whether the reaction system can approach thermodynamic equilibrium or whether the product distribution is governed predominantly by the rates of the competitive reactions. In general, higher reaction temperatures and longer residence times favor more extensive rearrangements and redistributions. The products may also be determined by steric factors that can apply, for example, to reactions catalyzed by acidic sites within zeolite cavities or by interlamellar species in montmorillonites.

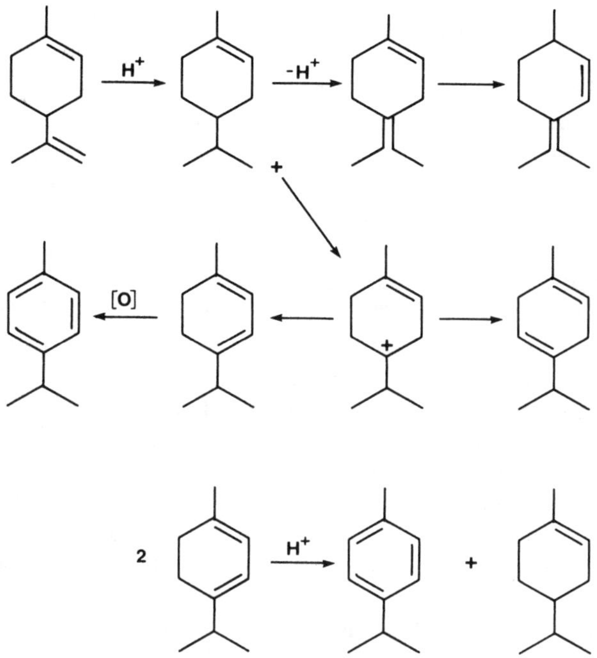

Figure 5.19 Some products from the acid-catalyzed isomerization of limonene.

The alkyl carbenium ions undergo isomerization by a series of 1,2-shifts of hydrogen, alkyl, or aryl groups, the process favoring the formation of the most stable carbenium ion, that is, the ion with the highest degree of alkyl substitution or allylic, benzylic, or other conjugation. In alicyclic systems, changes in ring size may also occur to reduce ring-strain and may involve distant unsaturated centers which can participate in the formation

Table 5.1   Correlation between Nature of Reaction and Catalyst Acidity[a]

Hammett Acidity of the Catalytic Centers[b]	Dehydration	Isomerization			
		cis–trans	1-2 C = C	1-*n* C = C	Skeletal
4.75 to 0.82	+	−	−	−	−
0.82 to − 4.04	+	+	+	−	−
− 4.04 to − 6.63	+	+	+	+	−
Less than − 6.63	+	+	+	+	+

[a]*Source:* J. P. Damon, B. Damon, and J. M. Bonnier, *J. Chem. Soc., Farad. I.*, **73**, 372 (1977).
[b]Indicators: 4-dimethylaminotriphenylcarbinol (pK_a + 4.75); tris(4-methoxylphenyl)carbinol (pK_a + 0.82); tris(4-methylphenyl)carbinol (pK_a − 4.04); triphenylcarbinol (pK_a − 6.63).

of transannular nonclassical carbenium ion intermediates (Figure 5.14, **7** and **9**).

Early studies of the isomerization of the three butenes on aluminosilicates showed that the *sec*-butyl carbenium ion (Figure 5.14, **8**) was a common intermediate [96]. This has been confirmed in recent tracer studies of labeled butenes [97] and by semiempirical quantum chemical calculations [98]. The formation of the carbenium ion on silica–aluminas is preceded by the formation of a π-bonded adsorption complex between the alkene and surface bridged-hydroxyl groups [98], whereas in zeolites, the π-bonded complex may be formed with the exchangeable cations [99].

Bronsted acid-catalyzed isomerization reactions are sensitive to the nature of any exchangeable cations and to the water content of the reaction system. For example, the proportion of *cis-trans* isomerization of 2-butene by zeolites, compared to skeletal isomerization in the presence of excess water, increases on substitution of exchangeable Na^+ with Ca^{2+} ions [100]. Moderate hydration in sodium zeolites can increase the rate of isomerization of butene, as the water molecules can act as bridges between the proton-donating zeolite hydroxyl groups and the alkene [101].

The carbenium ions generated on the mineral surface may alkylate other adsorbed, nucleophilic species, such as amines, and aromatic and unsaturated compounds [102]. The polymerization of vinylic monomers is a special case of the latter reaction. Other examples can be found in the dimerization or trimerization of unsaturated fatty acids and esters. The oligomerization of natural unsaturated fatty acids, such as oleic acid or "tall oil" acids, in the presence of acidic clay minerals at 200 to 250°C is often used for the preparation of high-molecular-weight acids for resin production [e.g., 103]. The fatty acid "dimer" products comprise a mixture of linear saturated, unsaturated, and mid-chain monomethyl-branched monomeric acids, acyclic dimers and trimers, plus alicyclic and aromatic

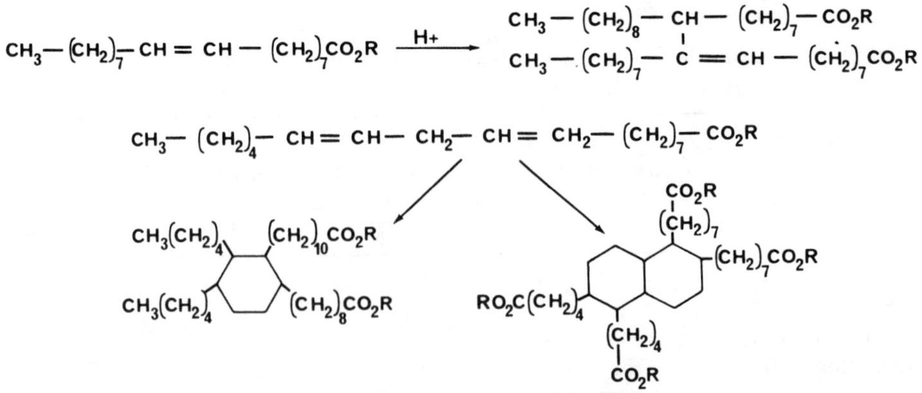

Figure 5.20 Some products from the acid-catalyzed dimerization of unsaturated fatty acids.

Table 5.2   Structural Composition of Dimerized Fatty Acids[a]

Structure	Linoleic	Tall Oil	Oleic and Elaidic
Linear	5	15	40
Monocyclic aromatic	25	20	5
Nonaromatic	30	50	50
Polycyclic	40	15	5

[a]Source: D. H. McMahon and E. P. Crowell, *J. Amer. Oil Chem. Soc.*, 51, 522 (1974).

materials (see Figure 5.20). The relative proportions of these products are shown in Table 5.2 [104].

The mineral-catalyzed alkylation of aromatics has already been demonstrated in the cyclization reactions of the styrene-derived carbenium ions. Some other examples include the condensation of carbonyl compounds, such as acetone, with aromatics to form carbinols [87]; the alkylation of aromatics by alkyl halides or alcohols [105,106]; the acylation of aromatics with carboxylic acids to form ketones [105]; the conversion of volatile unsaturated impurities in aromatic hydrocarbons to less volatile condensates by treatment with acidic minerals [107]; and the removal of contaminants from phenol prepared by decomposition of cumene hydroperoxide. Acid clays can also catalyze the condensation of alkylanilines and formaldehyde and the isomerization of resultant *bis-N,N'*-(arylamino)methanes to *bis*(aminoaryl)methanes [108] (see Figure 5.21). This reaction is an important stage in some of the dye-forming reactions discussed in Section 5.3.

Acidic montmorillonites can also catalyze the alkylation of aromatics by alkenes [109]. This may proceed via extensive isomerization, as in the cyclization of 2-phenylpent-2-ene to 1,1-dimethylindane [110]. Montmorillonites and zeolites containing exchangeable transition metal ions, such as a Cu^{2+}-montmorillonite, can have a surface acidity equivalent to that of aluminium chloride. These materials can catalyze the low-temperature alkylation of aromatic compounds [111] and the cracking of aliphatic hydrocarbons [112].

The carbenium ions may undergo intermolecular skeletal redistribution by transalkylation and depolymerization "cracking" reactions that are the reverse of the alkylation and polymerization reactions. These may also be accompanied by isomerization, as in the formation of isobutene from cracking of poly(1-butene). The cracking reactions usually occur at temperatures in excess of 200 to 300°C, unless the organic material is a polymer with a lower ceiling temperature. Carbenium ion, and radical, intermediates have been postulated for the geological transformation of organic materials into petroleum. These reactions are commonly believed to be catalyzed by aluminosilicate clay minerals codeposited in the organic sediments [113–115]. The type of clay mineral present may determine the nature of the petroleum deposit formed. (Kaolins promote the formation of

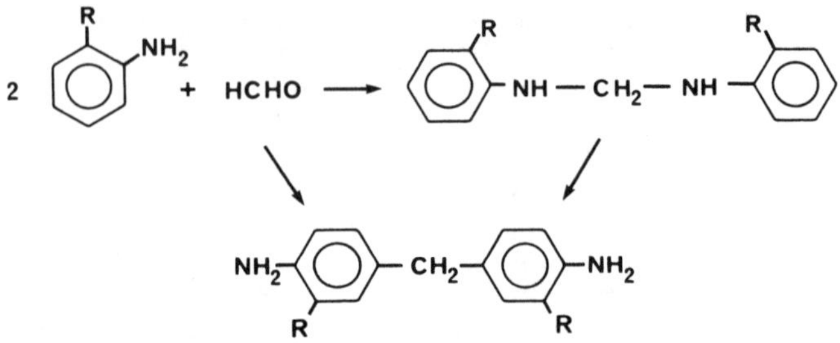

Figure 5.21

light petroleum and natural gas, while bentonites produce heavy petroleums [116].) The state of hydration of the deposit may also affect the course of the transformation. Long-chain carboxylic acids, for example, when heated at 250°C under pressure with aqueous montmorillonite, undergo radical-mediated decarboxylation, cleavage, and hydrogen abstraction to form low-molecular-weight n-alkanes, typical of those found in "young" deposits. In contrast, the reactions on dry montmorillonite yield carbenium ion-mediated products, such as branched-chain alkenes and alkanes, typical of those found in "old" deposits [114].

Hydrogen-transfer reactions may also occur between adsorbed species, resulting in an acid-catalyzed hydrogenation or dehydrogenation. Although alkanes can undergo hydride abstraction to form carbenium ions, this abstraction may also occur by "protonation" of a saturated carbon, followed by elimination of molecular hydrogen, a process that is similar to that catalyzed by the "super acids" [117]. However, the formation of carbenium ions from saturated materials adsorbed on aluminosilicates is more likely to arise by hydride exchange with other carbenium ions.

Most of the mineral-catalyzed reactions occurring above 100 to 150°C are accompanied by the formation of polymeric carbonaceous residues on the catalyst surface. This material can act as a source or sink for hydrogen and is a potential source of protons for Bronsted acid-catalyzed reactions on Lewis acidic mineral surfaces. Many reactions are found to occur at low temperatures only after the mineral has been activated by reaction at higher temperatures. This pretreatment results in the formation of a partially carbonaceous surface. Excessive "coke" production, however, can inhibit catalysis by blocking the mineral surface sites.

Another source of protons, which may induce Bronsted acid-catalyzed isomerization reactions on Lewis acidic oxide surfaces, is chemisorbed water molecules which become active at elevated temperatures. Water on an alumina surface dissociates at temperatures above 150°C [99,118], and this dissociated water can poison the surface acidic and basic sites involved in the formation of π-alkenyl anions from alkenes. The residual surface hydroxyl groups of alumina have significant Bronsted activity [99], for example, isomerization and deuterium exchange between 1,1-

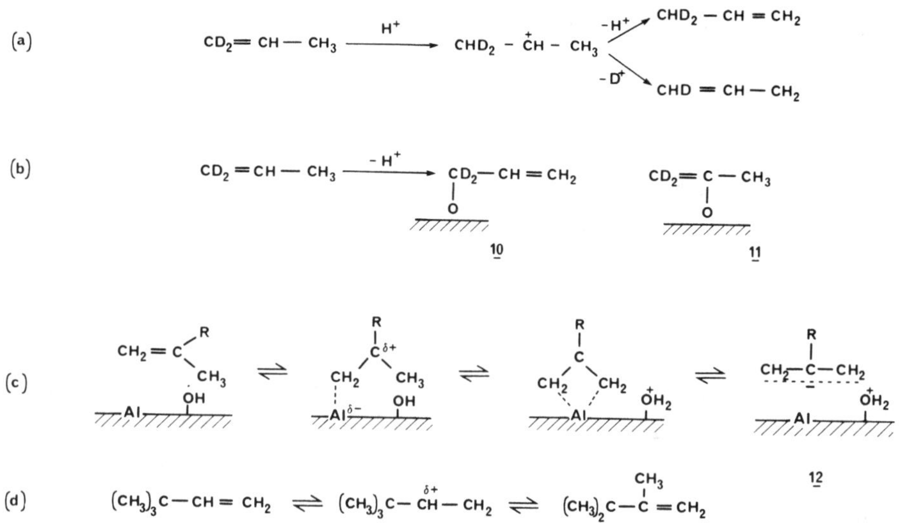

Figure 5.22   Reactions of allylic hydrocarbons on alumina. (a) Associative adsorption and exchange via carbenium ionic intermediates. (b) Dissociative adsorption, to form σ-bonded anionic species. (c) Dissociative adsorption, to form π-allylic carbanions. (d)

dideuteropropene and H_2O, characteristic of a carbenium ion intermediate, can occur at temperatures near 150°C. However, adsorption at temperatures near 30°C appears to involve the formation of σ-bonded propen-1-yl (10) and propen-2-yl (11) intermediate species, or of π-allylic carbanions (12).

The isomerization of alkenes on these surfaces can occur by two processes, an associative process involving proton transfer with some adsorbed protonic species to form a carbenium ion (Figure 5.22a) or a dissociative process involving the loss of a proton and the formation of an allylic carbanion adsorbed at a Lewis site (Figure 5.22c) [118,119]; the latter reaction is related to those occurring in the alumina-catalyzed dehydration of alcohols. Both modes may occur on the one catalyst surface under different reaction conditions; the reaction temperature appears to be a major factor.

Isomerization of hydrocarbon chains containing nucleophilic heteroatoms, such as oxygen or nitrogen, can occur readily on mineral surfaces. The isomerization of oxiranes (alkylene oxides) to carbonyl compounds is a common example of this class of reactions. Thus, butylene oxide adsorbed on Bronsted acidic surfaces is converted to butyraldehyde, with crotyl alcohol and methyl ethyl ketone as minor byproducts [120]. Isomerization of oxiranes may also occur on Lewis acidic surfaces but with the obvious exception of ethylene oxide, the products from these reactions are usually allylic alcohols formed via allylic carbanion intermediates. The carvomenthene oxide, for example, is converted by a strong Bronsted acid, such as silica–alumina, to a mixture of aldehyde and ketones, while reaction on a

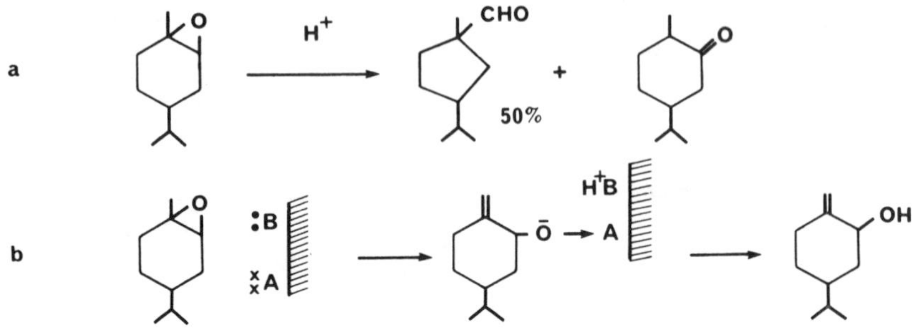

Figure 5.23

Lewis acid, such as titania–zirconia, produces a mixture of allylic alcohols [121] (see Figure 5.23). A commercially important example of the isomerization of epoxides is the acid clay-catalyzed degradation and detoxification of insecticides, such as dieldrin, which are converted to biologically inactive ketones when compounded with these supposedly chemically inert diluents [122].

5.2.5 Condensation and Hydrolysis

Acidic mineral surfaces can catalyze a wide variety of hydrolytic and condensation reactions (see Figure 5.24). Examples of the condensation reactions include the esterification of sebacic acid by octanol [123]; the transesterification of dimethyl terephthalate with ethylene glycol, as one method for synthesis of poly(ethylene terephthalate) [124]; the reaction between tetrahydrofuran and acetic anhydride to form poly(butylene glycol) diacetates [125]; and a related reaction, the step-growth polymerization of ethylene oxide to a poly(ethylene glycol).

The widespread use of potentially acidic minerals as fillers in plastics compositions, or as "inert" diluents or carriers for herbicides, pesticides, and other biologically active materials, can result in degradative hydrolytic reactions during compounding of the mixture or in subsequent storage. Soil surface-catalyzed hydrolysis is also an important mechanism for

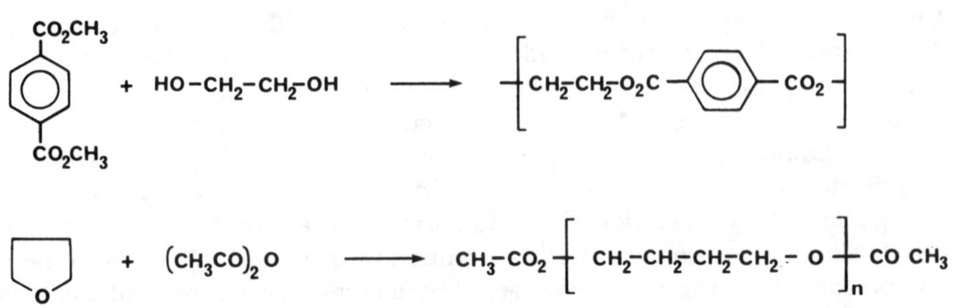

Figure 5.24

the detoxification of agricultural spray residues and can complement the microbiological degradation of these materials. Clay mineral-catalyzed hydrolysis and transesterification reactions are also important in the transformation of organic sediments [126,127] and in the catalysis of peptide bond formation between amino acids adsorbed on clay mineral surfaces [128]. Other geological organic transformations, such as the prebiotic formation of nucleotides and polypeptides, in which the clay mineral surfaces may have acted as templates, are described in Chapter 6.

Many of the pesticides used in agriculture are organophosphate derivatives; two examples are parathion and malathion. These thiophosphates are particularly susceptible to hydrolysis when adsorbed on the surface of aluminosilicate clay minerals, which are common constituents of most soils. The hydrolysis may be preceded by rearrangement of the thioester and occurs by reaction with ligand water molecules coordinated with the exchangeable cations on the mineral surfaces [129,130]. Minerals with adsorbed copper ions are potent catalysts for the hydrolysis of pyridone organophosphate pesticides, the reaction being promoted by the formation of a copper chelate [131] (see Figure 5.25a). Other agricultural chemicals that are hydrolyzed on clay mineral surfaces include the chlorotriazine herbicides (see Figure 5.25b) and the chlorinated aromatic and alicyclic pesticides [132]. Acidic clay minerals such as attapulgite or kaolin were used at one time as diluents and wetting agents in pesticide dusts and sprays, but their use was found to result in loss of activity on storage. A dieldrin–attapulgite mixture, for example, lost 98% of its activity after storage for one week at 65°C, a temperature typical of that reached in railway boxcars in summer [122]. This loss of activity on storage can be overcome by the use of neutralized clays.

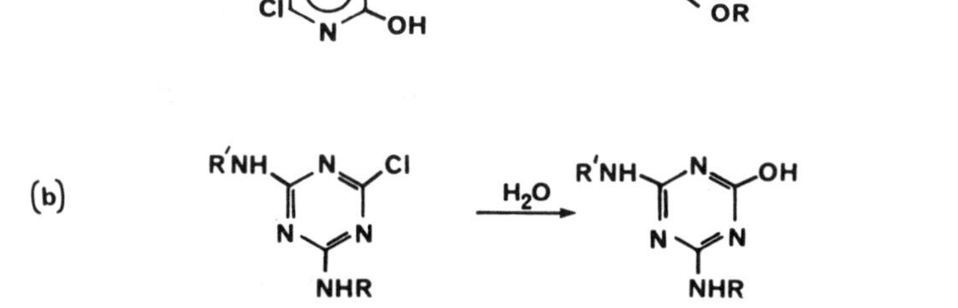

Figure 5.25 Surface-catalyzed hydrolysis of some agricultural chemicals. (a) Pyridone orthophosphates. (b) Chlorotriazines.

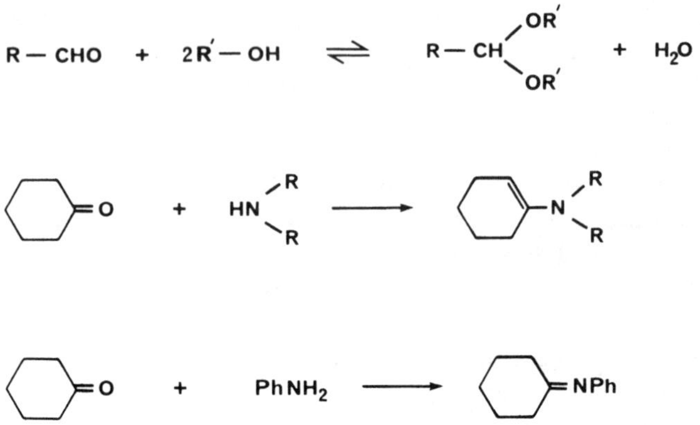

$$R-CHO \;+\; 2R'-OH \;\rightleftharpoons\; R-CH\big(OR'\big)_2 \;+\; H_2O$$

Figure 5.26   Examples of the surface-catalyzed condensation of carbonyl compounds with alcohols and amines.

The clay minerals can catalyze the hydrolysis and degradation of polypeptides (polyamides) and nucleic acids. Kaolinitic sediments, for example, can cause the hydrolysis of adsorbed purines, a reaction that normally requires catalysis by strong acids [133]. Clay minerals, such as montmorillonite, have been found to adsorb and degrade a variety of neutral drugs such as the digitalis glycoside, digoxin [134]. The adsorption and hydrolytic degradation of the active ingredients is an important consideration in the formulation or coadministration of clay-containing drugs. Clay minerals can also catalyze the hydrolysis or disproportionation of chlorination agents, such as Chloramine T [135]. The hydrolysis of hydrogen cyanide adsorbed on montmorillonite has been reported to yield formamide [136] and amino acids [137], the latter reaction being a possible prebiotic source of these materials.

The hydrolysis of adsorbed vinyl ethers to alcohols and aldehydes has already been noted. The acidic mineral surfaces can also catalyze the condensation between these products, or between other alcohols and aldehydes or ketones, to form acetals or ketals [138]; the analogous reaction between ketones and amines can be used for the production of ketimines [139,140] (see Figure 5.26). These condensation reactions are reversible, but the yields may be improved by the use of an acidic catalyst, such as silica–gel, which can absorb the coproduct water. The hydrolysis of aliphatic esters may also be catalyzed by aqueous aluminosilicates, but kinetics studies [141] have shown that the catalysis is effectively confined to acidic sites which, in the dry minerals, have Hammett acidities, $H_0 < -3.0$.

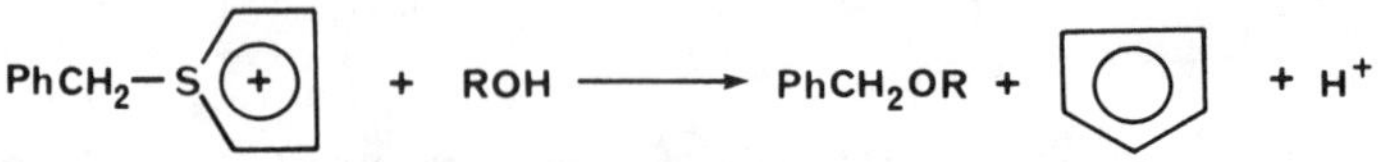

Figure 5.27

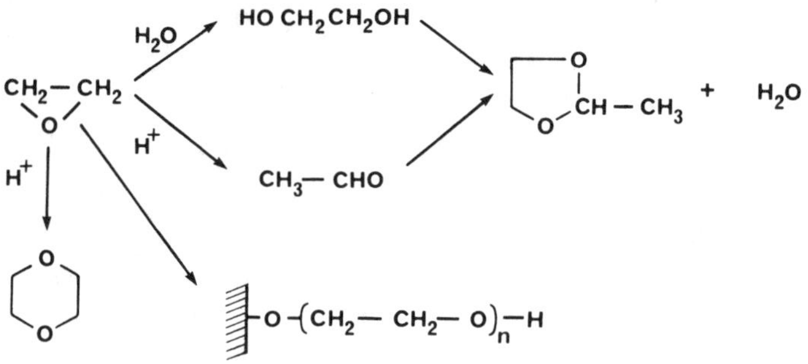

Figure 5.28

Acidic minerals can catalyze the condensation, at elevated temperatures, of alcohols with phenols to form alkyl aryl ethers; for example, phenol and methanol are converted to anisole when passed over a silica–alumina at 200°C [142]. Alkyl or cycloalkyl benzyl ethers can also be obtained in good yield by the reaction, at 150°C, between the respective alcohols and a S-benzyltetrahydrothiophenium montmorillonite intercalation complex, the reaction occurring in the interlamellar spaces [143] (see Figure 5.27).

Heterocyclic species, such as lactams, lactones, or oxiranes (epoxides), may undergo acid-catalyzed hydrolysis or condensation when contacted or adsorbed on mineral surfaces. Oxiranes may be converted by acidic minerals into glycols or polyethers, with side reactions which include the isomerization of the oxirane into a carbonyl derivative, which can then undergo conversion to an acetal. For example, ethylene oxide can be converted into ethylene glycol, poly(ethylene oxide)s, 1,4-dioxan, acetaldehyde, and 2-methyl-1,3-dioxolane [144] (see Figure 5.28). If an alcohol is present, the products will include the corresponding alkoxyethanol [144,145], while mineral surface hydroxyl groups may be similarly incorporated with the formation of poly(alkylene oxide)-grafted surfaces [146].

Related examples of the mineral-catalyzed reactions of cyclic ethers include the polymerization of styrene oxide by zeolites [38], the montmorillonite-catalyzed reaction between cyclohexene oxide and a secondary amine to form a 1,2-hydroxyamine [147], the condensation between furan and propylamine on zeolites to form N-propylpyrrolidine [148], the montmorillonite-catalyzed copolymerization of ethylene oxide and tetrahydrofuran to produce prepolymeric intermediates for polyurethanes [149], and the conversion of δ-lactones to polyesters [150] (see Figure 5.29).

Acidic mineral surfaces can catalyze the redistribution and grafting of siloxanes. Cyclic octamethyltetrasiloxane is converted by aluminosilicates into linear poly(dimethylsiloxane), the catalytic centers being Bronsted acidic water coordinated with exchangeable Al^{3+} ions [151]. In these reactions, adsorbed water can also act as a transfer or terminating agent, while reaction with surface hydroxyl groups, of kaolin, for example,

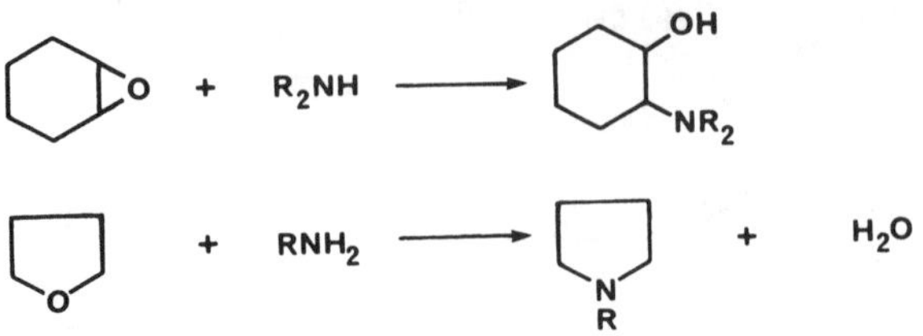

Figure 5.29   Some examples of the acid-catalyzed amination of cyclic oxides.

results in the formation of mineral–siloxane graft polymers [152].

The intermediate carbenium ion formed in the ring-opening of the adsorbed oxiranes can initiate the cationic polymerization of monomers such as styrene, and the propagating cations can, in turn, react with the oxirane; for example, styrene can be copolymerized with epichlorohydrin by use of an Al-montmorillonite catalyst [153,154] (see Figure 5.30).

Mineral fillers can assist in the formation of polyurethanes from isocyanates and polyols. Hydrated fillers, such as gypsum, can provide water for reaction with the isocyanates [155], while the surface hydroxyl groups of silica and silicate glasses can catalyze the condensation reactions [156]. Zeolites can be used as carriers for amine catalysts and can act as cocatalysts in one-pot polyurethane sealant compositions. The catalyst, triethylene diamine, for example, is absorbed in the zeolite and the resultant complex mixed with the isocyanate and hydroxyl terminated prepolymers. The mixture can be stored and is stable until contacted with moisture, which displaces the amine catalyst from its protective zeolite cage, thus initiating the curing reactions. Zeolites can also assist in other polycondensation reactions, such as polyesterifications, by absorbing the byproduct, water [157].

Another series of condensation reactions, which may be catalyzed by dry mineral surfaces, is involved in the formation of adsorbed "petrochor," or "smell of earth." This material is desorbed when soil is moistened and is responsible for the "smell" of rain which, in the arid areas of the Indian subcontinent and elsewhere, signals the end of a drought. Petrochor is a complex mixture of oxidized and nitrated phenolic and alicyclic com-

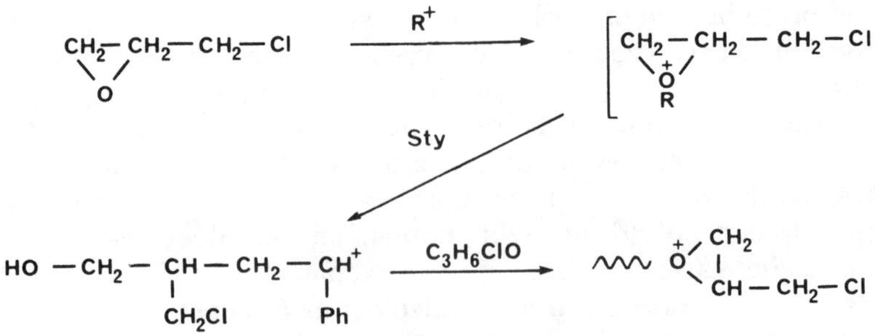

Figure 5.30

pounds. It is apparently formed when organic vapors, for example, terpenes from foliage, are adsorbed from the atmosphere on to the surfaces of dry soils or minerals, such as powdered basalt or kaolinized granites, and then exposed to sunlight [158]. The mechanisms involved are still unknown, and the extent of involvement by the mineral surfaces is obscured by the presence of some of the more volatile petrochor species in the atmosphere, as normal trace constituents, during sunny weather [159].

5.3 REACTIONS VIA RADICAL-CATIONIC INTERMEDIATES

Radical-cations can be formed by the transfer of electrons from electron-rich centers of adsorbed organic molecules to electron-deficient centers or species that may be present on a variety of mineral surfaces. The electron-donating centers may be the π-electron clouds of aromatic or unsaturated groups or the lone-pair electrons of nitrogen, oxygen, or sulfur. The oxidation process can result in the formation of a charge-transfer complex, in which the transferred electron is shared between the mineral center and the organic species, or the transfer may be complete, resulting in the formation of a radical-cation.

In homogeneous systems, the electron acceptor may be another organic molecule, such as tetracyanoethylene, which may form a charge-transfer complex with the donor molecule, or an oxidant, such as oxygen (see Figure 5.31), in a strongly acidic environment, which may be a solution in a strong mineral acid, such as sulfuric acid. The electron transfer between organic molecules may also be promoted by the presence of Bronsted acids [160]. Dissociation of the charge-transfer complexes, with the formation of organic radical-cations, is also promoted by visible or ultraviolet irradiation. Radical-cations can be formed via charge-transfer complexes with very strong Lewis acids, such as aluminum chloride in rigorously aprotic media.

Charge-transfer complexes can also be formed between the Lewis acidic sites on alumina or silica–alumina surfaces and coordinated aromatic com-

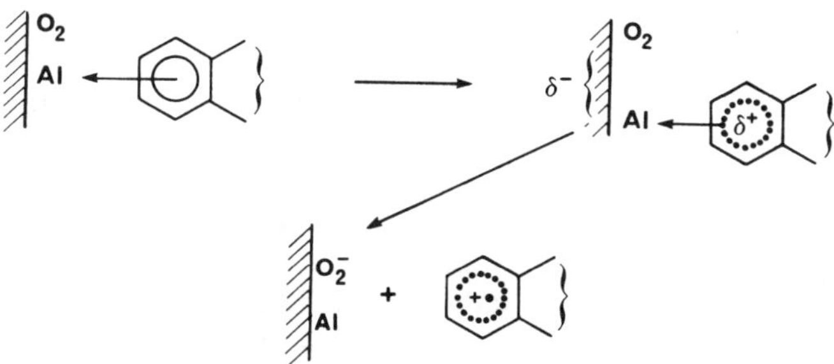

Figure 5.31 One mechanism for the formation of aromatic radical-cations on an oxygenated Lewis acidic surface.

pounds that have low ionization potentials [161]; the charge-transfer complexes may be capable of dissociation, to form free radical-cations. Electron transfer between the aromatic hydrocarbon, perylene, and Al^{3+} ions in an aluminum-doped silica gel has been shown by the appearance of a six-line signal from a second paramagnetic species in the esr spectrum of the adsorbed perylene radical-cation. This is believed to arise from electrons trapped either on, or in the vicinity of, isolated $^{27}Al^{3+}$ ions [162].

The nature of the oxidizing species involved in the generation of radical-cations on mineral surfaces is often uncertain. The potential electron acceptors may include transition metal ions at surface sites, or within the mineral lattice; adsorbed oxygen species, such as chemisorbed molecular oxygen, or oxygen radical-anions; exposed, coordination deficient metal ions; and lattice defects. The lattice defects may be formed during the activation of the mineral, for example, by outgassing at high temperatures under vacuum for the removal of adsorbed oxygen or water from the mineral surface. This process can also result in the elimination of molecular hydrogen from metal oxides or aluminosilicates and the formation of surface superoxide radical-anions, O_2^- [163]. The formation of radical-cations by reaction between chemisorbed oxygen and aromatic hydrocarbons has been established [164], and the counterions associated with the adsorbed radical-cationic species are probably O_2^-. The radical-cations and the oxygen species may be present as clusters on the mineral surfaces [165].

Alumina and silica–alumina samples can contain powerful electron donor and electron acceptor centers in close proximity [166], and these can promote electron transfer between adsorbed species. Electrons may also be donated by the mineral to electron-deficient species, such as tetracyanoethylene or quinones, to form the corresponding radical-anions [167].

The radical-cations are generally very reactive and, unless they are resonance stabilized or sterically hindered, they can participate in reactions that include cleavage into radical and cationic fragments; dimerization, with the formation of dications; reaction with anions to form radicals; or reaction with olefins to form polymers, with propagation by either the radical or cationic centers of the monomer–radical–cation adduct. The radical-cations may also undergo further oxidation in order to form diamagnetic dications, or disproportionate to form a mixture of neutral and dicationic species.

Many of the reactions of absorbed organic molecules, now known to involve cationic intermediates and to be catalyzed by surface Bronsted acid species, were originally thought to involve radical-cationic intermediates generated by electron transfer to surface Lewis acid species. This supposition arose from studies of silica–alumina cracking catalysts, and similar materials, in which the catalytic activity could be correlated with the activity for oxidation of perylene and other polynuclear aromatic molecules. Some correlation was also found between the respective numbers of Lewis acid sites and oxidizing species apparently present on the mineral surfaces. It was assumed that the surface species involved in oxidation of hydrocarbons were the same as those that promoted other reactions, such as polymerization or isomerization of adsorbed organic molecules. The relative importance of Bronsted and Lewis acidic centers in mineral catalysis was the subject of considerable controversy during the 1960's, the main protagonists being E. Hirschler (Bronsted acidity) and W. Keith Hall (Lewis acidity); their respective positions have been summarized in references 168–170.

Evidence supporting the concept of Lewis acid activity included the following observations.

(i) Infrared spectroscopic studies of pyridine and other amines adsorbed on catalytically active mineral surfaces showed that the dominant species were Lewis acid coordinated adducts, with little or none of the Bronsted acid protonated adducts. In these studies, the catalyst samples were usually activated by prior outgassing at high temperatures, which would have resulted in extensive dehydroxylation of the mineral surfaces.

(ii) The spectroscopic identification of the chemisorbed polynuclear hydrocarbon species as cation-radicals, similar to those readily generated by oxidation in concentrated sulfuric acid and similar media. Bronsted acids would have been expected to only generate cationic species.

(iii) The oxidation of triphenylmethane on silica–alumina and dry clay aluminosilicate surfaces, to form triphenylcarbenium ions that could be recovered as triphenylcarbinol after addition of water to the adduct. The carbenium ions were thought to form by hydride abstraction from the triphenylmethane by the Lewis acid species, although the fate of the hydride ions was never established.

(iv) The facile oxidation of a variety of aromatic amines in anaerobic conditions to form highly colored radical-cations.

(v) The decolorization of the stable free radical, diphenylpicrylhydrazyl (DPPH), supposedly by electron transfer to surface Lewis acid species.

The concept of Bronsted activity was supported by the following observations.

(i) Molecular oxygen can form thermally stable surface adducts with a variety of oxygen-containing mineral surfaces. These are known to be powerful oxidants. Activation of the mineral in a hydrogen atmosphere, or pretreatment with reductants, usually resulted in a drastic reduction in oxidizing potential of the catalyst, without a corresponding reduction in activity for other reactions, such as the polymerization or isomerization of alkenes. This showed that the oxidizing and the catalytic species were not identical.

(ii) When isolated and identified, the products from mineral-catalyzed

$$Ph_3CH \xrightarrow{\ H^+\ } Ph_2CH^+ \ + \ PhH$$

$$Ph_3CH \ + \ PhCH_2^+ \longrightarrow Ph_3C^+ \ + \ Ph_2CH_2$$

Figure 5.32

polymerization or isomerization reactions corresponded to those expected from a Bronsted-catalyzed process. These studies included the use of deuterium and tritium labeled minerals to demonstrate the involvement of surface hydroxylic species in the catalytic process.

(iii)   The catalytic activity of the mineral in the polymerization and isomerization reactions was also correlated with the measurable proton acidity of its surface.

(iv)   Analogies could be drawn with the catalysis of Friedel–Crafts and similar reactions by Lewis acids, such as aluminium chloride, in which the true catalytic species are frequently protonic species derived, for example, from trace amounts of adsorbed water. These Bronsted acid species may be several orders of magnitude greater in catalytic activity than pure aluminium chloride.

One of the most important examples of the power of Bronsted acid-initiated reactions of adsorbed species was eventually provided by Wu and Hall [171]. They showed that the "oxidation" of triphenylmethane was catalyzed by Bronsted acid species, and that the reaction did not involve adsorbed oxygen, as proposed by Hirschler and co-workers, but instead a retroalkylation, followed by hydride abstraction from triphenylmethane by the diphenylcarbenium ions (see Figure 5.32). Earlier workers had failed to detect the benzene and diphenylmethane coproducts.

5.3.1 Polymerization

The polymerization of styrene and its derivatives by dried, acidic aluminosilicate minerals, such as kaolinite, montmorillonite, or attapulgite, was originally thought to involve a radical-cationic initiation process [10] (see Figure 5.33). This hypothesis was based on the then-current view that Lewis acidic sites were the dominant reactive centers on anhydrous aluminosilicate minerals and was supported by the observation of a semiquantitative correlation between the polymerization activity of a variety of ion-exchanged minerals and Lewis acid-related activities for oxidation of aromatic amines and decolorization of DPPH [172]. The initial styrene radical-cations were thought to undergo rapid dimerization to form dications; these could propagate by a cationic mechanism and, because of the frequency of proton transfer to monomer, yield a polymer virtually indistinguishable from that initiated by a Bronsted acid.

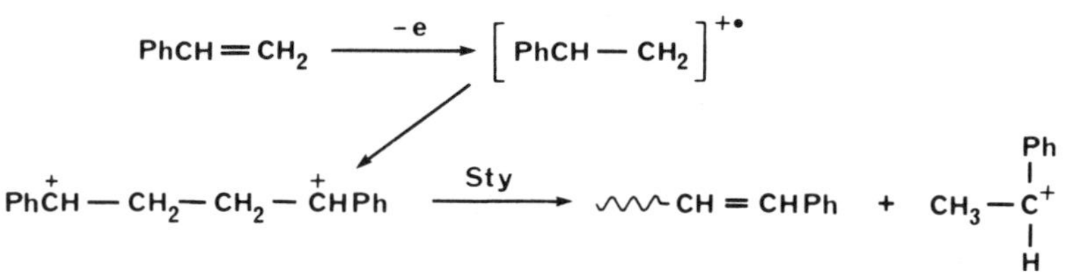

$$PhCH = CH_2 \xrightarrow{-e} \left[PhCH - CH_2 \right]^{+\bullet}$$

$$\overset{+}{PhCH} - CH_2 - CH_2 - \overset{+}{CH}Ph \xrightarrow{Sty} \text{\small\textbackslash\textbackslash\textbackslash} CH = CHPh \;+\; CH_3 - \overset{Ph}{\underset{H}{\overset{|}{\underset{|}{C}}}}{}^{+}$$

Figure 5.33

Further evidence for an electron-transfer-initiated polymerization was found in the enhanced activity of lamellar silicates containing exchangeable cobalt ions, compared to that of their sodium or calcium-exchanged forms. Cobalt vermiculite, for example, could initiate the polymerization of styrene, whereas the calcium and sodium forms of the mineral were inactive [10]. However, reexamination of the styrene oligomers produced in the cobalt vermiculite-catalyzed polymerization indicated that the reaction was initiated by Bronsted acidic species, probably residual water molecules coordinated with the Co^{2+} ions [19]. The dimeric fractions from this reaction and from that initiated by aluminum kaolinite consisted of a mixture of 1,3-diphenylbut-1-ene and isomeric 1-methyl-3-phenylindanes; related moieties comprised the end-groups of the higher oligomers and polymeric fractions [22]. No 1,4-diphenylbutane or phenyltetralin derivatives, which would be indicative of dimerized radical-cation intermediates, were detected in the products from these reactions.

Some evidence for possible radical-cationic intermediates in the Lewis acid-catalyzed reactions may be found in the esr observation of trapped radicals which are formed when styrene or α-methylstyrene is polymerized in the presence of 2,4,6-tri-t-butylnitrosobenzene and solutions of $AlCl_3$ or $BF_3 \cdot Et_2O$ [173] (Figure 5.34). However, the products from reactions in the absence of radical traps correspond to those formed by H^+, or $AlCl_2{}^+$, cation-initiated processes [5], indicating that the trapping agent itself is probably involved in the anomalous initiation.

5.3.2 Dye-Forming Reactions

Acidic clay aluminosilicates are capable of rapidly converting a variety of colorless, reduced, or leuco-dyestuffs into their oxidized, colored forms when solutions of leucodyes are contacted with the minerals [174]. These reactions are the basis of the widely used carbonless copying papers and duplicating materials. They probably also occur in the preparation of the ancient pigment, Maya Blue, an indigo–attapulgite complex [175]. Other dye-forming reactions, such as the oxidation of benzidine or N,N,N',N'-tetramethylbenzidine, have been used for the identification of bentonitic clay minerals [176,177]. These oxidations, like those of the polynuclear hydrocarbons, involve the formation of radical-cationic intermediates or products.

Montmorillonite contacted with a solution of benzidine dihydrochloride a $n\,PhCH = CH_2$ + $Ar - N = O$ $\xrightarrow{AlCl_3}$ $\sim\!\!\sim\!\!\sim CH_2 - \overset{\overset{\displaystyle H}{|}}{\underset{\underset{\displaystyle Ph}{|}}{C}} - \overset{\overset{\displaystyle O^\bullet}{|}}{N} - Ar$

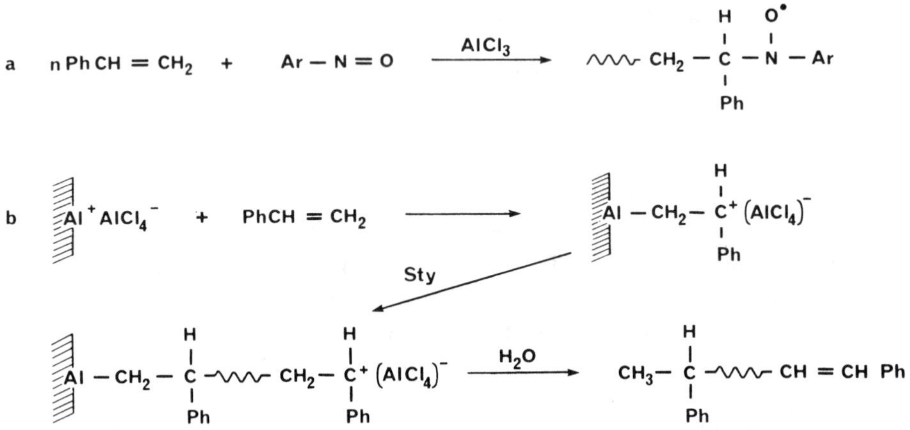

b $\equiv Al^+ AlCl_4^-$ + $PhCH = CH_2$ $\longrightarrow$ $\equiv Al - CH_2 - \overset{\overset{\displaystyle H}{|}}{\underset{\underset{\displaystyle Ph}{|}}{C^+}} \left(AlCl_4\right)^-$

$\xleftarrow{\quad Sty \quad}$

$\equiv Al - CH_2 - \overset{\overset{\displaystyle H}{|}}{\underset{\underset{\displaystyle Ph}{|}}{C}} \sim\!\!\sim\!\!\sim CH_2 - \overset{\overset{\displaystyle H}{|}}{\underset{\underset{\displaystyle Ph}{|}}{C^+}} \left(AlCl_4\right)^-$ $\xrightarrow{H_2O}$ $CH_3 - \overset{\overset{\displaystyle H}{|}}{\underset{\underset{\displaystyle Ph}{|}}{C}} \sim\!\!\sim\!\!\sim CH = CH\ Ph$

Figure 5.34 Polymerization of styrene by anhydrous aluminum chloride. (a) By a radical mechanism, in the presence of t-butyl nitroxide. (b) By a cationic mechanism, initiated by aprotic ionization of Al_2Cl_6.

rapidly develops a blue coloration, which changes through green to yellow as the complex is dried and as the surface acidity of the mineral increases [178,179]. The blue-colored species consists of the radical-cation, or possibly of a charge-transfer dimer, the electronic structure of which is analogous to that of quinhydrone. The oxidation can be inhibited by pretreatment of the mineral with hydrazine or polyphosphates [176].

The Benzidine Blue (see Figure 5.35) is formed by electron transfer between the adsorbed amine and an acceptor species within the mineral lattice. The adsorption and oxidation processes are initially reversible and are believed to occur by the sequence of equilibrated reactions [180,181]. Most of the oxidation sites occur on the interlamellar surfaces. The electron acceptors have been shown to be Fe^{3+} substituent ions located in the octahedral layer, electron exchange occurring across the aluminosilicate boundary [180]. The resultant Fe^{2+} ions can be regenerated by aerial oxidation. Freshly formed Benzidine Blue can revert to benzidine when the complex is extracted with ethanol [182]. On standing or on exposure to oxygen, further oxidation and condensation reactions occur, resulting in a permanent coloration.

The yellow-colored species, which is formed when the blue complex is dried, was originally thought to be a benzidine radical dication. It is now known to be the dication, formed by disproportionation of the radical-cation in the increasingly acidic environment that develops on dehydration of the mineral [183]. This reaction is largely reversible, and the blue color of the radical-cation reappears on moistening of the yellow complex.

Oxidation of benzidine to Benzidine Blue can occur on most other clay minerals, with the possible exception of talc. The development of color on nonswelling aluminosilicates, or on the magnesium silicate, hectorite, is often slower than on montmorillonite [176]. In the case of iron-free hectori-

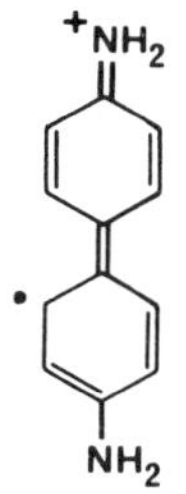

Figure 5.35 Benzidine radical-cation; one canonical structure.

tes, color development is dependent on aerial oxidation of the base [183], but Fe^{3+}-containing hectorites can rapidly convert benzidine to its radical-cation complex [184]. The development of a stable blue color is only possible when the radical-cation is adsorbed on, and can form a π-complex with, the silicate oxygen sheets; Benzidine Blue formed in solution is only transient and rapidly decomposes to brown-colored products [182].

Other benzidine derivatives, such as tetramethylbenzidine or diamino-stilbene, are also readily oxidized to colored species when contacted with acidic minerals. In the case of N,N,N',N'-tetramethylbenzidine, the greenish color of the radical-cation does not alter on dehydration of the complex. The bulky methyl substituents prevent the close approach of the nitrogen centers of adjacent molecules necessary for the electronic transfer process which yields the dication [185]. Other diphenyl derivatives that are isoelectronic with benzidine, such as 4,4'-dihydroxydiphenyl or its dimethyl ether, may also be oxidized to form green-colored quinonoid surface complexes with montmorillonite [186]. Aniline is slowly oxidized on montmorillonite, with formation of brownish materials; the probable oxidant in this case is atmospheric oxygen [183]. However, photoirradiation of aniline adsorbed on silica–alumina results in the formation of the aniline radical-cation. This can dimerize, forming benzidine, and undergo further oxidation to yield Benzidine Blue [187]. Photoassisted oxidation of phenazine to its radical-cation can occur on silica–alumina or magnesia surfaces [188].

The carbonless copying papers consist essentially of paper coated with a mixture of the acidic clay mineral, an acidic resinous binder, and microcapsules containing the color-forming leuco-dyes dissolved in an oily solvent system; the walls of the microcapsules serve to insulate the color-formers from the acidic mineral. Pressure from a pencil tip or the impact of a typeface causes the microcapsules to rupture, releasing the leuco-dye solution which is then adsorbed on, and reacts with, the mineral surfaces to form the colored imprint. Accidental pressure or creasing will also cause colored marks on these papers. An alternative system, which requires contact between two separate sheets of paper, is frequently used in docket books and stationery forms. In this system, the capsules containing the leuco-dye are coated on the reverse side of the original sheet, and

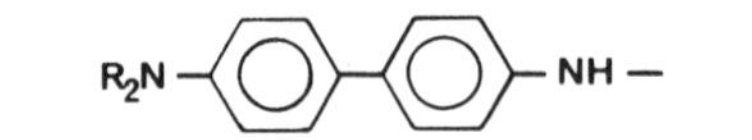

Figure 5.36 (a) Color-forming dialkylamino-biphenylamino and (b) dialkylaminobenzeneazoanilino functional groups.

the acidic mineral on the underlying duplicate sheet; pressure causes the capsules to rupture and transfer the color-forming solution to the mineral-containing layer on the duplicate sheet.

Benzidine is not suitable for use in carbonless papers, but derivatives, such as tetramethylbenzidine, and related compounds containing dialkyl-aminobiphenylamino (Figure 5.36a) or dialkylaminobenzeneazoanilino (Figure 5.36b) groups are used commercially as color-formers. Full color development by oxidation of the benzidine analogues is a slow process, and other color-formers which give a rapid reaction, but yield colors more susceptible to fading, are sometimes used in conjunction with the slower-acting, permanent color-formers. The most important of these rapid color-formers are lactones related to leuco-Crystal Violet lactone or Malachite Green o-carboxylic acid lactone. When one of these lactonic leuco-dyes is adsorbed on a mineral surface, the lactone ring opens to produce the colored zwitterionic form of the dyestuff (see Figure 5.37). This reaction results from the polarizing influence of the mineral surfaces [189] and is promoted by acidic species [190].

A wide variety of color-formers that can provide stable impressions with hues ranging across the spectrum from red to violet have been patented; some examples are briefly reviewed in references 191 and 192. A wide range of acidic minerals and activating treatments have also been patented for use in production of carbonless copying papers. The minerals are usually acid-treated bentonites, attapulgite, kaolin, or silica–aluminas. The activity of these may be enhanced by exchange with Fe^{3+} or Ni^{2+} ions [190] or by the inclusion of zinc chloride [193].

The acidic minerals can catalyze the oxidation of triphenylmethane (rosaniline) leuco-dyes (see Figure 5.38). The oxidation of the tris-

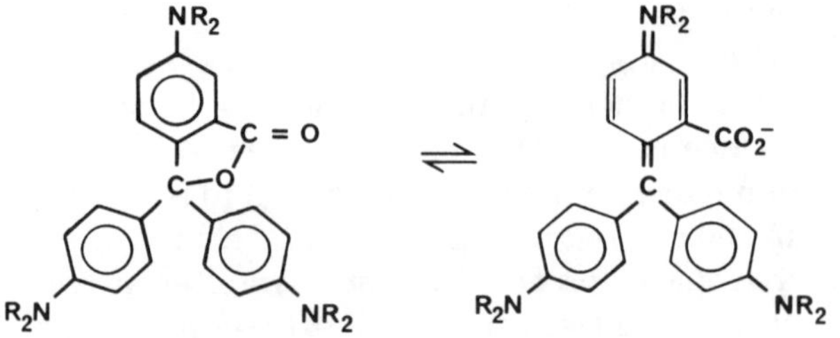

Figure 5.37 Formation of the colored zwitterionic form of a triphenylmethane lactone leuco-dye.

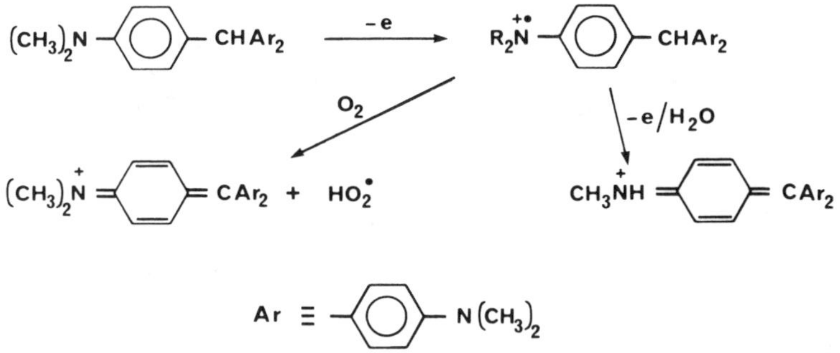

Figure 5.38   One mechanism for the oxidation of triphenylmethane leuco-dyes.

(dimethylaminophenyl)methane (leuco-Crystal Violet) is believed to involve an initial electron abstraction from one of the nitrogen lone-pairs with the formation of a radical-cation that can transfer a hydrogen atom (or electron plus proton) to a surface receptor. The hydrogen acceptor is usually oxygen adsorbed on the mineral surface. The reaction may become autocatalytic as the $HO_2^{\bullet}$ and HO_2^{-} species produced are efficient oxidants. Alkylamino derivatives of triphenylcarbinol are also used as leuco-dyes; these undergo protonation and eliminate water when adsorbed on acidic surfaces.

Leuco-derivatives of the cationic thiazine dyestuff, Methylene Blue, related compounds, and their phenoxazine analogues, are also used as color formers [194] (see Figure 5.39). These leuco-dyes usually consist of *N*-acyl derivatives, which can undergo hydrolysis and oxidation when adsorbed on an acidic mineral surface; *N*-benzoyl-leuco-Methylene Blue,

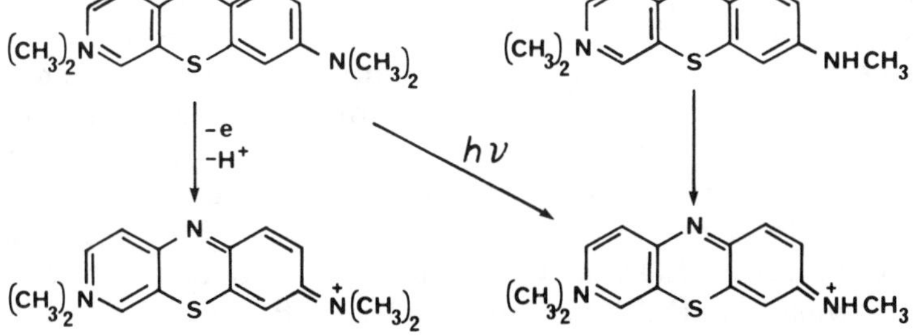

Figure 5.39   Some reactions involved in the surface-catalyzed color formation from benzoyl-leuco-Methylene Blue.

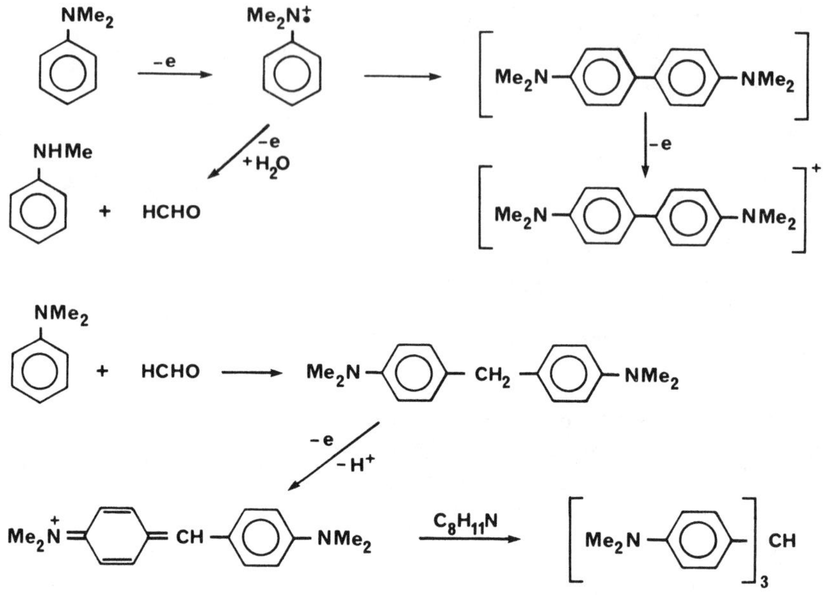

Figure 5.40   One reaction sequence postulated for the formation of leuco-Crystal Violet and tetramethylbenzidine radical-cationic species by the oxidation of dimethylaniline on Kaolinite or montmorillonite surfaces.

for example, would form the dye cation and benzoic acid. However, the color development process is more complex than implied by this simple reaction sequence. The initial products from the reaction on silica gel or on attapulgite, for example, can consist of a mixture of colorless mono-, di-, and trimethyl derivatives of N-benzoyl-leucothionine plus their oxidized derivatives, the mono-, di-, and tri-methylthiazines, as well as Methylene Blue (tetramethylthiazine) itself [195]. The demethylation of the leuco-dyes and/or their oxidized forms are reported to occur by a photoassisted radical process, although the reaction could involve electron-transfer mediated demethylation processes, similar to those which occur during the oxidation of other dimethylamino-substituted aromatic compounds, including the triphenylmethane leuco-dyes.

Aromatic amines with low ionization potentials, for example, dimethyl-aniline or o-methoxyaniline, can be oxidized on montmorillonite or other acidic mineral surfaces to form colored complexes. The oxidation products of dimethylaniline (DMA) have been examined in some detail and have been shown to include the tetramethylbenzidine (TMB) radical-cation complex and the triphenylmethane dyes, Crystal Violet and Methyl Violet (see Figure 5.40). DMA reacts with montmorillonite to form a green-colored complex that may be a derivative of TMB [196], although this has not been proven conclusively. DMA is dimerized to TMB via the radical-cation in electrochemical syntheses [197], and a similar electron-transfer initiated reaction may occur with the intercalated amine.

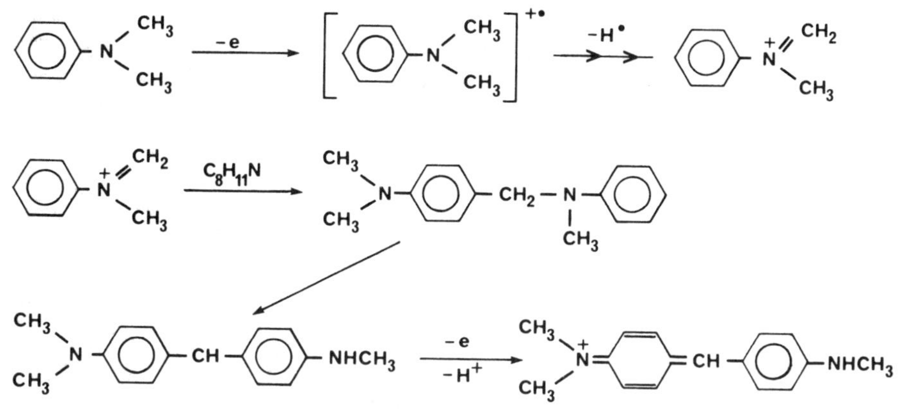

Figure 5.41 The probable mechanism for oxidation of dimethylaniline on Laponite® and on other acidic surfaces.

The oxidation of DMA adsorbed on kaolin apparently takes a different course, as the products from the greenish-yellow complex include Crystal Violet plus related species [172,196]. Oxidation of DMA with oxidants such as cupric sulfate is used for the commercial manufacture of Crystal Violet. The dyestuff is also formed by the oxidation of DMA in the presence of aluminum chloride [198].

The oxidation of DMA intercalated in Laponite®, a synthetic iron-free hectorite, has been shown to be complex [199]. It apparently involves the formation of an intermediate, a *bis*(anilino)methane, which can yield a blue quinoid cation on oxidation in an acidic environment; this monocation forms a yellow dication in strongly acidic environments. Methyl Violet is formed by condensation of the quinoid cation with DMA, followed by oxidation of the resultant leuco-base. The oxidation of DMA on montmorillonite may also involve the formation of the dianilinomethane as well as, or instead of, tetramethylbenzidine.

The dye-forming oxidation of DMA on Laponite® was thought to involve the formation of formaldehyde as a transient intermediate [199]; this is a dubious assumption. More probably, the mineral-catalyzed condensation involves an intermediate methyliminium species (Figure 5.41), formed from the initial DMA-mineral charge-transfer, a process analogous to that proposed for the oxidation of DMA in homogeneous systems [200]. The predominant formation of Crystal Violet requires the presence of excess DMA, to swamp the methylaniline and its derivatives which are formed as byproducts in the initial oxidation and condensation reactions. The composition of the products formed in the restricted environment of the montmorillonite interlamellar space probably differs from those formed in the oxidation of DMA adsorbed on the exterior surface of a kaolinite. However, detailed quantitative analyses of the products formed in the DMA-mineral reactions have not been reported. Triphenylmethane

dyestuffs may also be produced by the montmorillonite-catalyzed condensation of DMA and benzaldehyde [189]. Triphenylamine can be oxidized to
form a red-colored complex when adsorbed on the exterior surfaces of dry
montmorillonite [201]. The complex is believed to contain N,N,N',N'-
tetraphenylbenzidine, formed by dimerization of the triphenylamine
radical-cations.

5.3.3 Metal Arenes

Transition-metal ion-exchanged smectites are capable of forming
charge-transfer complexes with a variety of aromatic hydrocarbons
(arenes). The most notable of these complex-forming smectites is Cu^{2+}-
montmorillonite. This mineral can adsorb benzene to form two different
types of complex: a green-colored "Type I" complex and a reddish-brown
"Type II" complex [202]. The Type I complexes are formed by the
hydrated mineral and contain Cu^{2+} ions coordinated with the edge of the
benzene rings, which retain their planar structure and aromaticity.

The Type II complexes are formed by dehydration of the mineral in the
presence of the arene, which enables electron transfer to occur between the
bare Cu^{2+} ions and the aromatic rings, forming arene radical-cations.
These radical-cations are not free, but are bound to the Cu^+ ions in the form
of charge-transfer complexes. This is evident from the esr spectrum of the
benzene complex, which shows a single sharp resonance signal, arising
from the radical-cation. The expected hyperfine splitting by the ring protons is lost in the exchange-narrowing of the resonance caused by the rapid
interchange of electrons between the arene and its associated metal ion
[203]. A significant proportion of the radical-cations formed in the charge-
transfer process apparently undergo disproportionation, or further oxidation to form the diamagnetic dications, as the number of Cu^{2+} ions reduced
exceeds the number of radicals or radical-cations produced (see Figure
5.42).

Type II complexes can also be formed with thiophene, naphthalene,
anthracene, and biphenyl [203]. Methyl substituents can interfere with the
formation of the charge-transfer complexes, and the methylarenes tend to
form only Type I complexes [204]. In contrast, anisole can form a Type II
complex with Cu^{2+} montmorillonite [205].

The Type II complexes can revert to Type I species on rehydration in an
inert atmosphere. Rehydration in the presence of moist air can result in the
polymerization of the arene, the benzene complex, for example, forming
poly(phenylene) [206]. The polymerization may involve radical intermediates, as metal arenes of this type can react with oxygen, either to form
radicals or free radical-cations [29]. Aging of the anisole complex results in
formation of dimethoxybiphenyl through the dimerization and deprotonation of the radical-cation [207]; other arene complexes may also form oligomers on storage. The oxidative polymerization of benzene to

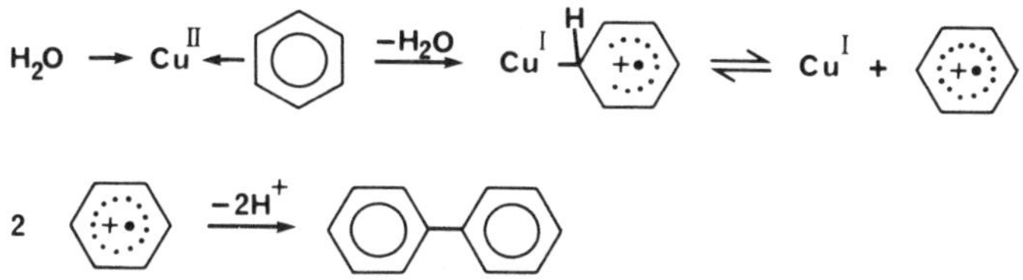

Figure 5.42 The formation and dimerization of a metal arene in a benzene–Cu-mont-morillonite charge-transfer complex.

poly(p-phenylene) by an $AlCl_3$–$CuCl_2$ mixture, which also involves forma-tion of radical-cationic intermediates, probably involves a stepwise cat-ionic propagation process [208]; the chemistry of this system would be relevant to that of the intercalated arene complexes.

Arene charge-transfer complexes can also be formed with montmoril-lonites or hectorites containing Ag^+, Fe^{3+}, or VO_2^+ ions. These complexes are not formed with smectites, such as saponite, in which the anionic charge results predominantly from isomorphous substitution in the tetra-hedral layers. In the case of the Fe^{3+} smectites, charge transfer only occurs with freshly prepared mineral specimens, which contain hydrated Fe^{3+} ions. These hydrated ions, like those of Al^{3+}, undergo hydrolysis on aging, in order to form oxy-cations which are ineffective as oxidants. Aging of Fe^{3+}-exchanged smectites also reduces their ability to oxidize benzidine to Benzidine Blue [184].

The formation of charge-transfer complexes can often enhance the chemical reactivity of adsorbed species. Dry montmorillonite, for example, forms a purple charge-transfer complex with pyridine, which can readily hydrolyze in the presence of water vapor, to give the ammonium enolic salt of glutacondialdehyde [209]. The complexes formed between alkylarenes and alumina are also readily oxidized, toluene, for example, being con-verted to chemisorbed benzoate [210].

5.4 REACTIONS VIA RADICAL INTERMEDIATES

Organic radicals are the reactive intermediates in a wide range of reac-tions that include the polymerization of monomers and the oxidation and degradation of polymers. Many of these reactions can be initiated, inhib-ited, promoted, or otherwise modified by the presence of mineral fillers and pigments. Organic radical species may be generated, for example, by the Bronsted acid-catalyzed breakdown of an alkyl peroxide, by redox reaction between peroxides and a mineral surface, or by photolytic processes involving the mineral crystal lattice. Radicals may also be produced by thermomechanical processes during the compounding of filled polymer compositions. Radical species may be stabilized by adsorption on one class of surface, yet destroyed by adsorption on another surface. These various reactions are of considerable interest because of the effects of mineral-catalyzed radical degradation reactions on the service life of many surface coatings and plastics compositions.

We propose to discuss these reactions under four headings.

(i)   Photolysis and oxidation. This section will be largely concerned with the photoinitiated oxidation reactions occurring on the surface of titania pigments, but will also include other radical-forming reactions that involve electron transfer with mineral surfaces.

(ii)   Peroxide breakdown, which will be principally concerned with the Bronsted acid-catalyzed homolytic cleavage of peroxides, and some related examples of radical-mediated reactions.

(iii)   Radical stabilization and destruction, which will include a brief survey of some mineral-radical physical and chemical interactions.

(iv)   Mechanochemical polymerization and grafting.

5.4.1   Photolysis and Oxidation

The potential of some mineral pigments and fillers to act as oxidation catalysts has been discussed in earlier chapters; the titanias are among the most reactive of these materials and can act as both photoactive and thermal oxidants.

The principal oxidizing species formed by actinic irradiation of titania surfaces have been shown to include adsorbed O_2^- radical-anions and $HO_2^{\bullet}$ and $HO^{\bullet}$ radicals. These species are produced by the reactions of photogenerated excitons or electron-hole pairs with adsorbed oxygen molecules or surface hydroxylic species. Other photogenerated oxidants may include atomic and singlet molecular oxygen, hydroperoxide anions, and hydrogen peroxide. Atomic hydrogen has also been proposed as a potential initiating species in both the oxygenation and reduction of organic materials. These oxidizing species can also be produced by the near-ultraviolet irradiation of zinc oxide pigments and may be formed on other metal oxides, such as magnesia. However, photocatalyzed oxidation of adsorbed organic materials on minerals other than titania or zinc oxide may require irradiation with light having wavelengths significantly shorter than 290 nm, the atmospheric absorption limit of actinic solar radiation. Superoxide and oxygen radical anions and related forms of chemisorbed oxygen may also be formed in the thermal treatment of minerals. Most nonphotolytic radical oxidation reactions that occur on mineral fillers and pigments below 300°C appear to involve some form of adsorbed oxygen and an adjacent electron-deficient surface species.

The actual mechanisms for the photooxidation of most types of organic compounds are still obscure. This uncertainty is reflected in the number of different oxidizing species that have been proposed as intermediates in the

$$2 \; RCH_2OH \; + \; O_2 \; \xrightarrow[h\nu]{TiO_2} \; 2 \; RCHO \; + \; 2 \; H_2O$$

$$2 \; R_2CHOH \; + \; O_2 \; \longrightarrow \; 2 \; R_2CO \; + \; 2 \; H_2O$$

Figure 5.43 Photooxidation of alcohols by titania; the conventional, simplistic stoichiometry.

titania-catalyzed photooxidations. Esr examination of the titania–organic adsorbate photolytic reactions that occur near room temperature fails to show organic radical species [211]. The intermediate organic species produced in the reactions are usually strongly chemisorbed; the desorbed products are often complex and can be formed by more than one process.

The oxidation of alcohols is one of the few mineral–organic photoassisted reactions that have been extensively investigated. When primary or secondary alcohols are irradiated in the presence of anatase or rutile titanias in an inert atmosphere, dehydrogenation of the alcohol can occur, with the respective formation of aldehydes and ketones and the reduction of lattice Ti^{IV} ions to the Ti^{III} state [212]. Regeneration of the Ti^{IV} state can occur in the presence of oxygen and the overall reaction can be summarized by the idealized equations shown in Figure 5.43. The rate of formation of

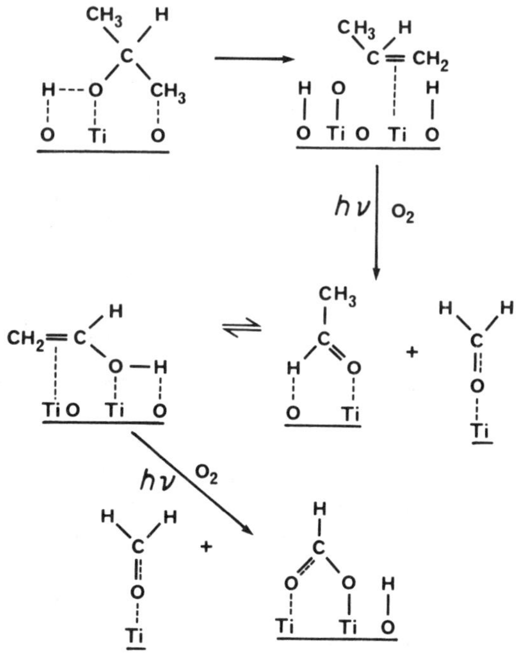

Figure 5.44 Photooxidation of alcohols on titania surfaces; reaction at an oxidizing surface site. (Figures 5.44 and 5.45 taken, with permission, from R. I. Bickley and R. M. K. Jayanty, *Farad. Discuss., Chem. Soc.*, no. 58, p. 194; copyright © Royal Society of Chemistry, London, 1974.)

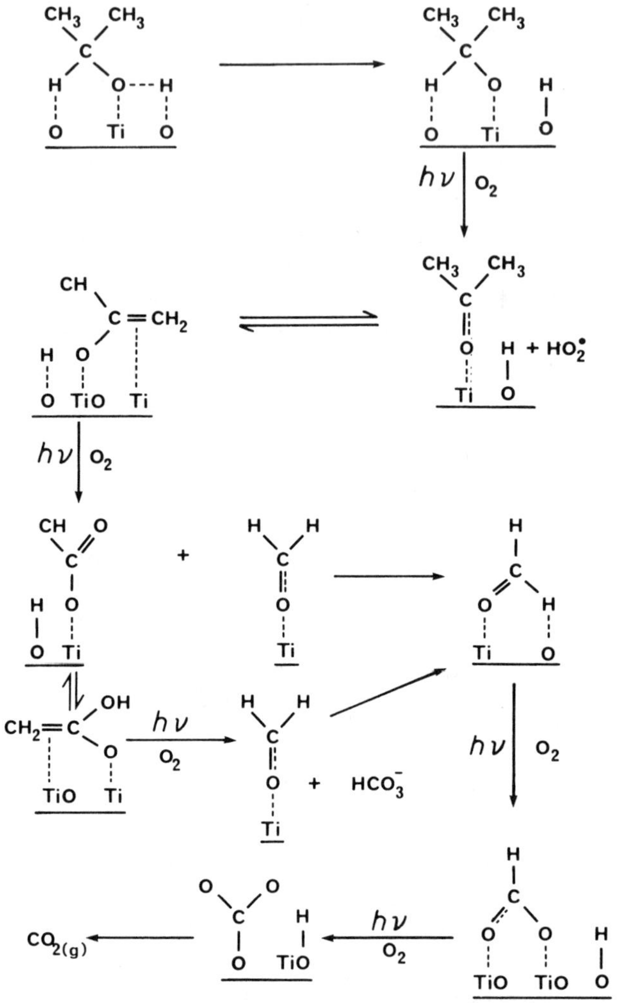

Figure 5.45   Photoassisted dehydration and oxidation of alcohols on titania.

acetone from isopropanol by this reaction has been proposed as one means for measuring the photoactivity of commercial titania pigments [213,214]. The same oxidizing species are believed to be involved in both the oxidation of the alcohol and the degradation of titania-pigmented plastics or surface-coating compositions.

The oxidations are more complex than these equations suggest. A mixture of products may be formed, the proportions of which may vary according to the state of hydration of the titania surface and its degree of saturation with the alcohol.

Bickley and co-workers originally proposed that the oxidation of adsorbed isopropanol on titania at ambient temperatures involved a non-

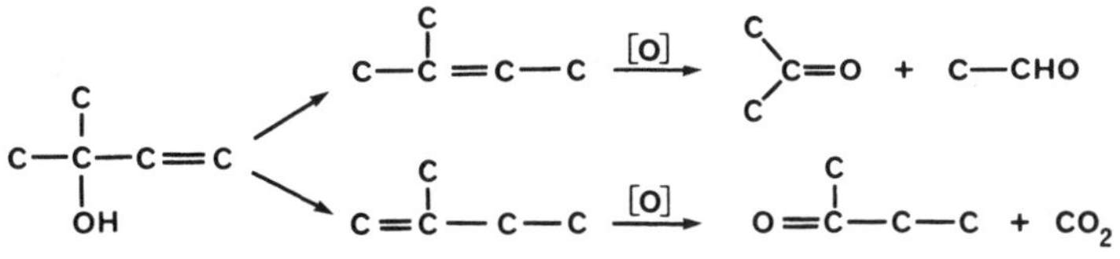

Figure 5.46

chain reaction with adsorbed O_2^- radical-anions or $HO_2^\bullet$ radicals as the primary oxidants and with $HO^\bullet$ radicals and H_2O_2 as intermediate oxidants [215]; part of their proposed scheme is shown in Figure 2.6. Other products of the oxidation include propene, acetaldehyde, formic acid (chemisorbed formate) and, ultimately, carbon dioxide. Oxidation of the alcohols was found to require the simultaneous action of the titania surface, irradiation, and adsorbed oxygen to yield appreciable amounts of carbonyl products. Bickley and co-workers subsequently proposed that the alcohol could be adsorbed on two different surface sites, one of which promoted photoassisted dehydration, followed by oxidation (Figure 5.44), and the other promoted photooxidation (Figure 5.45). The latter reaction involves the formation of a surface alkoxide, and the subsequent abstraction of α- and β- hydrogens, with the formation of an enol that may either desorb or undergo further oxidation [216]. The presence of excess alcohol or water favors desorption of the enol and limited oxidation of the alcohol to the corresponding aldehyde or ketone [212].

Teichner and co-workers [217] proposed that the photooxidation of adsorbed primary, secondary, and tertiary alcohol vapors on titania (anatase at 95°C) was generally preceded by dehydration of the alcohols to alkenes; these were then oxidized by photogenerated atomic oxygen. The relative ease of oxidation of a series of isomeric methylbutanols was found to be secondary > tertiary > primary. The products of the oxidations were:

(i)   From the tertiary alcohol (2-methylbutan-2-ol); acetone, acetaldehyde and butan-2-one (see Figure 5.46).

(ii) From the secondary alcohol (3-methylbutan-2-ol); acetaldehyde and acetone plus smaller amounts of 3-methyl-butan-2-one and 2-methylpropanal (see Figure 5.47).

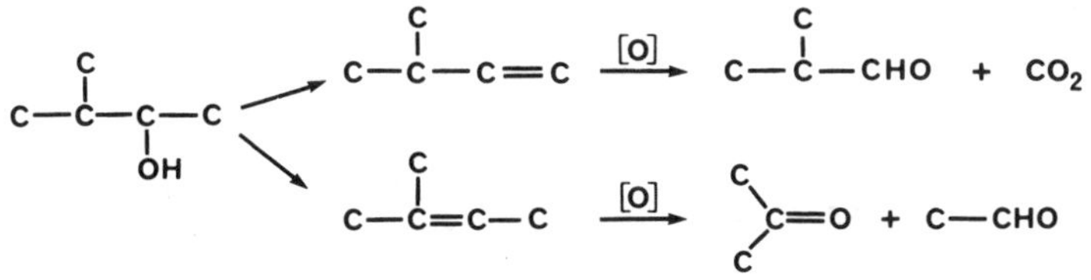

Figure 5.47

> (iii)   From the primary alcohol (3-methylbutan-1-ol); 3-methylbutanal plus lesser amounts of 2-methylpropanal, acetone, and acetaldehyde (see Figure 5.48). In this case, Teichner proposed that the dominant reaction was the direct oxidation of the alcohol, without prior dehydration to an alkene.

of t-butanol [219] had previously shown that the acetone and isobutene products were formed by different pathways, although some proportion of the acetone could have been formed by oxidation of the isobutene.

These reactions have been reinvestigated by Cunningham and co-workers [218], using both vapor–solid and liquid–solid reaction systems. They proposed that the dominant reaction was the photodehydrogenation (oxidation) of the alcohol, rather than a dehydration–oxidation reaction sequence. The oxidizing species were assumed to be adjacent, photo-generated surface hydroxyl and hydroperoxyl radicals and the abstraction of the skeletal hydrogens was accompanied by α-β cleavage reactions; the detailed mechanisms for these reactions were not discussed. Cunningham was unable to directly observe any alkene intermediates in these reactions, or the rapid oxidation of added alkenes at rates comparable with those of the alcohol oxidation reactions. Flash-photolysis studies of the oxidation

The photooxidation of alkanes on titania results in the formation of aldehydes and ketones. Most of the published studies have originated from Teichner and his colleagues, who originally proposed that the oxidation involves the insertion of atomic oxygen into the hydrocarbon molecules, with the formation of an alcohol; this then undergoes dehydration and oxidation to form the observed products [220]. The intermediate alcohols were not detected in these studies. This was ascribed to their much greater reactivity. The intermediate alkenes were only observed in the absence of gaseous oxygen [221]. More recently, Teichner has invoked a second oxidizing species, derived from lattice oxygen anions, to explain the high selectivity of the oxidation of 3-methylbutan-1-ol to 2-methylpropanal, via the hypothetical intermediate, 3-methylbut-1-ene [221]. The second oxidizing species were believed to be O^-, O_3^-, O_3^{3-}, or (presumably) O_2^- radical-anions, formed by the reactions of adsorbed oxygen with the photogenerated excitons [222].

Although alkenes have been assumed to be intermediates in the oxidation of alcohols, little is known of their own photocatalytic oxidation. The oxidation of adsorbed alkenes is often extensive, yielding CO_2 as the major product. The oxidation of propene may yield ethanal (acetaldehyde), propanal, propenal (acrolein), acetone, and propylene oxide. The last-named species may be the initial product in the oxidation reaction of propene by the activated oxygen species [223].

The proposed intermediacy of atomic oxygen in the oxidation of alkanes or alkenes is not in accord with the reactions of authentic atomic oxygen with polymers. In these reactions, atomic oxygen is observed to abstract hydrogen with the formation of hydroxyl radicals, and of organic radicals

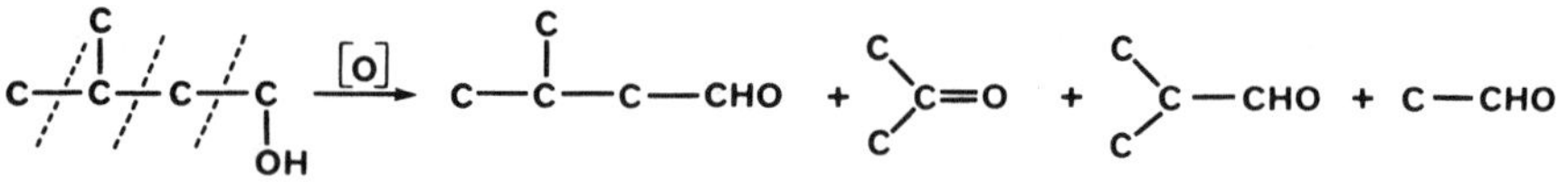

Figure 5.48

that react with oxygen to form peroxyl species [224]. Although Teichner and colleagues were unable to detect hydroperoxides or peroxides in the desorbed oxidation products, the products from the photooxidation of alcohols, alkenes or alkanes on titania surfaces resemble those which would be formed by hydrogen abstraction from the organic molecules in the presence of oxygen and surface species which would be capable of reducing the intermediate peroxyl radicals [e.g., 225], or alternatively, under conditions which would favor rapid decomposition of the intermediate hydroperoxides. The photoassisted electrooxidation of hydrocarbons, which has mechanistic similarities with the photooxidation process on hydrated titania surfaces, has been shown to involve the formation of hydroxylated organic intermediates [226].

The photooxidation of hydrocarbons by titania appears to involve an initial abstraction of hydrogen by trapped holes or bound hydroxyl radicals. The resultant hydrocarbyl radicals can react with oxygen to form adsorbed peroxyl radicals, or they can undergo dimerization, cleavage, or disproportionation to form other hydrocarbon skeletons [e.g., 227,228]. The peroxyl radicals can decompose on the mineral surface to form carbonyl derivatives, or react with the lattice oxygens to form adsorbed alkoxides, or else they can be involved in an autocatalytic oxidation process. In all, the photooxidation of a polymer may involve more than forty kinetically distinct competitive or consecutive reactions.

The photooxidation of methanol adsorbed on titania results in the formation of chemisorbed formate ions and eventually, carbon dioxide via adsorbed carbonate; methyl formate may be formed by esterification of the bound formate by the excess methanol. The oxidation of higher primary alcohols also results in the formation of chemisorbed carboxylate ions [229]. Aminoacids adsorbed on titania can be photooxidized, the products apparently being formed by an initial abstraction of an α-hydrogen atom, followed by decarboxylation, deamination, or cleavage [230]. Aromatic aminoacids, such as tyrosine, may also form phenolic dimers [231]. Titanias can also oxidize reduced, colorless forms of some dyestuffs, for example, Crystal Violet, to their colored forms. This reaction has been used as a means for estimating the photoactivity of the pigments [232].

Titania may act as a photoreductant in the absence of oxygen, provided the surface is not completely outgassed, and still retains bound surface hydroxyl groups. Acetylene can be reduced to form a mixture of methane, ethylene, ethane, and propene. The photoreduction of ethylene also yields these materials plus butene. The reactions probably occur by the addition of a hydrogen atom (or of an electron plus a proton) to the adsorbed alkene, followed by dimerization or disproportionation of the resultant radicals

[233]. The reaction of completely outgassed titania with acetylene or ethylene yields benzene or butene, respectively. Titania can also cause the photoreduction and bleaching of Methylene Blue [234]. The photoreduction of nitrogen by titania can yield ammonia and hydrazine [235], while photolysis of incompletely outgassed titania in the absence of nitrogen, alkenes, or other reducible species, results in the formation of molecular hydrogen; these inorganic reductions were outlined in Chapter 2.

Titania is an important component of many high-temperature thermal oxidation catalysts, but the photolytic activity of titanias can be usefully employed for oxidations at low temperatures, for example, in the formulation of self-cleaning exterior paints, mentioned in Chapter 2; the photodegradation of disposable poly(ethylene terphthalate) films which have zero strength after 400 hr sunlight exposure [236]; the removal of organic material from waste water by titania-assisted photooxidation [237]; the synthesis of acetone by the photooxidation of isobutane [238]; and photographic processes employing the titania-catalyzed photopolymerization of vinylic or acrylic monomers [239]. In the latter example, radicals produced by the oxidation or reduction cause the polymerization or cross-linking of the illuminated areas of the composition, the insoluble image being developed by washing and removal of the unexposed, soluble material.

Oxidizing species may be present on the surfaces of other mineral pigments and fillers. For example, adsorbed butene is oxidized to butadiene on α-ferric oxide at near-ambient temperatures [240] and on silica containing superoxide radical-anions [241]; these adsorbed oxygen species are readily formed when silica, or numerous other oxides, are heated in air. No oxygenated products are obtained from the reaction between butene and silica containing preformed oxygen radical-anions, whereas these are produced in the reaction of butene and oxygen on ultraviolet-irradiated silica [241]. The oxidation of arenes by Cu-montmorillonite has been described in Section 5.3.3. Cu-montmorillonite and freshly prepared Fe^{3+} smectites can oxidize hydroquinone to quinone and polymeric quinones; the formation of polymeric quinones appears to be a characteristic of the oxidation on a silicate surface [242].

Aluminosilicates can also oxidize organic materials via radical intermediates at low temperatures, although the reactions appear to involve chemisorbed oxygen species and/or variable valence transition metal ions. Zeolites, for example, can oxidize contacted α-methylstyrene to acetophenone [243] at room temperature, and montmorillonite can catalyze the autoxidation of acetaldehyde to acetic acid, the necessary oxygen being obtained from adsorbed water or chemisorbed oxygen [244]. Oxidative degradation of aminoacids such as tyrosine can be catalyzed by clay minerals, and in the case of tyrosine, the L-isomer is apparently degraded more rapidly than the D-isomer [231]. Fly-ash, the finely divided, airborne siliceous material produced from burning powdered coal, can promote the photooxidation of those polycyclic hydrocarbons which, like fluorene, contain benzylic hydrogens [245].

The formation of radical species by oxidative decarboxylation of long-chain aliphatic acids has been invoked to account for the formation of longer chain alkanes when these acids are heated with clay minerals in the presence of water. Behenic acid ($C_{21}H_{43} \cdot CO_2H$), for example, forms a mixture of C17 to C29 n-alkanes, similar to that formed by heating the acid with a radical initiator [114]. These products are markedly different from those expected from reactions involving carbenium ion intermediates and are believed to account for the smooth distribution of homologous n-alkanes and fatty acids in petroleum and petroleum reservoir waters.

Although the titania-catalyzed photooxidation reactions described above have dealt with relatively simple organic molecules, similar processes can occur in the photooxidation and degradation of polymers in contact with titania surfaces. The intermediates in these reactions are probably hydroperoxides and their derivative or precursor radicals. The degradation mechanisms of the oxidized polymers are probably similar to those occurring in bulk polymers and probably involve similar intermediate species [224,246–248].

The phenomenon of "chalking", or the erosion of the organic matrix, on outdoor exposure of titania-pigmented surface coating materials has been described in Chapter 2. Many of the exterior surface coating compositions are based on "air-drying" alkyds that are derived from polyunsaturated fatty acids, such as linoleic or linolenic acids. These species undergo autoxidation in the presence of oxygen, with the formation of radical intermediates which can also react with the unsaturated chains, causing the resin film to become cross-linked and insoluble, or to "dry." These reactions are catalyzed by the presence of transition metal driers, such as cobalt and manganese naphthenates. Because the driers are retained in the coating, autoxidation can continue long after the film has dried, which results in the deterioration of the film and its eventual failure. Incorporation of a photoactive titania into such a composition only hastens this process, as the oxidizing species formed on the pigment just supplement those formed in the autoxidation.

It is possible, by irradiation with light of different wavelengths, to differentiate between the ultraviolet-initiated degradation of the polymers and the titania photoassisted oxidation processes. In the case of anatase-pigmented materials, the latter reactions occur most readily when the composite is irradiated at wavelengths between 375 and 395 nm [249]. The actual reactions occurring between the pigment-derived oxidizing species and the matrix are still open to conjecture. Whether the oxidants are radicals or active oxygen species, the reactions almost certainly involve hydrogen abstraction by the oxidant(s), followed by autoxidation with atmospheric oxygen [249]; singlet oxygen can react with allylic centers to form hydroperoxides directly [250].

Titania or zinc oxide-catalyzed photooxidations of most other polymers probably also involve the formation of radical intermediates and the eventual degradation of material in the vicinity of the pigment surface to form

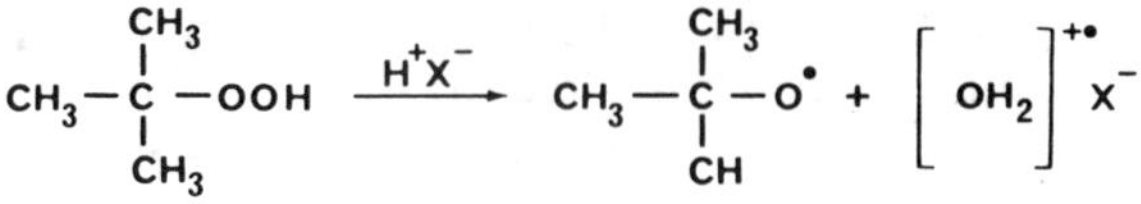

$$CH_3-\underset{\underset{CH_3}{|}}{\overset{\overset{CH_3}{|}}{C}}-OOH \xrightarrow{H^+X^-} CH_3-\underset{\underset{CH}{|}}{\overset{\overset{CH_3}{|}}{C}}-O^\bullet + \left[OH_2\right]^{+\bullet} X^-$$

Figure 5.49

volatile or extractable low-molecular-weight species. Polyolefins, such as polyethylene or polypropylene, undergo chain scission and formation of carbonyl compounds [251–253]; titania can also cause thermal oxidative degradation during melt-compounding of their composites. The reduced thermal stability of polypropylenes in the presence of titania may be partly due to the enhanced oxidative degradation of the phenolic stabilizers used in the commercial polymer [254].

Titania pigments can accelerate the photodehydrochlorination of poly-(vinyl chloride) and the chain scission of nylon polyamides [255]. Acrylic polymers may also be photodegraded by titania, either by oxidation or, in the case of methacrylates, by radical depolymerization. Conversely, titania can catalyze the radical polymerization of methyl methacrylate [256] and accelerate the "drying" of surface-coating formulations [257]. Titanias can also promote the benzoin ether photoinitiated polymerization of acrylic formulations by the transfer of excitation energy from the pigment to the initiator [258].

5.4.2 Peroxide Breakdown

The ability of acidic minerals, such as kaolinites and montmorillonites, to catalyze the heterolytic decomposition of peroxides, principally alky-laryl peroxides and hydroperoxides, has been discussed in Section 5.2.2. These minerals can also accelerate the rate of production of radicals from alkyl peroxides and hydroperoxides. This increase results from the greatly increased rate of decomposition of the protonated forms of the peroxides, a significant proportion of which decompose by homolytic fission. In the case of t-butyl hydroperoxide, approximately 20 to 30% of the material may decompose by the homolytic pathway (see Figure 5.49).

The alkoxy radicals produced in these reactions can initiate the polymerization of acrylic monomers and promote the autoxidation of unsaturated esters and hydrocarbons [61,259]. Similar reactions are catalyzed by Bronsted and Lewis acids in homogeneous media [259,260]. The homolytic decomposition of peroxides is accelerated by complex formation with other small, highly polar ions such as Li^+ (Figure 5.50a) [261]. This mode of decomposition is in contrast with the Haber-Weiss mechanism for hydroperoxide cleavage by variable valence ions (Figure 5.50b).

Spontaneous polymerization of acrylic monomers has been reported to occur on the surfaces of various montmorillonites, for example, hydroxyethyl and hydroxypropyl methacrylates [262] and methyl methacrylate.

a) $$ROOH + Li^+ \longrightarrow \left[ROOH \cdot Li \right]^+$$

$$\left[ROOH \cdot Li \right]^+ + ROOH \longrightarrow RO^\bullet + RO_2^\bullet + \left(H_2O \cdot Li^+ \right)$$

b) $$ROOH + M^{n+} \longrightarrow RO^\bullet + HO^- + M^{(n+1)+}$$

$$ROOH + M^{(n+1)+} \longrightarrow RO_2^\bullet + H^+ + M^{n+}$$

Figure 5.50 Metal cation-catalyzed decomposition of hydroperoxides. (a) By polarizing ions. (b) By variable-valence ions.

In the first two examples, the mechanism was originally believed to involve a transfer of electrons from Fe^{2+} ions contained in the mineral lattice, but it is equally possible that the initiators were adventitious peroxides, with the acidic mineral surface or variable-valence species acting as coinitiators. Zeolites can also catalyze the decomposition of aroyl and alkyl peroxides and hydroperoxides by a redox mechanism with zeolite donor or acceptor species to produce chemisorbed radical species [263]; these reactions are discussed in the next section.

Although the acid-catalyzed decomposition of alkyl peroxides can be used to initiate the polymerization of monomers at low temperatures, this reaction is obtained at the expense of initiator efficiency. These reactions are still important in plastics technology, for example, in the use of peroxyketal-cured styrene–polyester formulations, where an acidic filler can reduce the pot-life and increase the cure-time through surface-catalyzed reactions [264]. The accelerated decomposition of peroxides, such as cumyl peroxide, with reduced initiator efficiency, can also occur on titanias and in the presence of titanium compounds, such as titanate esters (or some titanate ester–mineral adducts) [265].

Other reactions have been reported that involve the mineral-catalyzed formation of radical species, although the modes of formation of the radicals are not described.

Montmorillonite can catalyze the degradative oxidation of intercalated alkylammonium ions via radical intermediates, ethylammonium ions, for example, decomposing to form methane via a transient carbonyl species; these reactions are in addition to the hydrolysis and redistribution reactions described in Section 5.2.3 and were thought to have been promoted by the mineral surface field [266]. We believe that it is more probable that the degradative reactions involve radical intermediates, produced by elec-

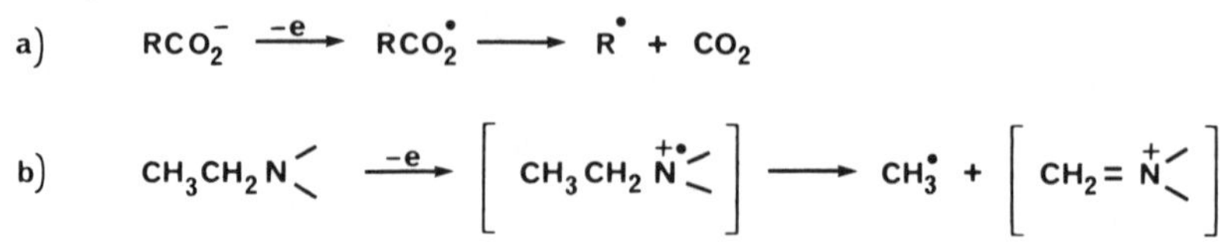

a) $\quad RCO_2^- \xrightarrow{-e} RCO_2^\bullet \longrightarrow R^\bullet + CO_2$

b) $\quad CH_3CH_2N\big\langle \xrightarrow{-e} \left[CH_3CH_2 \overset{+\bullet}{N}\big\langle \right] \longrightarrow CH_3^\bullet + \left[CH_2 = \overset{+}{N}\big\langle \right]$

Figure 5.51 Some mechanisms for the oxidative degradation of intercalated species. (a) Carboxylic acids. (b) Amines.

tron transfer between the amine and an oxidizing species, such as octahedral layer Fe^{III} substituent ions (see Figure 5.51). The putative radical intermediates in the behenic acid redistribution reactions (p. 233) may have been produced via a similar transfer between the carboxylate ions and lattice Fe^{III}.

Acidic montmorillonites can catalyze the polymerization of butenes and butadiene via carbonium ion intermediates. Friedlander found that a neutral sodium montmorillonite could also polymerize these monomers, apparently via a radical process [32]; the initiation mechanism is still obscure. Natural clays frequently contain species that have an esr signal centered near the $g = 2.00$ value typical of organic radicals. This signal is partly due to Fe^{3+}, but also results from the presence of humic acid, which is readily oxidized under mild conditions to form stable radical species [267].

5.4.3   Radical Stabilization and Destruction

Radicals can interact with mineral surfaces in a number of ways. These may include physical entrapment, either of an externally generated radical or of an organic radical generated *in situ*, as in the example of the stable, bound oxymethylene radicals formed on irradiation of a methoxy–silica [268] (see Figure 5.52). Alternatively, the radicals may react with the mineral surfaces and be destroyed by electron transfer with some surface species or else become adsorbed at Lewis or Bronsted acid sites. These complexed radicals may then undergo further reactions.

Physical stabilization of radicals can occur when the adsorbed or grafted radicals are isolated so that the normal processes of radical decay, by combination, disproportionation, or electron transfer with other species cannot occur. These stabilized radicals are not necessarily unreactive; they may combine with oxygen to form new radical species, and many are able to initiate the polymerization of vinylic monomers through the formation of propagating radicals. The propagating radicals are likely to have lower stability than the physically entrapped radicals because of the greater mobility of their active centers. Isolated stable radicals can be produced by γ-irradiation of the mineral, by decomposition of cationic or anionic initiators adsorbed on ion-exchange sites on the mineral surfaces [269], and by the decomposition of conventional azo or peroxide initiators in zeolitic cavities [270] or in microporous silicas or other oxide gels [271].

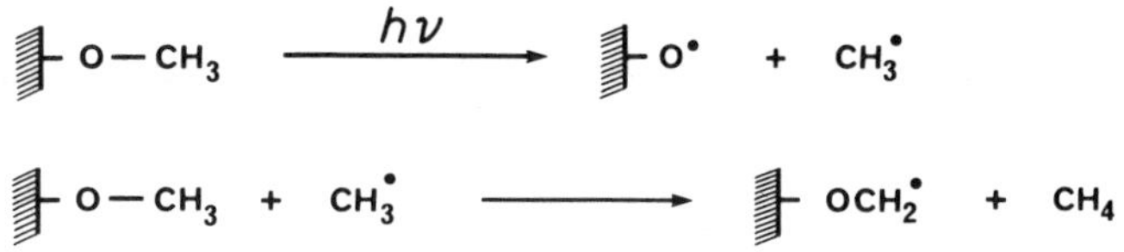

$$\text{⧄O}-CH_3 \xrightarrow{\;\;h\nu\;\;} \text{⧄O}^{\bullet} + CH_3^{\bullet}$$

$$\text{⧄O}-CH_3 + CH_3^{\bullet} \longrightarrow \text{⧄OCH}_2^{\bullet} + CH_4$$

Figure 5.52   Formation of stable radicals species on a methoxy–silica surface.

Photolysis of butyl methacrylate on the surface of an activated alumina results in the formation of a methacrylate radical stabilized by adsorption on Lewis acidic sites [272]. The adsorbed radical can react with oxygen to form peroxy radicals, which are also stabilized, or with excess butyl methacrylate or with vinyl acetate to form the respective propagating radicals; the propagating radicals lose stability as they escape from the confines of the mineral surface. The trapped, stabilized radicals can be displaced from the alumina surface by adsorbed bases.

The peroxy radicals formed by the reaction of atmospheric oxygen with zeolite-stabilized poly(butyl methacrylate) radicals (formed by γ-irradiation of the adsorbed monomer) may have limited stability as repetitive 1,5-hydrogen shift reactions can result in the formation of free monomer molecules; the recombination of these with other radicals can cause the active centers to move beyond the stabilizing confines of the zeolite cavities [270].

Stabilized radicals can be formed in synthetic sodium zeolites by reaction between peroxides or hydroperoxides with electron acceptor (Z_a) or donor species (Z_b) in the zeolite lattice [263] (Figure 5.53). Peroxides, such as cumyl peroxide or t-butyl hydroperoxide, can react with electron acceptor centers to form radical-cationic complexes that decompose, forming peroxy radicals and a zeolitic radical-anionic species. These lattice radical species are capable of electron transfer to tetracyanoethylene, which is converted to its radical-anion; the peroxides do not react with tetracyanoethylene in the absence of the zeolite. Decomposition of these peroxides in a Co-exchanged zeolite, a donor species, results in the formation of a radical-anionic complex that decomposes forming oxyradicals and a zeolitic radical-cation; the latter can accept an electron from diphenylamine, forming its radical-cation. Benzoyl peroxide decomposes via the radical-anionic route in sodium zeolites, with the formation of stabilized benzoyloxy radicals which, unlike free benzoyloxy radicals, do not undergo decarboxylation.

In many systems, the interactions between radical species and mineral surfaces can only be indirectly evaluated from kinetics measurements, for example, of polymerization processes. Details of these interactions have only been observed in the esr spectra of a few highly stable radical species, such as the nitroxides frequently used as spin-probes.

Di-t-butyl nitroxide (DTBN) is adsorbed on silica, silica–alumina, alumina, and decationized zeolites at two distinct types of stabilizing sites [273]. In one, the DTBN is hydrogen-bonded to surface hydroxyl groups,

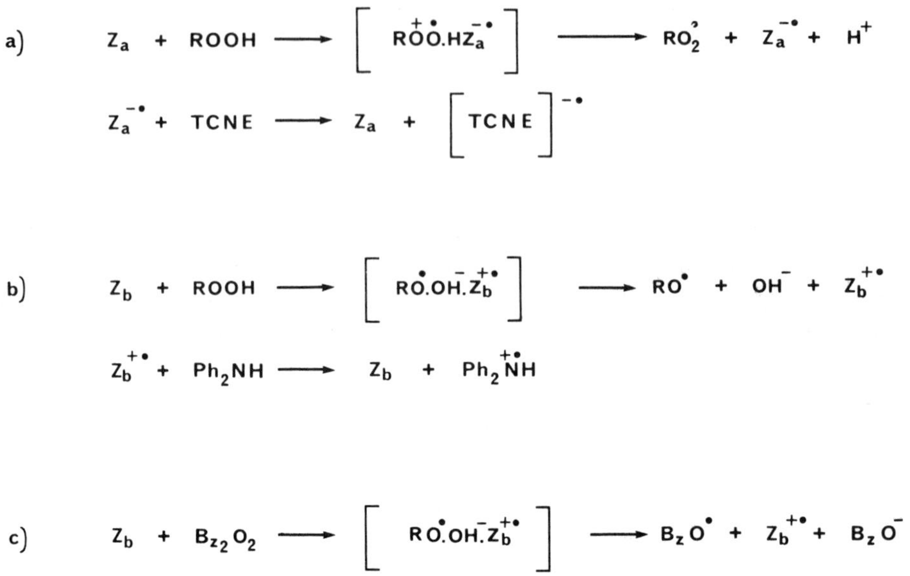

a) $\quad Z_a + ROOH \longrightarrow \left[R\overset{+}{O}\overset{\bullet}{O}.H\overset{-\bullet}{Z_a} \right] \longrightarrow RO_2^{\circ} + Z_a^{-\bullet} + H^+$

$\quad Z_a^{-\bullet} + TCNE \longrightarrow Z_a + \left[TCNE \right]^{-\bullet}$

b) $\quad Z_b + ROOH \longrightarrow \left[R\overset{\bullet}{O}.\overset{-}{O}H.\overset{+\bullet}{Z_b} \right] \longrightarrow R\overset{\bullet}{O} + OH^- + Z_b^{+\bullet}$

$\quad Z_b^{+\bullet} + Ph_2NH \longrightarrow Z_b + Ph_2\overset{+\bullet}{N}H$

c) $\quad Z_b + Bz_2O_2 \longrightarrow \left[R\overset{\bullet}{O}.\overset{-}{O}H.\overset{+\bullet}{Z_b} \right] \longrightarrow Bz\overset{\bullet}{O} + Z_b^{+\bullet} + BzO^-$

Figure 5.53 Some modes of decomposition of peroxides in zeolite cavities. (a) Hydroperoxides on zeolite electron-acceptor sites, to form reduction sites. (b) Hydroperoxides on zeolite electron-donor sites, to form oxidation sites. (c) Formation of stabilized benzoyloxy radicals.

and in the other, to coordination-deficient Al ions, when these are present. Adsorption at the latter site is shown by the appearance of hyperfine splitting of the nitroxide triplet esr spectrum, which results from electronic interaction between the radicals and $^{27}Al^{3+}$ ions. In contrast to the adsorption at these two types of site, adsorption on Bronsted acid sites results in the disappearance of the esr signal of the radical, probably as a result of bimolecular disproportionation of the mobile protonated radicals. The esr spectrum of another nitroxide, tetramethylpiperidinyloxyl, when adsorbed on silica–alumina, can be used to differentiate between adsorption on Bronsted (esr triplet) and Lewis (esr multiplet) sites [274].

The partial decolorization of the stable radical diphenylpicrylhydrazyl (DPPH) following adsorption on acidic clay minerals was originally thought to result from electron transfer to a second DPPH molecule via surface Lewis acid sites or other mineral electron-acceptor species. The overall redox reaction in the presence of adsorbed water resulted in the formation of diphenylpicrylhydrazine (DPPH·H) and quinonoid products [275]. Reinvestigation of the reaction of DPPH on dried aluminosilicate and on acidified titania surfaces [276] showed that reactions on both types of surface yielded similar products. These reactions involve an initial, reversible adsorption followed by the slow disproportionation between protonated and unprotonated adsorbed DPPH molecules, which gives DPPH·H and an intermediate species which undergoes irreversible reaction with adsorbed water to form a magenta-colored product, believed to be an indoaniline (see Figure 5.54).

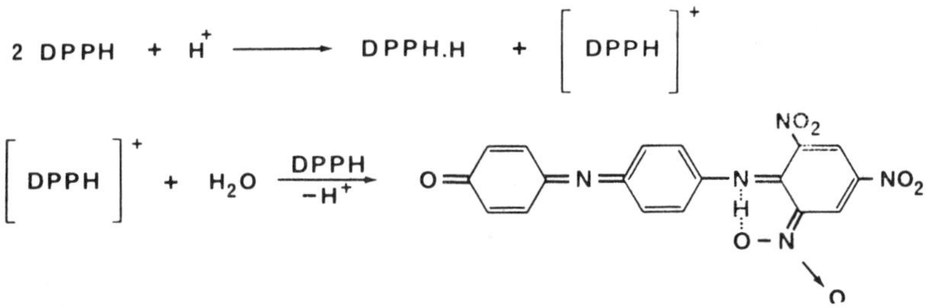

$$2\ DPPH + H^+ \longrightarrow DPPH\cdot H + \left[DPPH\right]^+$$

$$\left[DPPH\right]^+ + H_2O \xrightarrow[-H^+]{DPPH} \cdots$$

Figure 5.54   Reactions involved in the irreversible hydrolytic disproportionation of DPPH on acid surfaces.

The decomposition of nitroso-*t*-butane is promoted by electron-donor species which occur on the surface of magnesia, and to a lesser extent on alumina, but not on silica [277]. Decomposition by irradiation, or on magnesia surfaces, results in the formation of DTBN, but the reaction on alumina yields a different radical product, apparently due to the conversion of the intermediate *t*-butyl radicals to isobutyl radicals on the electron-accepting mineral surface, a process which is otherwise inherently unfavorable. Photooxidation of nitroxide radicals can occur on zinc oxide surfaces [278] and probably on titanias (see Figure 5.55).

The decomposition of azo compounds on silica can differ significantly from the decomposition in homogeneous media. Hydrogen bond formation between the hydroxylic surface and the azo group can significantly reduce the activation energy for decomposition, thus increasing the rate. However, the rates of recombination of the geminate radicals may also be increased, thereby reducing the initiator efficiency. This appears to occur in the accelerated decomposition of azobisisobutyronitrile on hydrated titanias [265] and in the decomposition of azocumene on silica [279]. In the latter reaction, the yields of cumene and α-methylstyrene-derived products, formed by disproportionation of the geminate cumyl radicals, are considerably greater than those formed in homogeneous solution (see Figure 5.56). Decomposition of the adsorbed azocumene also results in the formation of α-*o* and α-*p* cumyl radical dimers which undergo aromatization on the mineral surface. In homogeneous media, these short-lived dimers are not aromatized, but revert to cumyl radicals, and thence to the stable products, dicumyl, cumene, and α-methylstyrene.

$$(CH_3)_3C-N=O \longrightarrow (CH_3)_3C^\bullet + NO$$

$$(CH_3)_3C-N=O + (CH_3)_3C^\bullet \xrightarrow{MgO} (CH_3)_3C-\underset{\underset{O^\bullet}{|}}{N}-C(CH_3)_3$$

$$(CH_3)_3C-N=O + (CH_3)_3C^\bullet \xrightarrow{Al_2O_3} (CH_3)_3C-\underset{\underset{O^\bullet}{|}}{N}-CH_2-CH(CH_3)_2$$

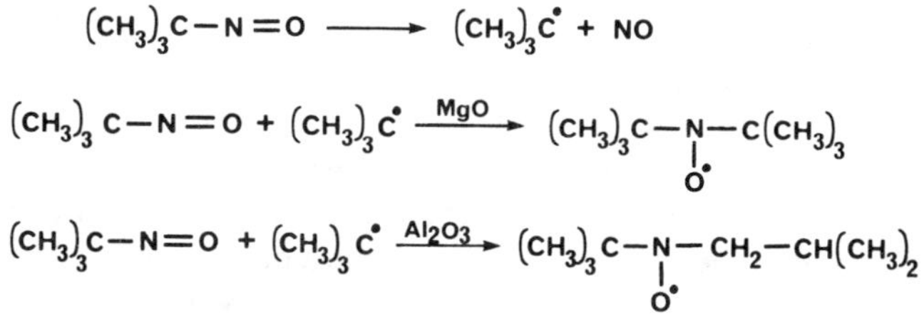

Figure 5.55   Surface-catalyzed transformations of *t*-butyl nitroxide.

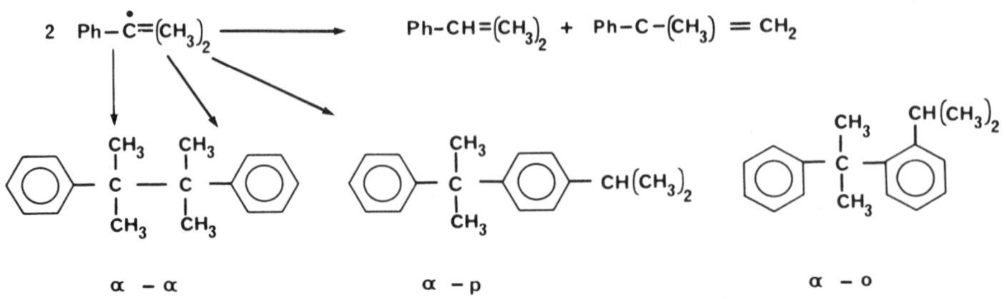

$$2 \; Ph-\overset{\bullet}{C}=(CH_3)_2 \longrightarrow Ph-CH=(CH_3)_2 \; + \; Ph-C-(CH_3)=CH_2$$

$$\alpha - \alpha \qquad\qquad \alpha - p \qquad\qquad \alpha - o$$

Figure 5.56 The aromatic products from the dimerization and disproportionation of cumyl radicals on mineral surfaces.

The radical-initiated polymerization of acrylic monomers can be inhibited by the presence of acidic aluminosilicate minerals such as dry attapulgite, montmorillonite, or kaolinite, whereas the neutral mineral, talc, has little effect [280]. This inhibition has been ascribed to electron transfer between the propagating radicals and coordination-deficient Al^{3+} ions, and the formation of nonpropagating carbonium ions. This process is analogous to the oxidation of triphenylmethyl radicals to triphenylcarbonium ions on silica–alumina or alumina surfaces. In contrast, very high polymerization rates were observed for acrylic monomers in aqueous clay-containing emulsions, the acceleration in this case being due to the low rate for the radical termination processes resulting from the presence of the mineral surfaces [269].

Polymerization of acrylic monomers is generally accelerated by the presence of finely divided silicas. This has been ascribed to increased initiation efficiency, in the case of azobisisobutyronitrile initiator on an aerosil [281], and to stabilization of the propagating radical by adsorption on the silica surface. This reduces the rate of bimolecular termination, but has a lesser effect on the rate of propagation, as monomer can still rapidly diffuse to the radical centers [282]. The presence of silica also decreases the termination rate in the polymerization of vinyl acetate and increases the average molecular weight of the polymer [283].

The effects of aerosil or methylated aerosil on the absolute rates of propagation, transfer, and termination reactions in the radiation-initiated polymerization of adsorbed vinyl acetate have been studied [284]. The rate of propagation of the adsorbed monomer was found to be 100 to 200 times less than that in the bulk monomer, as was the rate of transfer to monomer; the relative rate of chain transfer to monomer was slightly higher in the case of adsorbed monomer on aerosil. The termination process on methylated aerosil is a bimolecular process, and the rate of this reaction was five orders of magnitude less than the corresponding reaction in bulk monomer. In consequence, the average molecular weight of the surface polymerized material is considerably higher than that of the bulk polymer. Acidic, coated titanias have been reported to increase the rate of polymerization of methyl methacrylate compared to rates in the presence of basic titanias

[285]. In this case the acceleration was ascribed to the formation of layers of adsorbed monomer that allowed increased rates of propagation.

5.4.4   Mechanochemical Polymerization and Grafting

Polymers may be grafted, that is, chemically bound to mineral fillers by a number of processes. Those already described in Chapter 4 include the condensation of a polymer, having active groups, with reactive mineral surface species, usually hydroxyl groups; the copolymerization of polymeric propagating radicals with chemisorbed unsaturated organic species; and the initiation of polymerization by chemisorbed azo or peroxide initiators. Grafted polymers can also be obtained by mechanochemical processes that can occur when minerals are ground or subjected to ultrasonic vibration in the presence of monomers or polymers, or by irradiation of a mixture of mineral and monomer by ionizing radiation.

When mineral particles are fractured, for example, by grinding in a ball-mill, the freshly formed surfaces contain active species that can, for example, react with organic materials, such as alcohols and hydrocarbons; initiate the polymerization of vinylic monomers; and adsorb oxygen or nitrogen, converting these molecules to electronically excited states [286]. Much of the published material on mechanochemical processes in mineral –organic systems has originated from Russian workers. Reference should be made to translated review articles for details of the chemistry involved [287,288]. The initial species formed by interaction between organic molecules and the freshly cleaved mineral surfaces may be radicals, ion-radicals [284], or, in the case of unsaturated molecules, activated π- or σ-bonded species [290].

The location of the active sites can be observed (by electron microscopy) by the "decoration" produced on exposure of freshly cleaved salt or mineral surfaces to styrene vapor [291]. These appear to be accumulations of charged point defects on the cleavage steps of the crystalline lattice. The activity of the cleaved salt surfaces for the polymerization of styrene increases with increasing polarity of the ions, namely, $KCl < NaCl < CaCO_3 < SiO_2 < LiF < CaF_2$. Decoration also occurs on cleaved amorphous glasses.

Cleavage of silicates or silica results in the formation of silyl and silyloxy radicals. Although most of these can undergo vicinal recombination, approximately 1% of the cleaved bonds remain as isolated, stabilized radicals [292]. The silyl radicals can react with oxygen to give adsorbed silylperoxyl radicals; the latter are oxidizing species and, like the silyl and silyloxy radicals, can also initiate the polymerization of adsorbed monomers (see Figure 5.57). Milling of alumina also generates radicals that can be observed by esr spectroscopy. These mechanochemical radicals can react with ethylene at low temperatures to form observable propagating radicals [293]. Milling of alumina in the presence of ethylene at elevated

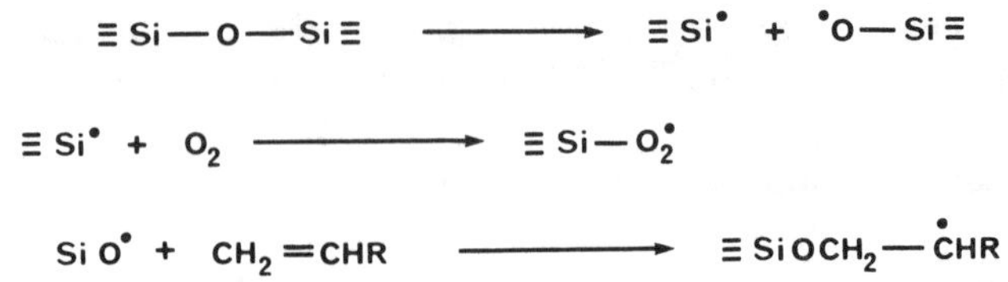

$$\equiv Si-O-Si \equiv \quad \longrightarrow \quad \equiv Si^{\bullet} + {}^{\bullet}O-Si \equiv$$

$$\equiv Si^{\bullet} + O_2 \quad \longrightarrow \quad \equiv Si-O_2^{\bullet}$$

$$Si\,O^{\bullet} + CH_2 = CHR \quad \longrightarrow \quad \equiv Si\,OCH_2-\overset{\bullet}{C}HR$$

Figure 5.57   Mechanochemical formation and reactions of silyl radicals.

temperatures produced low-density polyethylene, half of which was grafted to the mineral surface; the molecular weight of the nongrafted fraction was approximately 10,000 daltons. Similar treatment in the presence of polypropylene yielded an atactic oligomer having a molecular weight of about 400 daltons. The lower molecular weight results from the greater probability of the polypropylene radical undergoing chain transfer [293].

The formation of radicals and grafted species can also occur during ultrasonic-induced cavitation at a liquid–solid interface. The ultrasonic treatment of a mixture of kaolin and iodoform in carbon tetrachloride, for example, results in the liberation of iodine and the formation of a diiodomethyl-grafted kaolin surface [294] (see Figure 5.58).

The mechanochemical polymerization of styrene on cleaved metal oxide surfaces appears to involve radicals as the propagating species, but the polymerization on salts, including calcium carbonate and barium sulfate, apparently involves a radical-anionic process [295]. The proportion of polymer grafted to the mineral surface is dependent on the amount of water or oxygen present because oxygen inhibits radical-initiated polymerization [290], while water can inhibit the formation of polymer grafts [296].

Grinding or melt-compounding of polymers can result in the formation of radical species by mechanical cleavage of molecular bonds [297]. The formation of polymer radicals is promoted by the presence of fillers, which increase the melt viscosity of the composites. The radicals produced may graft to the surfaces of reactive fillers, like those of the carbon blacks [298], or of the unsaturated monomer-modified fillers described in Chapter 4. Polymer radicals can also be produced by abstraction reactions between cleaved minerals and polymers. These, in turn, may form grafts with other centers on the cleaved surfaces.

The γ-irradiation of minerals can result in the formation of persistent reactive surface species. These species can initiate the graft polymerization of monomers added after the irradiation. In the reaction of styrene with preirradiated silica gel [299], approximately 10% of the polymer formed was found to be grafted to the silica surface. The polymer had a bimodal molecular-weight distribution, containing a high-molecular-weight fraction, which was shown to result from a radical polymerization process, and a low-molecular-weight fraction that resulted from a cationic polymerization process. In another study, the proportion of styrene grafted to silica gel during irradiation was found to be greatest when the

$$CHI_3 \rightleftharpoons CH\overset{\bullet}{I_2} + \overset{\bullet}{I}$$

$$CH\overset{\bullet}{I_2} \xrightarrow{\text{CLAY}} \equiv Si-CHI_2$$

Figure 5.58

amount of styrene adsorbed corresponded to a surface monolayer coverage. The grafted polymer had a higher average molecular-weight than the nongrafted polymer, although the difference decreased with increasing styrene content of the irradiated material [300]. Simultaneous formation of grafted and nongrafted polymer was observed during the irradiation of silica gel containing adsorbed vinyl acetate or acrylonitrile, the grafted poly(vinyl acetate) also having a higher molecular weight than the ungrafted material [283].

Irradiation of unfilled polymers can result in both cleavage and cross-linking of polymer chains. This nonthermal cross-linking process can be used, for example, in the manufacture of heat-shrink tubing and the improvement of low-density polyethylene cable insulation. The extent of cross-linking can be greatly increased by the introduction of small proportions of finely divided fillers. The addition of 0.2% v/v of silica to polyethylene, for example, results in a 60% increase in the gel (cross-linked polyethylene) content after 5 Mrad of γ-irradiation [301]. The rate of irradiation-initiated polymerization of methyl methacrylate is also increased by the addition of a mineral filler, such as kaolin, with the formation of grafted and nongrafted polymer by radical propagation processes [302].

γ-Irradiation is often used to initiate the radical polymerization of monomers intercalated in montmorillonites and other lamellar minerals. The initiation sites can occur both on the interlamellar and exterior surfaces of the minerals, but, unless the intercalated monomer molecules are chemisorbed to specific sites, or have a regular intermolecular alignment that favors a rapid propagation process, much of the intercalated monomer can diffuse to, and be preferentially polymerized on, the exterior surfaces where the propagating radicals are less subject to steric restraint.

5.5 REACTIONS VIA ANIONIC OR POLARIZED INTERMEDIATES

Many mineral fillers are basic oxides or carbonates, such as zinc oxide, magnesia, or calcite, or they are amphoteric, containing both potential acidic and basic surface species, like those of alumina or titania. Basic surface species are also present on some aluminosilicates and other minerals that have a predominantly acidic character. Although strongly basic mineral derivatives can be prepared, for example, by doping metal oxides such as magnesia with sodium, the common mineral fillers and pigments do not normally develop highly basic surface sites which complement the highly acidic sites present on the surfaces of zeolites and silica–aluminas.

While most of the mineral-catalyzed reactions described so far have involved cationic or radical intermediates, many others can involve anionic intermediates. Some examples of the latter group, which may be formed by dissociative adsorption on Lewis acidic surfaces, have already been described in Section 5.2. Some other examples of reactions involving anionic intermediates, or of concerted reactions involving coordinated intermediates, are described below.

Basic sites on amorphous silica–aluminas [303] or silica–magnesias [304] are believed to be associated with hexacoordinated aluminum–hydroxy or magnesium–hydroxy groups, while the acidic sites contain tetracoordinated metal–hydroxy groups in a tetrahedral silica network. Acidic and basic species can coexist in close proximity on mineral surfaces, and many catalytic reactions are dependent on this surface duality. Silica and other nonbasic oxides can also develop basic sites when activated at high temperatures. These basic species are thought to be O_2^- radical-anions [305], or oxide ionic defect centers [306].

Anionic derivatives may be formed from adsorbed organic molecules by a one-electron transfer, usually from surface O_2^- radical-anions, to yield organic radical-anions, or by hydride-abstraction (two-electron transfer). These two types of reaction may occur at different sites. They can be differentiated by selective poisoning and by titration using Hammett indicators (for basic species) and tetracyanoethylene (for reductants) [306,307]. Strongly basic oxide surfaces can be prepared by sodium-doping of magnesia; these materials can have sites with Hammett basicity values, $H_0 \geqslant 35$, and are capable of abstracting hydride ions from triphenylmethane ($pK_a = 33$) and of isomerizing 1-alkenes to 2-alkenes [307]. The *cis–trans* isomerization of 2-butenes on alumina and zinc oxide surfaces proceeds without loss or exchange of the hydrogens at the 2- or 3-positions; this indicates that the reaction probably proceeds via an allylic anionic intermediate [308].

The isomerization of alkenes via carbanionic intermediates on alumina, zinc oxide, and magnesia surfaces implies that strongly basic sites can be formed during activation by outgassing in vacuum, although the formation of the intermediate allylic anions actually involves the cooperative interaction of adjacent Lewis and Bronsted acidic centers, or of a Lewis acid such as an Al^{3+} ion, and an adjacent O^{2-} anion (see Figure 5.22). The effective surface Lewis acid sites may be electron-deficient O^{2-} anions associated with coordination-deficient Al^{3+} ions rather than the cations themselves [309].

The formation of strongly basic O_2^- or other oxygen radical-anions on mineral surfaces usually involves activation by irradiation, or by calcination, followed by cooling in the presence of oxygen. Freshly calcined calcium oxide, for example, adsorbs oxygen at room temperature to form O_2^- radical-anions [310]. These species can also be formed on alumina that has been dried at 300°C [311]. They can be produced by an organic adsorbate-sensitized process that does not involve mineral surface one-electron transfer processes, but rather hydride abstraction from the sensitizer, for example, pyridine, propene, or some other unsaturated species, followed by reaction of the organic carbanionic intermediate with atmospheric oxygen [312]. The resultant organic radical can form peroxides or dimers, repetition of the process with the latter yielding polymeric species.

Basic pigments, such as zinc oxide or zinc-doped titanias, can react with the ester groups of alkyd resins that are used in many surface coating formulations. These reactions, while promoting the dispersion of the pigments, also can result in extensive saponification and deleterious changes in rheology of the composition, with the eventual formation of metal carboxylate gels. Studies of the interaction between zinc oxide and poly(tetramethylene adipate) and the model hemiester, monoethyl succinate, have shown that metal–ligand bonds are formed between the polymer and the mineral; these bonding species include zinc carboxylate and zinc-hydroxo-carboxylate complexes [313] (see Figure 5.59). The coordinative cross-linking of low-molecular-weight polyesters by zinc oxide, or by basic zinc or magnesium acylates, is used for the thickening of the monomer–resin mixtures used in reinforced sheet molding compositions [314].

Zinc oxide is used as both a pigment and a vulcanization-accelerating additive in sulfur-vulcanized elastomers. The initial stages of the vulcanization process involve the formation of a zinc mercaptide, which can react with sulfur and with the unsaturated elastomer to introduce polysulfide cross-links; the vulcanization reaction is a radical-mediated process [315]. Poly(vinyl chloride) can also be cross-linked by reaction with magnesia and a triazinedithiol, the intermediate mercaptide presumably displacing reactive chloride from the polymer chains [316]. Basic oxides can be used to reduce the thermal creep of fluorocarbon (Viton) elastomers that occurs with the onset of oxidative degradation at temperatures above 200°C [317].

Basic minerals can catalyze the ring-opening polymerization of epoxides and lactones. Ethylene oxide is polymerized on the surface of stron-

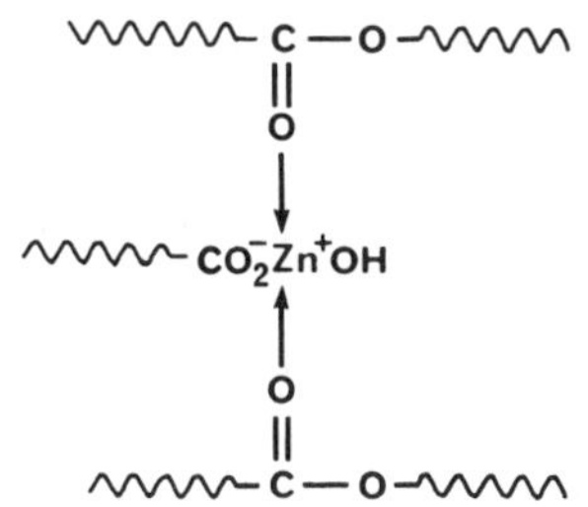

Figure 5.59 Thickening of polyesters by zinc oxide or basic zinc carboxylates. Structure of the complexed zinc salt.

tium carbonate that has been activated by adsorbed water [318]; similar reactions may occur on the surface of calcite. Zinc oxide can catalyze the condensation of epoxides with polycarboxylic acids, the active species being a surface zinc carboxylate [319]. Calcined hydrotalcite, a magnesia–alumina composition, can catalyze the polymerization of β-propiolactone. This reaction occurs on basic sites having Hammett basicity, $H_0 > 15$; adsorbed water or carbon dioxide can deactivate the basic catalytic centers [320]. In contrast, basic mineral pigments may retard the acid-catalyzed curing of urea–formaldehyde and amino resin polymers; alumina-coated titania, for example, may retard the cure of some stoving enamels containing amino resins [321].

The adsorption of alcohols on alumina and the subsequent formation of surface alkoxide species has been mentioned in Section 5.2.3. The adsorption of alcohols on titania involves the formation of surface alcoholates and displacement of surface water. On heating, these complexes dissociate; primary alcohols yield alkoxides, while secondary and tertiary alcohols yield carbonium ions and alkenes [322]. Decomposition of the primary alkoxides can occur when the complexes are heated at $300°C$, and in the case of the ethoxide adduct, the products are ethylene, diethyl ether, and isomeric butenes [323]. The dehydration of surface alkoxides is a concerted process, involving the abstraction of β-hydrogen by surface oxide or bridged-hydroxyl groups [324], and probably involving the formation of a six-membered cyclic intermediate charge-transfer complex. Similar cyclic intermediates are probably involved in the alkylation of alkoxides (ether formation) [324] and in the alkylation or elimination reactions of alkylamines or alkylarylamines on alumina at 200 to $300°C$ [325,326].

The decomposition of titania surface methoxides is a special case as the adsorbed species lack β-hydrogens, and therefore cannot undergo monomolecular dehydration like the other alcohols, although anatase methoxide can form dimethyl ether at temperatures above $300°C$. This reaction is accompanied by the formation of a surface methyltitanium species that can initiate Ziegler–Natta catalyzed oligomerization of olefins; the methyltitanium species formed on anatase having an environment similar to those of alkyltitanium surface groups that are formed by the alkylation of $TiCl_3$ [323]. Ethyltitanium species are also formed during the dehydration of ethanol on anatase; their reaction with the ethylene coproduct yields the butenes also obtained from the dehydration of ethanol over anatase (see Figure 5.60). In contrast, the decomposition of rutile surface methoxide groups occurs at $250°C$ and produces methane and coke, possibly via carbene intermediates.

The adsorption of acetone on a titania or alumina surface Lewis acid site, followed by the abstraction of one of the β-hydrogens by an adjacent O^{2-} ion, can result in the formation of surface enolate anions; these can react with other acetone molecules to form mesityl oxide [327]. Hexafluoroacetone is adsorbed on basic titania sites to form salts of

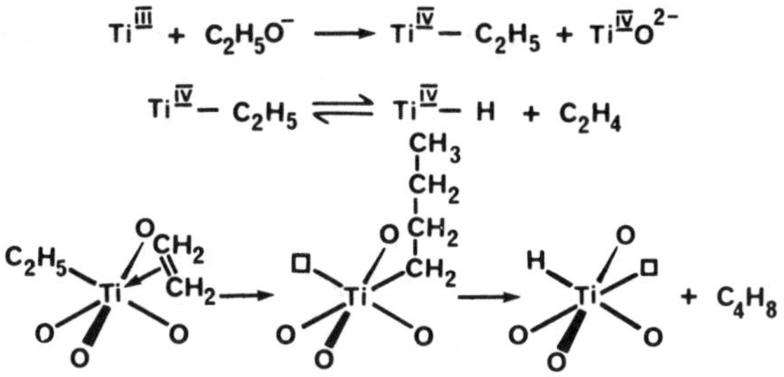

$$Ti^{III} + C_2H_5O^- \longrightarrow Ti^{IV}- C_2H_5 + Ti^{IV}O^{2-}$$

$$Ti^{IV}- C_2H_5 \rightleftharpoons Ti^{IV}- H + C_2H_4$$

Figure 5.60   Mechanism for the formation of butene on desorption of ethanol from anatase.

hexafluoropropan-2,2-diol that readily decompose forming trifluorome-thane and adsorbed fluoroacetate ions [328]. Adsorbed formate ions on titania surfaces decompose at 150 to 250°C, forming carbon monoxide; adsorbed acetate species are more stable. Anatase surface acetates only decompose at temperatures in excess of 350°C, the reaction forming a surface methyltitanium species. In contrast, acetic acid vapor undergoes a bimolecular decomposition on anatase surfaces at temperatures above 250°C, with the formation of acetone; adsorbed ketene is believed to be the key intermediate in this reaction [329] (see Figure 5.61).

Chemisorbed pyridine, ketones, and nitriles on alumina have been reported to form 2-pyridone, carboxylates, and amides, respectively. These reactions apparently involve paired acidic–basic sites and a nucleophilic interaction between the adsorbed species, coordinated with the Al^{3+} ions and basic, surface AlOH groups [330]. The hydrolysis of the nitriles, for example, involves displacement of a surface hydroxo ligand, followed by a nucleophilic attack of the coordinated nitrile by the liberated HO^- group; this reaction occurs at 120°C. The reaction with pyridine occurs at temperatures in excess of 300°C and, in the absence of air, involves the

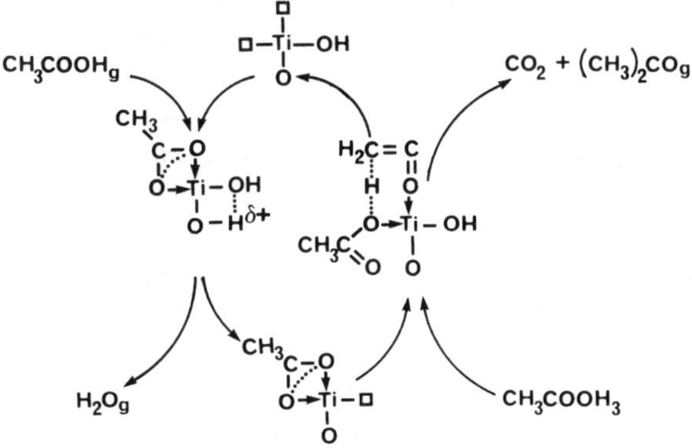

Figure 5.61   Catalytic cycle for the thermal generation of acetone from acetic acid vapor in the presence of anatase. [Reprinted, with permission, from F. Gonzalez, G. Munuera, and J. A. Prieto, *J. Chem. Soc., Farad. I*, **74**, 1526 (1978).]

elimination of molecular hydrogen from the intermediate hydroxylated adduct. Surface isopropoxides on alumina can reduce coadsorbed ketones in a mineral-catalyzed Meerwein–Ponndorf–Verley reaction. Alternatively, acetone can oxidize coadsorbed alcohols via the related Oppenauer process [331].

Alkanes, such as propane or butane, can exchange terminal methyl and methylene hydrogen atoms with deuterium over activated alumina; this is believed to involve the formation of carbanionic intermediates, the exchange occurring on particularly active surface O^{2-} or adsorbed O_2^- species [311]. The exchange reaction can be poisoned by treatment of the alumina with an alkene, which can form adsorbed π-allylic species on the exchange sites.

Acetylene can be adsorbed and slowly polymerized on the surface of activated alumina and acidic zeolites. The adsorption process appears to involve an interaction between the π-electrons of the acetylene molecules and the surface AlOH groups. The polymerization yields *trans*-polyacetylene, possibly via a Lewis acid-initiated process [332]. The aromatic acetylenes, 3-phenyl-1-propyne and 1-phenyl-1-propyne, are isomerized on zinc oxide, with the formation of phenylallene and propargyllic anionic intermediates [333]. Methylacetylene is adsorbed on zinc oxide, with dissociation, to form an allenic anion [334].

In many mineral-catalyzed reactions, the catalytic processes are uncertain and have occasionally been ascribed to the effects of polarization of the adsorbed molecules in the surface electric field of the minerals; examples already discussed in this chapter include the transformation of acids and ammonium salts intercalated in montmorillonites and the polymerization of butene on neutral clays. ^{13}C nmr studies on a variety of organic molecules adsorbed on low-iron zeolites and silica have shown that the surface field can induce chemical shifts generally comparable to those occurring on dissolution. In the case of butenes, polarization of the molecules is observed, redistribution of the electron cloud making the molecules more susceptible to specific attack by protons or other charged species [335].

The polymerization of methyl methacrylate or styrene on degassed silica, magnesia, or alumina has been reported to involve the formation of strained bidentate π-complexes that undergo homolysis and initiate radical polymerization processes. The mechanochemical polymerization of these monomers has been ascribed to a similar process that can occur on freshly formed surfaces [336].

REFERENCES

1   K. Tanabe, *Solid Acids and Bases,* Academic, New York, 1970.

2   I. M. Kolesnikov, *Russ. Chem. Rev.,* 44, 306 (1975).

3 H. P. Leftin, in G. A. Olah and R. Schleyer, Eds., *Carbonium Ions*, Vol. 1, Wiley-Interscience, New York, 1968.

4 V. P. Telichkun, Yu. I. Tarasevich, V. V. Goncharuk, F. D. Ovcharenko, *Proc. Acad. Sci. USSR, Dokl. Phys. Chem.*, **221**, 346 (1975).

5 A. Gandini and H. Cheradame, *Adv. Polym. Sci.*, **34/35**, 1 (1980).

6 P. H. Plesch, Ed., *Cationic Polymerization*, Pergamon, Oxford, 1963.

7 D. H. Solomon, J. D. Swift, and A. J. Murphy, *J. Macromol. Sci. Chem.*, **A5**, 587 (1971).

8 M. Biswas and N. C. Maity, *Adv. Polym. Sci.*, **31**, 48 (1979).

9 H. Hasegawa and T. Higashimura, *Polym. J.*, **11**, 737 (1979).

10 D. H. Solomon and M. J. Rosser, *J. Appl. Polym. Sci.*, **9**, 1261 (1965).

11 T. A. Kusnitskaya and I. K. Ostrovskaya, *Polym. Sci. USSR*, **A9**, 2510 (1967).

12 J. A. Bittles, A. K. Chaudhuri, and S. W. Benson, *J. Polym. Sci., Polym. Phys. Ed.*, **2**, 1221 (1964).

13 T. Matsumoto, I. Sakai, and M. Arihara, *Kobunshi Kagaru*, **26**, 378 (1969).

14 D. G. Hawthorne and D. H. Solomon, *Clays Clay Miner.*, **20**, 75 (1972).

15 A. G. Savkin, F. D. Ovcharenko, M. T. Bryk, and N. G. Vasilev, *Proc. Acad. Sci. USSR, Dokl. Phys. Chem.*, **247**, 663 (1979).

16 D. S. MacIver, R. C. Zabor, and P. H. Emmett, *J. Phys. Chem.*, **63**, 484 (1959).

17 J. J. Jurinak and D. H. Volman, *J. Phys. Chem.*, **63**, 1373 (1959).

18 J. A. Bittles, A. K. Chaudhuri, and S. W. Benson, *J. Polym. Sci., Polym. Phys. Ed.*, **2**, 1847 (1964).

19 D. G. Hawthorne (in preparation).

20 A. D. Kettley, *Stereospecific Polymerization*, Vol. 2, Arnold, London, 1967, p. 219.

21 S. A. Hassan, A-R. M. Mousa, M. Abdel-Khalik, A-A. Abdel-Azim, *J. Catalysis*, **53**, 175 (1978).

22 D. G. Hawthorne, S. R. Johns, and R. I. Willing (in preparation).

23 J. A. Bittles, A. K. Chaudhuri, and S. W. Benson, *J. Polym. Sci., Polym. Phys. Ed.*, **2**, 3201 (1964).

24 D. H. Solomon and H. H. Murray, *Clays Clay Miner.*, **20**, 135 (1972).

25 N. M. Seidov, A. A. Bakhshi-Zade, I. M. Chernikova, and Z. M. Melikova, *Azerb. Khim. Zh.*, 57 (1962).

26 K. Takshatake and H. Hasui, *Jap. Kokai* **78**, 21,149 (Feb. 27, 1978); *Chem. Abs.*, **89**, 42731 (1978).

27 Y. Kawakami, N. Toyoshima, and Y. Yamashita, *Chem. Lett.*, 13 (1980).

28 Kh. Dimitrov, Z. Popov, and Kh. Khaimov, *Geterogennyi Katal., Tr. Mezhdunar. Simp. 3rd*, 513 (1975); *Chem. Abs.* **90**, 104410 (1979).

29 J. M. Thomas, J. M. Adams, S. H. Graham, D. Tilak, and B. Tennakoon, in J. B. Goodenough and M. S. Whittingham, Eds., *Solid State Chemistry of Energy Conservation and Storage*, Adv. Chem. Ser. No. 163, American Chemical Society, New York, 1977, p. 299.

30 R. M. Barrer and A. T. T. Oei, *J. Catalysis*, **30**, 460 (1973); *J. Catalysis*, **34**, 19 (1974).

31 D. G. Hawthorne and G. F. Walker, *J. Catalysis*, **15**, 83 (1969).

32 H. Z. Friedlander, *J. Poly. Sci., Polym. Symp.*, No. 4, 1291 (1963).

33 D. G. Hawthorne, J. H. Hodgkin, B. C. Loft, and D. H. Solomon, *J. Macromol. Sci. Chem.*, **A8**, 649 (1974).

34 M. L. Poutsma, in J. A. Rabo, Ed., *Zeolite Chemistry and Catalysis*, Adv. Chem. Ser. No. 171, American Chemical Society, New York, 1976, p 437.

35 R. A. Rhein and J. S. Clarke, *Polymer*, **14**, 333 (1976).

36   C. J. Norton, *Ind. Eng. Chem., Proc. Des. Devel.,* **3**, 230 (1964).

37   Z. A. Lityaeva, I. I. Martsin, Yu. I. Tarasevich, R. V. Alekseeva, M. M. Kuvaeva, A. M. Mikhaikina, and L. M. Tsipenyuk, *Ukr. Khim. Zh.,* **38**, 1004 (1972).

38   W. H. Lang and W. O. Haag (Mobil Oil), Ger. Offen. 1,915,117 (October 23, 1969); *Chem. Abs.,* **72**, 3913 (1970).

39   J. H. Hodgkin, D. G. Hawthorne, J. D. Swift, and D. H. Solomon (C.S.I.R.O.), Ger. Offen. 2,206,371 (September 7, 1972).

40   H. Pezerat and I. Mantin, *Compt. Rend.,* **265**, 941 (1967).

41   H. Pezerat and M. Vallet, *Proceedings of the International Clay Conference* Madrid, 1972, Cons Super Inest Cient., Madrid, 1973, p. 683.

42   M. Vallet and H. Pezerat, *Bull. Group Fr. Argiles,* **24**, 89 (1972).

43   S. G. Burroughs (Penn. Industrial Chem.) U.S. Patent 2,455,225 (November 30, 1948).

44   E. A. Hauser and R. C. Kollman (National Lead), U.S. Patent 2,951,087 (August 30, 1960).

45   D. G. Hawthorne and D. H. Solomon, *J. Macromol. Sci. Chem.,* **A8**, 659 (1974).

46   S. E. Gebura (Interpace), U.S. Patent 3,573,944 (April 6, 1971).

47   Soc. Anomyme, Compaigne Fr. Raffinage, *Bull. Proprietary Industry,* No. 42 (1967).

48   V. T. Stannett, T. A. DuPlessis, and A. M. Goineau, *J. Appl. Polym. Sci.,* **16**, 2847 (1972).

49   D. H. Solomon (Georgia Kaolin), U.S. Patent 3,549,585 (December 22, 1970).

50   Stamicarbon N. V., Netherlands Application 6,516,208 (June 14, 1967); *Chem. Abs.,* **68**, 3327*b* (1968).

51   J. C. Nease (Engelhard), U.S. Patent 3,594,204 (June 20, 1971).

52   C. W. Umland III, A. M. Gessler, and W. C. Smith (Esso), U.S. Patent, 3,003,990 (October 10, 1961).

53   L. O. Amberg and W. D. Willis, *Proceedings of the International Rubber Conference,* 1959, American Chemical Society, Washington, 1959, p. 565.

54   L. J. Filar (Hercules), U.S. Patent 2,683,751 (July 13, 1974).

55   D. H. Solomon, J. D. Swift, G. O'Leary, and I. G. Treeby, *J. Macromol. Sci. Chem.,* **A5**, 995 (1971).

56   H. Kawasaki, K. Itadani, H. Hata, and F. Kusano, *Nippon Gomu Kyokaishi,* **44**, 452 (1971); *Chem. Abs.,* **75**, 49973 (1971).

57   I. M. Kolesnikov, G. M. Panchenkov, and T. I. Odinokova, *Deposited Doc., VINITI* 3490, 184 (1977); *Chem. Abs.,* **90**, 103549 (1979).

58   G. G. Joris (Allied Chem.), U.S. Patent 2,757,209 (July 31, 1956).

59   J. P. Fortuin and H. I. Waterman (Shell), U.S. Patent 2,671,809 (March 4, 1952); D. I. H. Jacobs, and B. H. M. Thompson (Hercules), U.S. Patent 2,801,262 (June 30, 1957).

60   E. L. Pollitzer, J. J. Louvan and G. E. Illingworth (Universal Oil), U.S. Patent 3,305,590 (February 21, 1967).

61   E. Rizzardo and D. H. Solomon, *J. Macromol. Sci. Chem.,* **A11**, 1697 (1977).

62   S. Sivaram, R. K. S. Singhal, and I. S. Bhardway, *Polym. Bull. (Berlin),* **3**, 27 (1980).

63   J. Chaussidon and R. Calvet, *J. Phys. Chem.,* **69**, 2265 (1965).

64   M. Frenkel and D. H. Solomon, *Clays Clay Miner.,* **25**, 463 (1977).

65   I. M. Eltantawy and P. W. Arnold, *Nature Phys. Sci.,* **244**, 144 (1973).

66   G. F. Walker, *Clay Miner.,* **7**, 111 (1967).

67   Y. Kitayama and A. Michishita, *J. Chem. Soc., Chem. Comm.,* 401 (1981).

68   Gulf Oil Canada, Netherlands Patent 77 06,072 (December 6, 1977); *Chem. Abs.,* **89**, 5923 (1978).

69   J. M. Adams, J. A. Ballantine, S. H. Graham, R. J. Laub, J. H. Purnell, P. I. Read, W. Y. M. Shaman, and J. M. Thomas, *J. Catalysis,* **58**, 238 (1979).

70 J. A. Ballatine, M. Davies, H. Purnell, M. Rayanakorn, J. M. Thomas, and K. J. Williams, *J. Chem. Soc., Chem. Comm.*, 427 (1981).

71 B. Durrand, R. Pelet, and J. J. Fripiat, *Clays Clay Miner.*, **20**, 21 (1972).

72 J. J. Fripiat and M. M. Lambert-Helsen, in W. M. Meier and J. B. Uytterhoeven, Eds., *Molecular Sieves*, Adv. Chem. Ser. No 121, American Chemical Society, New York, 1973, p. 518.

73 Kh. I. Areshidze, and B. S. Tsereteli, *Soobshch. Akad. Nauk Gruz. SSR*, **72**, 365 (1973); *Chem. Abs.*, **80**, 70387 (1973).

74 J. Take, T. Matsumoto, and Y. Yoneda, *Bull. Chem. Soc. Jap.*, **51**, 1612 (1978).

75 J. P. Damon, B. Damon, and J. M. Bonnier, *J. Chem. Soc., Farad. I*, **73**, 372 (1977).

76 B. Stuebner, H. Knoezinger, J. Conard, and J. J. Fripiat, *J. Phys. Chem.*, **82**, 1811 (1978).

77 O. Sieskind and G. Ourisson, *Compt Rend., Ser. C*, **274**, 2186 (1972).

78 A. K. Galwey, *J. Catalysis*, **19**, 330 (1970).

79 A. I. Maksimov, I. I. Levitskii, and Kh. M. Minachev, *Bull. Acad. Sci. USSR, Div. Chem. Sci.*, 965 (1979).

80 G. G. Ferrier, C. S. John, F. H. Leach, L. V. F. Kennedy, and J. K. Tyler, *J. Chem. Soc., Chem. Comm.*, 835 (1978).

81 A. K. Galwey, *J. Chem. Soc., Chem. Comm.*, 577 (1969).

82 M. C. Wilson and A. K. Galwey, *J. Chim. Phys.*, **73**, 441 (1976).

83 M. Sugioka, T. Kamonaka, and K. Aomura, *J. Catalysis*, **52**, 531 (1978).

84 J. A. Ballantine, M. Davies, J. H. Purnell, M. Rayanakorn, J. M. Thomas, and K. J. Williams, *J. Chem. Soc., Chem. Comm.* 8 (1981).

85 R. M. Barrer and S. Kravitz, in *Molecular Sieves*, Society of Chemical Industry, London, 1968, p. 95.

86 S. K. Verma and P. K. Deshpande, *Indian J. Technol.*, **16**, 223 (1978).

87 P. B. Venuto, E. N. Givens, L. A. Hamilton, and P. S. Landis, Preprints, 152th ACS Meeting, New York, 1966.

88 J. M. Bakke and J. Liashan (Bofors), Ger. Offen. 2,826,433 (January 4, 1979); *Chem. Abs.*, **90**, 121181 (1979).

89 M. Fermery, M. Elstner, and J. Zimmermann (Rheinische Braunkohl.), Ger. Offen. 2,720,749 (November 23, 1978); *Chem. Abs.* **90**, 171337 (1979).

90 J. R. Anderson, T. Mole, and V. Christov, *J. Catalysis*, **61**, 477 (1980).

91 W. M. Meier and J. B. Uytterhoeven, Eds., *Molecular Sieves*, Advances in Chemistry Series, No. 121 (1973).

92 M. I. Kuadzhe, *J. Appl. Chem. USSR*, **38**, 868 (1965); *Russ. J. Phys. Chem.*, **37**, 67 (1963).

93 M. Frenkel and L. Heller-Kallai, *Org. Geochem.*, **1**, 3 (1977).

94 M. Hino and K. Arata, *J. Chem. Soc., Chem. Comm.*, 1148 (1979).

95 Sh. B. Battalova, T. R. Mukitonova, and N. D. Pak, *Proc. Acad. Sci. USSR*, **242**, 412 (1978).

96 J. W. Hightower and W. K. Hall, *J. Amer. Chem. Soc.*, **89**, 778 (1967).

97 M. Misono, N. Tani, and Y. Yoneda, *J. Catalysis*, **55**, 314 (1978).

98 N. D. Chuvyekin and V. B. Kazanskii, *Kinet. Catal.*, **11**, 74 (1978).

99 C. S. John and A. Tada, *J. Chem. Soc., Farad. I*, **74**, 498 (1978).

100 E. A. Lombardo, G. A. Sill, and W. K. Hall, *J. Catalysis*, **22**, 54 (1971).

101 D. Freude, U. Lohse, D. Michel, H. Pfeifer, and H. J. Zaeber, *Z. Phys. Chem.*, **259**, 225 (1978).

102 P. B. Venuto, L. A. Hamilton, and P. S. Landis, *J. Catalysis*, **5**, 484 (1966).

103 B. Rabizynska, I. Mazgajska, W. Sieniaszko, and H. Staniak, Polish Patent 87,981 (November 30, 1976); *Chem. Abs.*, **90**, 104970 (1979).

104   D. H. McMahon and E. P. Crowell, *J. Amer. Oil Chem. Soc.*, **51**, 522 (1974).

105   J. Kumanotani and T. Kuwata, *J. Soc. Org. Chem. Jap.*, **16**, 11 (1958); C.S.I.R.O. translation.

106   J. Gosselek and W. W. Lotz, *Z. Naturforsch.*, **33b**, 346 (1978).

107   D. Ogawa, K. Iwayama, and M. Kihara (Toray Ind.), Jap. Kokai 78 23,935 (March 6, 1968).

108   L. K. Popov, M. B. Uskakova, and G. M. Tolmacheva, *J. Appl. Chem. USSR*, **43**, 2815 (1970).

109   M. Elstner, M. Fremery, and H. Hoever, Ger. Offen. 2,208,363 (February 13, 1973).

110   Kh. Dimitrov, S. Arnaudov, and M. Yankov, *God. Sofii, Univ. Khim. Fak.*, **69**, 125 (1979); *Chem. Abs.*, **90**, 121265 (1979).

111   W. W. Lotz and J. Gosselek, *Z. Naturforsch.*, **34b**, 121 (1979).

112   T. P. Goldstein (Mobil Oil), U.S. Patent 4,078,991 (March 14, 1978).

113   A Shimoyama and W. D. Johns, *Nature Phys. Sci.*, **232**, 140 (1971).

114   J. W. Jurg and E. Eisma, *Science*, **144**, 1451 (1964).

115   M. I. Kuadzhe, *Russ. Chem. Rev.*, **40**, 658 (1966).

116   V. R. Vlodarskaya, M. K. Kalinko, and G. I. Nosov, *Tr. Vses. Nauchno-Issled. Geologorazved. Neft. Inst.*, **205**, 128 (1979); *Chem. Abs.*, **91**, 159949 (1979).

117   G. A. Olah, G. K. S. Prakash, and J. Sommer, *Science*, **206**, 13 (1979).

118   C. S. John, L. V. F. Kennedy, and R. C. Paterson, *J. Chem. Res (M)*, 1141 (1978).

119   E. A. Irving, C. S. John, C. Kemball, A. J. Pearman, M. A. Day, and R. J. Sampson, *J. Catalysis*, **61**, 326 (1980).

120   K. Aomura, M. Nitta, and S. Matsumoto, *Hokkaido, Daigashu Kogakuba Kenkyu Hokoku*, 171 (1973); *Chem. Abs.*, **80**, 47361 (1974).

121   K. Arata, S. Akutogawa, K. Tanabe, *Bull. Chem. Soc. Jap.*, **51**, 2289 (1978).

122   F. M. Fowkes, H. A. Benesi, L. B. Ryland, W. M. Sawyer, K. D. Detling, E. S. Loeffler, F. B. Folchemer, M. R. Johnson, and Y. P. Sun, *J. Agr. Food Chem.*, **8**, 203 (1960).

123   G. N. Nosyreva, K. Ione, and V. I. Zheivot, *Kinet. Catal.*, **20**, 221 (1979).

124   H. Kawakami, S. Shiozaki, T. Chujo, and Y. Mitsuishi (Teijin), Jap. Kokai Tokkyo Koho 79 15,979 (February 6, 1979); *Chem. Abs.*, **90**, 205420 (1979).

125   H. Mueller, O. H. Huchler, and H. Hoffmann (BASF), Ger. Offen., 2,801,578 (July 19, 1979).

126   G. Eglinton, S. K. Haj Ibrahim, J. R. Maxwell, J. M. E. Quirke, G. J. Skau, J. J. Volkman, and A. M. K. Wardroper, *Phil. Trans. Roy. Soc. Lond.*, **293**, 69 (1979).

127   W. Van der Velden, G. J. F. Chittenden, and A. W. Schwarz, *Adv. Org. Geochem., Proc. Int. Meet. 6th*, 293 (1974).

128   A. W. Flegmann and H. Scholefield, *J. Mol. Evol.*, **12**, 101 (1978).

129   U. Mingelgrin, S. Yariv, and S. Saltzman, *J. Soil Sci. Soc. Amer.*, **42**, 519 (1978).

130   U. Mingelgrin and S. Saltzman, *Clays Clay Miner.*, **27**, 72 (1979).

131   M. M. Mortland and K. V. Raman, *J. Agr. Food. Chem.*, **15**, 163 (1967).

132   J. D. Russell, M. Cruz, and J. L. White, *Science*, **160**, 1340 (1968).

133   W. Van der Velden, G. J. F. Chittenden, and A. W. Schwartz, *Adv. Org. Geochem., Proc. Int. Meet. 6th* (1973); *Chem. Abs.*, **85**, 110996 (1978).

134   L. S. Porubean, G. S. Born, J. L. White, and S. L. Hem., *Pharm. Sci.*, **68**, 358 (1979).

135   P. Razdan and S. K. Srivastava, *Trans. Brit. Ceram. Soc.*, **78**, 37 (1979).

136   M. Cruz. A. Kaiser, P. G. Rouxhet, and J. J. Fripiat, *Clays Clay Miner.*, **22**, 417 (1974).

137   F. Aragon de la Cruz and C. Viton Barbolla, *An. Quim.*, **73**, 1107 (1977).

138   V. M. Thuy, P. Maitte, *Bull. Soc. Chim. Fr.*, 2558 (1976).

139   K. Taguchi and F. H. Westheimer, *J. Org. Chem.*, **36**, 1570 (1971).

140   T. Isshiki, T. Tomita, M. Abe, N. Takeda, and M. Miura, U.S. Patent 4,083,869 (April 1, 1978).

141   S. P. Walvekar and A. B. Halgeri, *J. Indian Chem. Soc.*, **50**, 246 (1973).

142   H. Imamura, K. Onisawa, Jap. Kokai 78 12,826 (February 4, 1978); *Chem. Abs.*, **89**, 23944 (1978).

143   J. Gosselek and I. Stahl, *Naturwiss.*, **66**, 110 (1979).

144   B. Burczyk and F. Kukla, *J. Catalysis*, **49**, 201 (1977).

145   H. Herold and R. Reinhold (Hoechst), German Patent 2,406,293 (May 22, 1975).

146   B. Casal-Piga, and E. Ruiz-Hitzky, *Proceedings of the Third European Clay Conference,* Nordic Society for Clay Research, Oslo 1977, p. 35.

147   K. Gladwin (Coalite), Ger. Offen. 2,824,976 (December 21, 1978).

148   K. Hatada, K. Fujita, and Y. Ono, *Bull. Chem. Soc. Jap.*, **51**, 2419 (1978).

149   H. E. Bellis (du Pont), U.S. Patent 4,127,513 (November 28, 1978).

150   V. F. Jenkins, D. C. Izzard (LaPorte), British Patent 1,272,572 (May 3, 1972).

151   N. N. Baglei and M. T. Bryk, *Polym. Sci. USSR*, **20**, 2777 (1979).

152   F. D. Ovcharenko, M. T. Bryk, N. N. Baglei, and N. G. Vasilev, *Proc. Acad. Sci. USSR, Dokl. Phys. Chem.*, **235**, 865 (1977).

153   E. Zh. Menligaziev, S. M. Serikbaeva, E. E. Ergozhin, Sh. B. Battalova, A. A. Likerova, and T. Chukenova, USSR Patent 495,328 (December 15, 1975); *Chem. Abs.* **84**, 90809 (1976).

154   E. Zh. Menligaziev, E. E. Ergozhin, K. Kh. Tastanov, S. M. Serikbaeva, Sh. B. Battalova, and A. A. Likerova, *Izv. Akad. Nauk Kaz. SSR, Ser. Khim.*, **26**, 42 (1976); *Chem. Abs.*, **85**, 94726 (1976).

155   F. Bovenden and P. Buesgen, Ger. Offen. 2,651,400 (May 18, 1978); *Chem. Abs.*, **89**, 25463 (1978).

156   L. S. Sheinina, T. E. Lipatova, T. I. Novikova, and Sh. G. Vengerovskaya, *Sint. Fiz.-Khim. Polim.*, **22**, 12 (1978); *Chem. Abs.*, **89**, 75795 (1978).

157   F. Bovenden and P. Buesgen, Ger. Offen. 2,658,121 (June 29, 1978); *Chem. Abs.*, **89**, 111255 (1978).

158   I. J. Bear and R. G. Thomas, *Nature*, **201**, 993 (1964).

159   I. J. Bear and Z. H. Kranz, *Aust. J. Chem.*, **22**, 2343 (1969).

160   K. Kalnins, M. M. Koton, N. G. Antonov, V. M. Svetlichnyi, V. V. Kudryavtsev, G. V. Lyubimova, and V. I. Blinova, *Proc. Acad. Sci. USSR, Dokl. Phys. Chem.*, **244**, 46 (1979).

161   G. N. Asmolov and O. V. Krylov, *Kinet. Catal.*, **19**, 807 (1978); *Kinet. Catal.*, **19**, 975 (1978).

162   B. R. T. Garrett, I. R. Leith, and J. J. Rooney, *J. Chem. Soc., Chem. Comm.*, 222 (1969).

163   F. Freund and H. Gentsch, *Ber. Deut. Keram. Ges.*, **44**, 51 (1967).

164   H. Ueda, *Bull. Chem. Soc. Jap.*, **43**, 319 (1970).

165   G. M. Muha, *J. Phys. Chem.*, **82**, 1843 (1978).

166   B. D. Flockhart, I. R. Leith, and R. C. Pink, *Trans. Farad. Soc.*, **66**, 469 (1979).

167   V. B. Golubev and V. I. Evreinov, *Russ. J. Phys. Chem.*, **40**, 1525 (1966).

168   E. Hirshler, *J. Catalysis*, **5**, 390 (1966); *J. Catalysis*, **5**, 196 (1966).

169   W. K. Hall and F. R. Dollish, *J. Colloid Interface Sci.*, **26**, 261 (1968); W. K. Hall and R. P. Porter, *J. Catalysis*, **5**, 544 (1966); *J. Catalysis*, **5**, 366 (1966).

170   B. D. Flockhart and R. C. Pink, *J. Catalysis*, **4**, 90 (1965); *J. Catalysis*, **8**, 293 (1967).

171   C. Y. Wu and W. K. Hall, *J. Catalysis*, **8**, 394 (1967).

172   D. H. Solomon, B. C. Loft, and J. D. Swift, *Clay Miner.*, **7**, 399 (1968).

173   K. Yamada, H. Tanaka, and H. Kawazura, *J. Polym., Sci., Polym. Chem. Ed.*, **14**, 517 (1976).

174   B. K. G. Theng, *Clays Clay Miner.*, **19**, 383 (1971).

175   R. Kleber and L. Masschelein-Kleiner, *Stud. Conserv.*, **12**, 41 (1967); *Chem. Abs.*, **68**, 22756 (1968).

176   D. H. Solomon, B. C. Loft, and J. D. Swift, *Clay Miner.*, **7**, 389 (1968).

177   S. B. Hendricks and L. T. Alexander, *J. Amer. Soc. Agron.*, **32**, 455 (1940).

178   N. Lahav, *Israel J. Chem.*, **10**, 925 (1972).

179   N. Lahav and S. Raziel, *Israel J. Chem.*, **9**, 683 (1971); *Israel J. Chem.*, **9**, 691 (1971).

180   D. T. B. Tennakoon, J. M. Thomas, and M. J. Tricker, *J. Chem. Soc., Dalton*, 2211 (1974).

181   D. T. B. Tennakoon, J. M. Thomas, and M. J. Tricker, *J. Chem. Soc., Dalton*, 2207 (1974).

182   S. Yariv, N. Lahav, and M. Lacher, *Clays Clay Miner.*, **24**, 51 (1976).

183   T. Furakawa and G. W. Brindley, *Clays Clay Miner.*, **21**, 279 (1973).

184   M. B. McBride, *Clays Clay Miner.*, **27**, 224 (1979).

185   D. T. B. Tennakoon and M. J. Tricker, *J. Chem. Soc., Dalton*, 1802 (1975).

186   Y. Matsunaga, *Bull. Chem. Soc. Jap.*, **45**, 770 (1972).

187   Yu. B. Vymorkov and E. I. Kotov, *Kinet. Catal.*, **16**, 413 (1975).

188   K. S. Seshadri and L. Petrakis, *J. Phys. Chem.*, **74**, 1317 (1970).

189   G. Kortuem and J. Vogel, *Chem. Ber.*, **93**, 706 (1961).

190   R. G. Bayless and D. J. Striley (National Cash Register), Fr. Demande 2,012,741 (March 20, 1970); *Chem. Abs.*, **74**, 32785 (1971).

191   E. N. Abrahart, *Chem. Ind.*, **583** (1973).

192   A. Hakusui, Y. Matsunaga, and K. Umehari, *Bull. Chem. Soc. Jap.*, **43**, 709 (1970).

193   B. W. Brockett (National Cash Register), U.S. Patent 3,516,845 (June 20, 1970).

194   G. A. Hemstock, B. S. Miller (Engelhard), U.S. Patent 3,565,653 (February 23, 1971).

195   H. A. Potts, A. H. Wood, and C. C. Cook, *J. Appl. Chem. Biotechnol.*, **22**, 651 (1972).

196   D. Krueger and F. Orberlies, *Chem. Ber.*, **74B**, 1711 (1941).

197   Z. Galus, R. M. White, F. S. Rowland, and R. N. Adams, *J. Amer. Chem. Soc.*, **84**, 2065 (1962).

198   H. Sato, *Bull. Chem. Soc. Jap.*, **38**, 1719 (1965).

199   E. F. Vasant and S. Yariv, *J. Chem. Soc., Farad. I*, **73**, 1815 (1977).

200   T. Nagai, T. Singaki, and H. Yamada, *Bull. Chem. Soc. Jap.*, **50**, 248 (1977).

201   M. J. Tricker, D. T. B. Tennakoon, J. M. Thomas, and J. Heald, *Clays Clay Miner.*, **23**, 77 (1975).

202   M. M. Mortland and T. J. Pinnavaia, *Nature Phys. Sci.*, **229**, 75 (1971).

203   J. P. Rupert, *J. Phys. Chem.*, **77**, 784 (1973).

204   T. J. Pinnavaia and M. M. Mortland, *J. Phys. Chem.*, **75**, 3857 (1971).

205   D. Fenn, M. M. Mortland, and T. J. Pinnavaia, *Clays Clay Miner.*, **21**, 315 (1973).

206   F. S. Stoessel, J. L. Guth, and R. Wey, *Clay Miner.*, **12**, 255 (1977).

207   T. J. Pinnavaia, P. L. Hall, S. S. Cady, and M. M. Mortland, *J. Phys. Chem.*, **78**, 994 (1974).

208 M. B. Jones, P. Kovacic, and R. F. Howe, *J. Polym. Sci., Polym. Chem. Ed.*, **19**, 235 (1981).

209 H. Lueddeke, H. Meyer, and A. Weiss, *Proceedings of the Third European Clay Conference*, Nordic Society for Clay Research, Oslo, 1977, p. 111.

210 M. Deflin, I. M. Eltantawy, and M. Baverez, *J. Catalysis*, **54**, 345 (1978).

211 R. I. Bickley, in M. W. Roberts and J. M. Thomas, Eds., *Chemical and Physical Properties of Solids and Surfaces*, Chemical Society, London, 1978, p. 119.

212 A. D. Buss, M. A. Malati, and R. Atkinson, *J. Oil Colour Chem. Assoc.*, **59**, 369 (1976).

213 R. B. Cundall, R. Rudham, and M. S. Salim, *J. Chem. Soc., Farad. I*, **72**, 1642 (1976).

214 R. B. Cundall, B. Hulme, R. Rudman, and M. S. Salim, *J. Oil Colour Chem. Assoc.*, **61**, 351 (1978).

215 R. I. Bickley, G. Munuera, and F. S. Stone, *J. Catalysis*, **31**, 398 (1973).

216 R. I. Bickley and R. K. M. Jayanty, *Farad. Disc. Chem. Soc.*, **58**, 194 (1974).

217 A. Walker, M. Formenti, P. Meriaudeau, and S. J. Teichner, *J. Catalysis*, **50**, 237 (1977).

218 J. Cunningham, B. K. Hodnett, and A. Walker, *Proc. R. Irish Acad.*, **77B**, 411 (1977).

219 J. Cunningham and P. Meriaudeau, *J. Chem. Soc., Farad. I*, **72**, 1499 (1976).

220 N. Djeghri, M. Formenti, F. Juillet, and S. J. Teichner, *Farad. Disc. Chem. Soc.*, **58**, 185 (1974).

221 N. Djeghri and S. J. Teichner, *J. Catalysis*, **62**, 99 (1980).

222 H. Courbon, M. Formenti, and P. Pichat, *J. Phys. Chem.*, **81**, 550 (1977).

223 P. Pichat, J. M. Herrmann, J. Disdier, and M-N. Mozzanega, *J. Phys. Chem.*, **83**, 3122 (1979).

224 B. Ranby and J. F. Rabek, *Photodegradation, Photo-oxidation, and Photostabilization of Polymers*, Wiley-Interscience, New York, 1975.

225 W. P. L. Carter, K. R. Darnall, R. A. Graham, A. M. Winer, and J. N. Pitts, Jr., *J. Phys. Chem.*, **83**, 2305 (1979).

226 S. L. Kaliguine, B. N. Shelimov, and V. B. Kazansky, *J. Catalysis*, **55**, 384 (1978).

227 V. B. Kazanskii, *Kinet. Catal.*, **18**, 31 (1977).

228 I. Izumi, W. W. Dunn, K. O. Wilbourn, F-R. F. Fan, and A. J. Bard, *J. Phys. Chem.*, **84**, 3207 (1980).

229 V. N. Filimonov, *Kinet. Catal.*, **7**, 451 (1966).

230 M. B. Patil and G. D. Kalyankar, *Marathwada Univ. J. Sci. Nat. Sci.*, **15**, 47 (1976); *Chem. Abs.*, **89**, 24751 (1978).

231 T. D. Thompson and A. Tsunashima, *Clays Clay Miner.*, **21**, 351 (1973).

232 W. A. Weyl and T. Foerland, *Ind. Eng. Chem.*, **42**, 257 (1950).

233 A. H. Boonstra and C. A. H. A. Mutsaers, *J. Phys. Chem.*, **79**, 2025 (1975).

234 A. V. Pamfilov, Ya. S. Mazurkevich, and E. P. Pahkomova, *Kinet. Catal.*, **10**, 752 (1969).

235 G. N. Schrauzer and T. D. Guth, *J. Amer. Chem. Soc.*, **99**, 7189 (1977).

236 H. P. Hermmerich, G. Winter, H. J. Rosenkranz, and P. Woditsch (Bayer), Ger. Offen. 2,436,260 (April 30, 1975).

237 L. C. Kinney and V. R. Ivanuski, *Robert A. Taft Water Res. Cent., Report* TWRC-13 (1969); *Chem. Abs.* **75**, 67258 (1977).

238 F. Juillet, S. Teichner, and M. Formenti, British Patent 1,331,084 (September 19, 1973).

239 R. W. Baxedale, G. W. Luckey, and H. C. Yutzy (Eastman Kodak), U.S. Patent 3,346,383 (October 10, 1967); *Chem. Abs.*, **68**, 35431 (1968).

240 M. C. Kung, W. H. Cheng, and H. H. King, *J. Phys. Chem.*, **83**, 1737 (1979).

241 M. Anpo, C. Yun, and Y. Kubokawa, *J. Catalysis*, **61**, 267 (1980).

242   T. D. Thompson and W. F. Moll, Jr., *Clays Clay Miner.*, **21**, 337 (1973).

243   W. K. Learman (Mobil Oil), U.S. Patent 3,341,603 (September 12, 1967).

244   I. M. Eltantawy and M. Baverez, *Clays Clay Miner.*, **26**, 285 (1978).

245   W. A. Korfmacher, D. F. S. Natusch, D. R. Taylor, G. Mamantov, and E. L. Wehry, *Science,* **207**, 763 (1980).

246   E. T. Denisov, *Russ. Chem. Rev.*, **47**, 572 (1978).

247   B. Ranby and J. F. Rabek, Eds., *Singlet Oxygen,* Wiley, New York, 1978.

248   J. F. Rabek, J. Lucki, and B. Ranby, *Euro. Polym. J.,* **15**, 1088 (1979); J. Lucki, B. Ranby, and J. F. Rabek, *Euro. Polym. J.,* **15**, 1111 (1979) and references cited therein.

249   H. G. Voelz, G. Kaempf, G. Fitzky, and A. Klaeren, *Org. Coat. Plast. Chem.,* **42**, 660 (1980).

250   S. S. Pappas and R. M. Fisher, *J. Paint Technol.,* **46**, 65 (1974).

251   A. Holmstroem, A. Andersson, and E. M. Soervik, *Euro. Polym. J.,* **13**, 483 (1977).

252   P. Laarenson, R. Arnaud, J. Lemaine, J. Quemnen, and G. Roche, *Euro. Polym. J.,* **14**, 129 (1978).

253   W. Hughes, *10th. FATIPEC Congress Book,* Verlag Chemie, 1970, p. 67.

254   C. W. Uzelmeier, *28th. Tech. Conf., Soc. Plast. Eng. Tech. Papers,* 29 (1970).

255   H. A. Taylor, W. C. Tinchner, and W. F. Hamner, *J. Appl. Polym. Sci.,* **14**, 141 (1970).

256   A. A. Degtyareva, A. A. Kachan, N. E. Tretyakov, and V. N. Filimov, *Teoret. Ekspr. Khim.,* **4**, 566 (1968).

257   S. Keifer, *DEFAZET (Dtsch. Farben-Z),* **28**, 364 (1974); *Chem. Abs.,* **82**, 100131 (1975).

258   S. P. Pappas and W. Kuehhirt, *J. Paint Technol.,* **47**, (610), 42 (Nov. 1975).

259   E. Rizzardo and D. H. Solomon, *J. Macromol. Sci. Chem.,* **A14**, 33 (1980).

260   L. V. Petrov, V. M. Solyanikov, and E. T. Denisov, *Bull. Acad. Sci. USSR, Div. Chem. Sci.,* **26**, 670 (1977).

261   L. V. Popov, V. M. Solyanikov, and Z. Ya. Petrova, *Proc. Acad. Sci. USSR, Dokl. Phys. Chem.,* **230**, 864 (1976).

262   D. H. Solomon and B. C. Loft, *J. Appl. Polym. Sci.,* **12**, 1253 (1968).

263   J. Rychly, L. Matisova–Rychla, and M. Lazar, *Coll. Czech. Chem. Comm.,* **40**, 865 (1975).

264   J. P. Cassoni, R. B. Gallagher, and V. R. Kamath, *Modern Plast.,* **56** (4), 78 (1979).

265   A. V. Sukin, M. A. Bulatov, V. I. Komonenko, B. V. Mitroifanov, and G. P. Shveikin, *Vysokomol Soedin., Ser. B,* **22**, 552 (1980).

266   J. J. Fripiat, *9th. Int. Congr. Soil Sci.,* **1**, 679 (1968).

267   P. L. Hall, *Clay Miner.,* **15**, 321 (1980).

268   E. Melamud, M. G. Reisner, and U. Garbatski, *J. Phys. Chem.,* **77**, 1023 (1973).

269   H. G. G. Dekking, *J. Appl. Polym. Sci.* **11**, 23 (1967).

270   J. Rychly and M. Lazar, *Euro. Polym. J.,* **8**, 711 (1972).

271   N. T. Kartel, N. N. Tsyba, and V. V. Strelko, *Adsorbtsiya Adsorbenty,* **6**, 21 (1978); *Chem. Abs.,* **91**, 63180 (1979).

272   V. B. Golubev, V. P. Galkin, B. A. Korolev, and E. V. Lunina, *Russ. J. Phys. Chem.,* **50**, 2291 (1976).

273   G. P. Lozos and B. M. Hoffman, *J. Phys. Chem.,* **78**, 2110 (1974).

274   A. K. Selivanovskii, V. B. Golubev, E. V. Lunina, and B. V. Strakhov, *Russ. J. Phys. Chem.,* **52**, 2811 (1978).

275   D. N. Misra, *Tetrahedron,* **26**, 5447 (1970).

276   D. G. Hawthorne and D. H. Solomon, *J. Macromol. Sci. Chem.,* **A6**, 661 (1972).

277   V. P. Oleshko, T. V. Bychkova, V. B. Golubev, E. V. Lunina, and L. I. Nekrasov, *Russ. J. Phys. Chem.*, **52**, 1062 (1978).

278   J. R. Harbour and M. L. Hair, *J. Colloid Interface Sci.*, **76**, 333 (1980).

279   J. E. Leffler and J. J. Zupancic, *J. Amer. Chem. Soc.*, **102**, 259 (1980).

280   D. H. Solomon and J. D. Swift, *J. Appl. Polym. Sci.*, **11**, 2567 (1967).

281   E. M. Morozova, T. R. Aslamazova, and V. I. Eliseeva, *Vysokomol. Soedin.*, *Ser. B*, **20**, 859 (1978).

282   T. R. Manley and B. Murray, *Euro. Polym. J.*, **8**, 1145 (1972).

283   M. A. Bruk, S. L. Mund, I. B. Aksman, and A. D. Abkin, *Polym. Sci., USSR*, **19**, 1932 (1977).

284   M. A. Bruk, S. A. Pablov, and A. D. Abkin, *Russ. Acad. Sci., Dokl. Phys. Chem.*, **245**, 254 (1977).

285   L. V. Kozlav and S. M. Alekseev, *Lakokras. Mater. Ikh Primen*, 12 (1978); *Chem. Abs.*, **89**, 110954 (1978).

286   R. E. Benson and J. E. Castle, *J. Phys. Chem.*, **62**, 840 (1958).

287   G. S. Khodakov, *Russ. Chem. Rev.*, **32**, 386 (1963).

288   V. V. Boldyrev and E. G. Avvakumov, *Russ. Chem. Rev.*, **40**, 847 (1971).

289   M. Jacovic, *Double Liason*, 83 (Oct., 1964).

290   G. A. Gorokhovskii, P. N. Logvinenko, and V. G. Kucher, *Poverkhn. Yavleniya Polim.*, 96 (1978); *Chem. Abs.*, **87**, 118166 (1977).

291   L. P. Yanova, V. A. Kuznetsov, A. E. Chalykh, and S. N. Tolstaya, *Russ. J. Phys. Chem.*, **51**, 851 (1977).

292   V. A. Radtsig and V. A. Khalif, *Kinet. Catal.*, **20**, 582 (1979).

293   N. Kurokawa, M. Tabata, and J. Sohma, *J. Appl. Polym. Sci.*, **25**, 1209 (1980).

294   E. V. Kukharskaya and Yu. I. Skorik, *Bull. Acad. Sci. USSR, Div. Chem. Sci.*, 1612 (1966).

295   V. A. Kargin, N. A. Plate, I. A. Litvinov, V. P. Shibaev, and E. G. Lurie, *J. Polym. Sci.*, **59**, S 56 (1962).

296   A. B. Taubman, L. P. Yanova, and G. S. Blyskosh, *Mekhanoemissiya Mekhanokhim. Tverd. Tel.*, 2nd, 254 (1969); *Chem. Abs.*, **82**, 156953 (1975).

297   J. Sohma and M. Sakaguchi, *Adv. Polym. Sci.*, **20**, 111 (1976).

298   F. G. Lavrushin, V. M. Goldberg, M. S. Akutin, and F. G. Gilimyanov, *Vysokomol. Soedin., Ser. B*, **20**, 367 (1978).

299   K. Fukano and E. Kageyama, *Toyo Soda Kenkyu Kokoku*, **21**, 57 (1977); *Chem. Abs.*, **88**, 38198 (1978).

300   Y. Kusama, A. Udagawa, and M. Takehisa, *J. Polym. Sci., Polym. Chem. Ed.*, **17**, 405 (1979).

301   W. J. Chappas and J. Silverman, Report 1978, CONF-780305-17; *Chem. Abs.*, **90**, 7023 (1979).

302   J. J. Beeson and K. G. Mayhan, *J. Appl. Polym. Sci.*, **16**, 2765 (1972).

303   K. Takimoto, *Bull Chem. Soc. Jap.*, **45**, 2391 (1972).

304   B. S. Girgis, A. M. Youssef, and M. N. Alaya, *Bull. Chem. Soc. Fr.*, Pt. 1, 63 (1979).

305   J. B. Peri, *J. Phys. Chem.*, **69**, 220 (1965).

306   K. Esumi and K. Meguro, *J. Colloid Interface Sci.*, **66**, 192 (1978).

307   J. Kijenshi, B. Zielinski, R. Zadrozny, and S. Malinowski, *J. Res. Inst. Catalysis, Hokkaido Univ.*, **27**, 145 (1979).

308   E. A. Lombardo, W. C. Connor, Jr., R. J. Madon, W. K. Hall, V. V. Kharlamov, and Kh. M. Minachev, *J. Catalysis*, **53**, 135 (1978).

309   J. Dewing, G. T. Monks, and B. Youll, *J. Catalysis,* **44,** 226 (1976).

310   Y. Ono, H. Takagiwa, and S. Fukuzumi, *Z. Phys. Chem., N. F.,* **115,** 51 (1979).

311   C. S. John, C. Kemball, E. A. Pearce, and A. J. Pearman, *J. Chem. Res. (M),* 4830 (1979).

312   E. Garrone, A. Zecchina, and F. S. Stone, *J. Catalysis,* **62,** 396 (1980).

313   A. Szilagyi, V. Izvekov, and I. Vancso–Szmercsanyi, *J. Polym. Sci., Polym. Chem. Ed.,* **18,** 2803 (1980).

314   M. Gruskiewicz and J. Collester, *35th. Tech. Conf. Reinf. Plast./Comp.,* Soc. Plast. Ind., Tech. Paper 7E (1980).

315   M. Porter, Vulcanization of Rubber, in S. Oae, Ed., *Organic Chemistry of Sulfur,* Plenum, New York, 1977.

316   K. Mori and Y. Nakamura, *Kobunshi Ronbunshu,* **35,** 193 (1978).

317   J. M. Barton, *Brit. Polym. J.,* **10,** 151 (1978).

318   K. S. Kazanskii, A. N. Tarasov, and S. G. Entelis, *Kinet. Catal.,* **19,** 470 (1978).

319   H. Aoi and I. Itoh (Mitsubushi Petrochem.), Jpn. Kokai Tokkyo Koho 79,157,523 (December 12, 1979).

320   T. Nakatsuka, H. Kawasaki, S. Yamashita, and S. Kohjiya, *Bull. Chem. Soc. Jap.,* **52,** 2449 (1979).

321   G. G. Parekh, *J. Paint Technol.,* **51,** 101 (1979).

322   I. Carrizosa and G. Munuera, *J. Catalysis,* **49,** 174 (1977); *J. Catalysis,* **49,** 189 (1977).

323   I. Carrizosa, G. Munuera, and S. Castanar, *J. Catalysis,* **49,** 265 (1977).

324   H. Knoezinger, *Angew. Chem. Int. Ed.,* **7,** 791 (1968).

325   P. Hogan and J. Pasek, *Coll. Czech. Chem. Comm.,* **38,** 1513 (1973).

326   P. Hogan and J. Pasek, *Coll. Czech. Chem. Comm.,* **39,** 3696 (1974).

327   D. M. Griffiths and C. H. Rochester, *J. Chem. Soc., Farad. I,* **74,** 403 (1978).

328   D. M. Griffiths and C. H. Rochester, *J. Chem. Soc., Farad. I,* **73,** 1913 (1977).

329   F. Gonzalez, G. Munuera, and J. A. Prieto, *J. Chem. Soc., Farad. I,* **74,** 1517 (1978).

330   H. Knoezinger, *Proc. 6th Int. Congress Catalysis, London,* 1976, Chemical Society, London, 1977, p. 183.

331   L. Horner and U. B. Kaps, *Liebigs Ann. Chem.,* 192 (1980)

332   J. Heaviside and P. J. Hendra, *J. Chem. Soc., Farad. I,* **74,** 2542 (1978).

333   T. T. Nguyen, J. C. Lavalley, J. Saussey, and N. Sheppard, *J. Catalysis,* **61,** 503 (1980).

334   Zh. L. Dykh, C. V. Bakubina, L. I. Lafer, V. I. Yakerson, A. M. Rubinshtein, A. M. Taber, and I. V. Kalechits, *Bull. Acad. Sci. USSR, Div. Chem. Sci.,* 2201 (1978).

335   D. Denney, V. M. Mastikhin, S. Namba, and J. Turkevich, *J. Phys. Chem.,* **82,** 1752 (1978).

336   G. A. Gorokhovskii, P. N. Logvinenko, and V. G. Kucher, *Poverkhn. Yavlenuya Polim.,* 96 (1976); *Chem. Abs.,* **87,** 118166 (1977).

ORIENTATION AND CONTROL OF ORGANIC REACTIONS BY MINERALS

The adsorption of organic molecules on mineral surfaces, and the various modes of bonding, have been discussed in Chapter 4. The regular crystalline order of many of the mineral surfaces and the disposition of the various specific binding sites can result in the adsorption of the organic molecules in fixed, sometimes known, orientations with respect to the mineral surface and neighboring adsorbate molecules. The intercalation complexes formed on adsorption of aliphatic amines or alcohols by montmorillonites or vermiculites, for example, can have structures in which the alkyl chains form regular, close-packed arrays in the interlamellar spaces with the axes of the chains inclined to the silicate surfaces (see Figure 1.8). In the case of the n-alkylammonium smectites, the angle of inclination is a function of the charge density of the silicate surface, with typical values in the range 55 to 70°. The apparent angle of inclination can be deduced from the basal spacings of the smectite and its intercalation complex and the Van der Waals dimensions of the alkyl chains [1,2].

The possibility of carrying out reactions on molecules adsorbed at interfaces in a given, fixed orientation has attracted the interest of scientists for many years. It offers a potent means for control of chemical transformation, with the prospect of obtaining products, formed from reaction at specific points on the adsorbed molecules, which may differ from those obtained from the reagents in the less restricted environment provided by a homogeneous medium. In the case of the clays, there is a more esoteric attraction. A number of theories on the origin of life and the formation of large, optically active constituents of living matter are based on the concept of concentration by adsorption of small, abiotic molecules on the mineral surfaces from the primordial seas, followed by stereospecific condensation to form more complex protobiotic species. In these various reactions, the mineral surface may act simply as a template in the concentration and organization of the organic molecules, or it may also act as a catalyst or initiator for the subsequent reactions.

In the previous chapters, we have introduced a number of reactions that are affected by, or are dependent on, the adsorption process. These include the Benzidine Blue reaction, in which the colored product is stabilized by adsorption on the silicate surfaces, and the transformations of intercalated amines, catalyzed by acidic water molecules that are activated by adsorption in the interlamellar spaces of smectites.

An analogy has been drawn between the potential for orientation and stereospecific reaction of adsorbed organic molecules on clay mineral surfaces and the stereospecific Ziegler–Natta polymerization of vinylic monomers by solid organometallic complexes, in which the surface of the catalyst is all-important in presenting the monomer to the propagating entities [3]. However, there are relatively few examples known in which the course of the reaction or the microstructure of a polymeric product is determined by some specific preorientation of the organic molecules on the surface of a clay or other mineral; this would appear to be an area of research that warrants more extensive attention.

We confine our discussion of the orientation and control of organic reactions by mineral surfaces largely to systems containing the clay minerals or silica–alumina compositions. The reactions will be discussed under the following headings:

   (i)   Transformation of simple molecules, to form low-molecular-weight products.

   (ii)   Template polymerization of vinylic or acrylic monomers.

   (iii)   Prebiotic syntheses, which includes the clay-catalyzed syntheses of chiral peptides, polypeptides, nucleotides, and carbohydrates.

6.1   TRANSFORMATION OF SIMPLE MOLECULES

In addition to the electronic effects that may arise from interaction between the adsorbed species and surface binding sites, the adsorption of organic molecules on minerals imposes steric constraints that can affect their reactivity. These steric effects may take several forms, which include (1) the shielding, by the mineral surface or by neighboring adsorbed molecules, of the bulk of a target molecule or specific groups from nonselective attack by reagents and (2) the preferential formation of particular geometrical isomers during addition or elimination reactions.

The adsorption of organic compounds in close-packed perpendicular arrays on a surface can force reagents to preferentially attack exposed portions of the molecules; these molecules are often the terminal or near-terminal bonds or groups. Aliphatic acids adsorbed on alumina, for example, show an increased degree of terminal chlorination compared to that in homogeneous systems; the yield of 8-chlorooctanoic acid is nearly doubled (to 33%) when octanoic acid is adsorbed on neutral alumina prior to reaction with chlorine [4].

Selective halogenation has also been reported in the bromination of alkylbenzenes and olefins in the presence of zeolites [5]. Toluene and ethylbenzene normally react with bromine, in nonpolar homogeneous media, to form bromoalkylbenzenes as the major products. Bromination occurs via radical attack on the alkyl chains. However, if the hydrocarbons are adsorbed in the pores of a Ca^{2+} or Fe^{3+} zeolite-X, the dominant products are p-alkylhalobenzenes, with bromination occurring on the aromatic ring. The predominance of the $para$ isomers was ascribed to shielding of the alkyl substituent group and $ortho$ positions of the benzene ring by the walls of the zeolitic cavities. The $ortho$–$para$ ratio of the products obtained using the Fe^{3+} zeolite was greater than that obtained using $FeCl_3$ as a catalyst in homogeneous media. [6]. Bromine is also strongly adsorbed by zeolites, and the adsorption adducts have been proposed as selective bromination reagents for the preferential attack of terminal or unbranched alkene groups or other substituents that are capable of entering the zeolite cavities [7]. Bromine adsorbed in a zeolite-5A, for example, was reported to react only with the styrene of a styrene–cyclohexene mixture.

The reported mechanisms for these selective bromination reactions have been criticized. In the reactions with alkylbenzenes, the preferential nuclear bromination probably reflects a change in the dominant reaction pathway from that of a radical-mediated attack on the alkyl substituent to that of a zeolite-promoted, or zeolite–bromine-byproduct-promoted, bromonium ion addition to the aromatic ring, instead of being solely due to steric effects arising from intercalation in the mineral. In the case of the zeolite–bromine adducts, if the openings in the zeolite lattice are small enough to prevent the entry of branched chain alkenes, they would probably be too small to permit desorption of the brominated products. The selective bromination of styrene, in admixture with cyclohexene, is more characteristic of a homogeneous radical-mediated bromination by desorbed bromine than of a steric-controlled ionic reaction [8].

Mutual steric effects resulting from adsorption at high surface loadings have been observed in the ozonation of adsorbed branched-chain esters. Oxidation of 3,7-dimethyloctyl acetate adsorbed on silica, for example, with ozone at 20°C gives good yields of the 7-ol (41%) and 7-norketo (27%) derivatives, whereas reaction in homogeneous media is less selective, providing respective 28% and 10% yields of these products. The specificity of attack at the 7-position of the adsorbed ester is increased further when the adsorbed ester is ozonized at −78°C [9].

A number of hydrocarbon hydrogenation or cracking petrochemical processes have been described that use various zeolites as "shape selective" catalysts. The selectivity in these systems may occur when only one part of the reactant molecule can pass through the catalyst pores to reach the internal catalytic sites or a clathrated coreagent; when only certain products, having sufficiently small dimensions, can diffuse out of the pores; or when steric restrictions within the cavities prevent the formation of the transition state species for some reactions, but not those leading to the desorbable products. Detailed discussion of these systems is beyond the scope of this book, but can be found, for example, in reference 5. Other template effects have been indicated, for example, in the fragmentation reactions that may occur during the high-temperature, high-vacuum desorption of long-chain alcohols and acids from kaolinite and montmorillonite (Section 5.2.3) [10] and in the transformation of long-chain alcohols into benzene on the surface of ferric oxide at moderate temperatures [11].

Adsorption of organic species can also affect the product distribution formed on irradiation of the complexes. γ-Irradiation of pentane or other hydrocarbons, for example, adsorbed as a thin surface layer on outgassed, finely divided silica or aluminosilicate, produces a higher yield of branched chain species than that obtained by irradiation of the bulk hydrocarbons [12]. This probably results from the stabilization of the cationic intermediates formed during the irradiation by interaction with the dehydroxylated mineral surfaces.

We have described in Section 1.7.4 some examples in which the "two-dimensional" interlamellar environment favors the intercalation of a particular cationic complex ion instead of some related interconvertable species which may be more stable in homogeneous media. If the two species differ, for example, in catalytic selectivity, then adsorption can indirectly influence the course of the catalyzed reactions.

$$[Rh(PPh_3)_3H_2]^+ \rightarrow [Rh(PPh_3)_3H] + H^+$$

This effect is apparent in the reactions of a rhodium cationic hydrogenation catalyst (shown above), the active form of which exists in equilibrium with its deprotonated derivative. The protonated form, which can be stabilized by intercalation in hectorite, is a good terminal hydrogenation catalyst, but a poor isomerization catalyst. However, the deprotonated species is effective for both hydrogenation and isomerization. Reduction of hexene, for example, using the methanol-swollen cationic rhodium–hectorite complex yields hexane, whereas reduction by the rhodium complex in homogeneous solution results in extensive internal isomerization [13].

In many reactions of adsorbed species, the products are found to be essentially similar to those formed in homogeneous media, contrary to simplistic predictions. In a number of cases this finding has provided a useful insight into the reaction mechanisms for the adsorbed species. The bromination of diethyl fumarate, for example, proceeds very slowly in homogeneous media of low polarity; however, when the ester is adsorbed on a hydroxylated silica, the bromination is very fast and highly stereospecific. Both reactions yield the same product, diethyl *meso*-2,3-

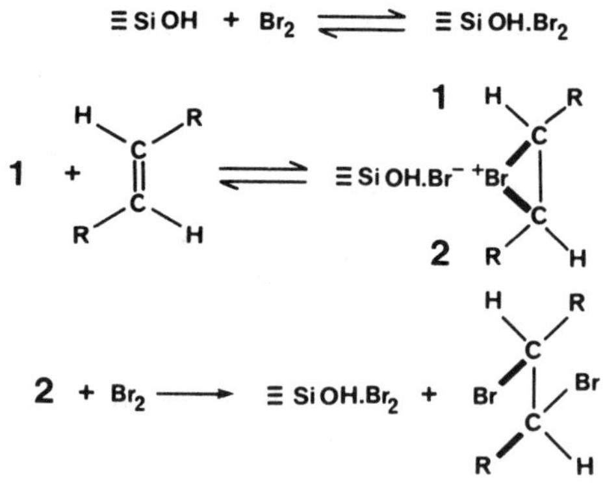

$$\equiv Si\,OH + Br_2 \rightleftharpoons \equiv Si\,OH.Br_2 \qquad \mathbf{1}$$

Figure 6.1 Bromination of diethyl fumarate adsorbed on a silica surface; R represents a $C_2H_5O{\cdot}CO{\cdot}$ group.

dibromosuccinate. This *trans* addition probably involves the formation of an intermediate bromonium ion. The reaction in the presence of the silica is believed to be promoted by the formation of an activated silanol–bromine complex (1). This can react with the unsaturated ester to form a bromonium–hydrogen-bonded bromide ion pair (2) (Figure 6.1) [14]. A similar activated complex may be involved in the aromatic bromination of alkylbenzenes in the presence of zeolites.

The mineral surface-catalyzed elimination of hydrogen halides from haloalkanes, or of carboxylic acids from esters, yields initial products that are formed predominantly by the *trans* elimination of a β-proton, similar to those formed in homogeneous media, rather than the expected *cis* elimination products that might have been expected from the reaction between an adsorbed organic species and acceptor sites on a planar surface [15]. The elimination of hydrogen bromide from bromoalkanes, such as 2-bromobutane, appears to proceed by a concerted (*anti*-E_2) mechanism over basic silica surfaces, which could require the two acceptor sites to be physically opposed, for example, on the walls of micropores (Figure 6.2); reaction over acidic silica occurs by the *syn* mode, and yields a greater proportion of the *cis* olefin [16]. The elimination reactions usually yield a mixture of geometrical isomers, as they may proceed, simultaneously and competitively, by both the concerted process and a nonconcerted carbenium ion-mediated (E_1) process; the initially formed alkenes also may undergo surface-catalyzed isomerization (Section 5.2.4).

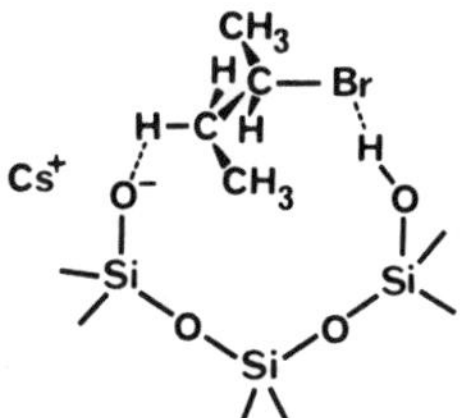

Figure 6.2 The intermediate complex postulated for the dehydrobromination of 2-bromobutene in a porous Cs-silica. (Reprinted, with permission, from M. Misono, T. Takizawa, and Y. Yoneda, *J. Catalysis*, 52, p. 404, copyright © Academic Press, New York, 1978.)

Elimination of water from alcohols adsorbed on the Lewis acidic surfaces of alumina, on the other hand, commonly yields products apparently corresponding to a *cis* (*syn*) elimination, although various studies have shown that these reactions occur by a concerted process, once again with the *anti* elimination of the *trans*-β-proton. These reactions are believed to proceed via an E_2-like, partial-carbenium ionic transition state, in which the hydroxyl group of the alcohol is bound to alumina surface hydroxyls, "preforming" the eliminated water, and in which the alkyl chain is inclined at an angle to the alumina surface. Sufficient vibrational freedom (rocking) of the alkyl chain occurs prior to complete scission of the C–O bond in order to allow the *trans*-β-proton to approach the surface basic receptor site (Figure 6.3) [17]. Steric restrictions are minimized, for molecules in this inclined configuration, when the adsorbed molecule has a conformation in which the more bulky substituents are eclipsed rather than opposed.

A similar E_2-like elimination mechanism could also occur in concerted dehydrohalogenations and related reactions. This obviates the necessity of reaction in hypothetical micropores to explain *anti* eliminations on planar surfaces. The *cis–trans* ratio of the initial products is determined by the preferred conformation of the adsorbed species and by the acidity or basicity of the catalyst surface, which influences the relative frequency of E_1 versus E_2 elimination. Strongly acidic or basic surfaces favor an E_1 elimination, with an ionic intermediate, and formation of the thermodynamically more stable *trans*-alkenes [15].

6.2   POLYMERIZATION OF ORIENTED VINYLIC MONOMERS

Although there have been relatively few reported examples of addition reactions between adsorbed species on mineral surfaces, in which the specific organization or orientation of the organic molecules prior to reaction has been demonstrated, there have been numerous reports of the stereospecific polymerization of molecules adsorbed on the external surfaces or in the interlamellar spaces of the clay minerals. These reactions include the chain polymerization of vinylic or acrylic monomers, described in this section, and the stereospecific step-growth condensation reactions, for example, of amino acids, which will be discussed in the following section.

The orientation of the monomer molecules can occur at three levels. In the lowest level, for monomers confined as an extended two-dimensional phase by intercalation between "inert" planar surfaces, or as an extended one-dimensional phase in a channel, the orientation of the molecules with respect to each other will be largely determined by their mutual interactions. At the next level, small numbers of molecules may exist in organized groups, which may result from their association with the randomly located interlamellar exchangeable cations, for example, but without extended organization of the adsorbed phase. Extended organization of the mono-

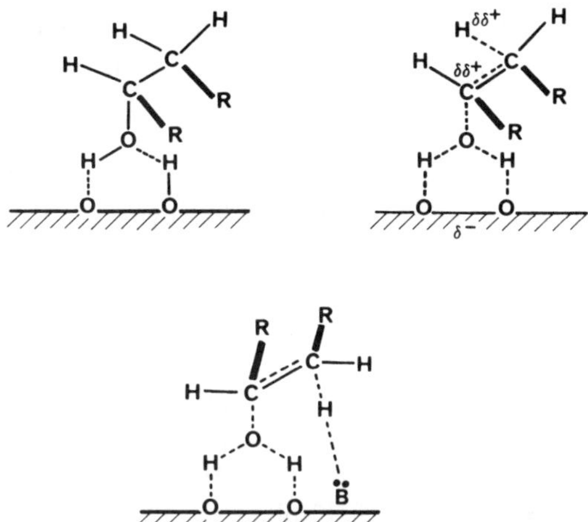

Figure 6.3 Mechanism for *trans*-elimination of water from alcohols on an alumina surface via a pseudocationic intermediate in a rocking mode of molecular vibration.

mer molecules occurs at the third level, and results from specific interactions with some regular array of lattice surface species. In the case of polymerization of vinylic (or acrylic) monomers adsorbed on clay mineral surfaces, effective organization of the molecules prior to reaction has only been found at the second level, although extended organization of monomers can occur in biological systems and in the template polymerization of monomers on synthetic polymer matrices [18].

Montmorillonites can intercalate vinylic monomers, such as styrene or vinylpyridine, or acrylic monomers, including acrylic and methacrylic acids, their esters, and related amides and nitriles, and the intercalated monomers polymerized *in situ* to form "insertion" polymers [19–22]. The polymerization may be initiated by mineral surface species, by irradiation of the mineral–monomer complex, or by added radical initiators. The resultant polymers may be extractable by solvents, but frequently their isolation requires dissolution of the mineral, for example, with concentrated hydrofluoric acid. In many cases, the insertion polymers are found to have properties and structures that differ significantly from those of similar polymers prepared in homogeneous media.

Although there have been many published examples of the application of the polymerization of clay mineral–monomer complexes for preparation of novel composite materials, structural details of the polymers formed have only been reported for a relatively few systems, such as the montmorillonite–acrylic ester or montmorillonite–acrylonitrile systems studied by Blumstein and co-workers.

Dry Na-montmorillonites can intercalate methyl methacrylate (MMA) to form complexes having basal spacings of 17.2 Å and containing the equivalent of two layers of monomer molecules lying parallel to the silicate surfaces. The bulk of the adsorbed ester is weakly held and can be removed by washing or by evaporation, leaving a complex that contains only a monolayer of intercalated monomer in each interlamellar region. This residual monomer can be polymerized, either by γ-irradiation or by thermal initiation using coadsorbed radical initiators [23]. The polymer formed has been shown to have a significantly higher proportion of isotactic triads (groups of three monomer units) than conventional radical-initiated poly-(methyl methacrylate) (PMMA), which contains mostly syndiotactic or atactic triads. The proportion of isotactic triads increases from 15 to 50% with increasing cation exchange capacity of the mineral and, in systems containing other ion-exchanged montmorillonites, increases with the polarizing power of the exchangeable cations [18,24]. Blumstein reported that the solution properties of isolated insertion PMMA did not differ noticeably from those of a conventional polymer of similar average molecular weight [25], although its stereoblock nature was reflected in peculiarities in the compression isotherm of polymer monolayers spread on water [21]. The polymers, while still intercalated in the mineral, have much greater thermal stability than the free polymers. This is ascribed to the reduced oxidative degradation, due to isolation from aerial oxygen, and to the hindrance of thermal motion and the preferential recombination of radicals formed by thermal decomposition of the intercalated polymer [26].

Blumstein has proposed that the monomer molecules in the monolayer MMA–mineral complex are present in one of two forms: as free monads, or independent molecules, adsorbed with their long axes laying parallel with, and weakly bound to, the silicate surface, but having a high degree of rotational and translational freedom within the two-dimensional interlamellar spaces; as isotactic diads that are bound, by ion–dipolar interactions, to the exchangeable cations (Figure 6.4). The PMMA formed consists of short stereosequences, as expected for a random combination of placements of isotactic diads and free monads, the proportions of which can be correlated with the cation exchange capacity of the mineral [24]. The rate of polymerization of monomer in the monolayer complex is at least one order of magnitude less than that of MMA in bulk. This has been ascribed to steric effects due to the reduced mobility of the monomer in the interlamellar monolayers which, in turn, is dependent on the monomer-ionic interactions; the stronger the interaction, the lower the monomer mobility and propagation rate [27].

Cross-linked insertion PMMA can be prepared by the monolayer intercalation, and subsequent polymerization, of MMA containing a *bis*-acrylic ester such as tetramethylene dimethacrylate. The insertion polymer, isolated after dissolution of the mineral lattice, consists of two-dimensional sheets that retain a memory of the mineral lattice [25,28]. The dimensions of the sheets, determined by light-scattering methods, correspond to the basal dimensions of the original montmorillonite particles. The solution properties differ significantly from those of both linear and branched-

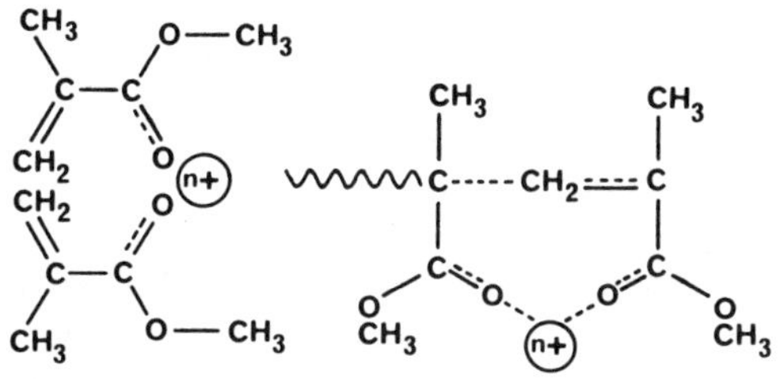

Figure 6.4 Preorientation and polymerization of intercalated methyl methacrylate. Organization of complexed monomers and the formation of isotactic diads.

chain polymers; this is most noticeable in the tendency of the polymer sheets to associate in book-like structures, containing ten or more layers, like the platelets of a coagulated smectite. These sheet polymer aggregates can be dispersed in strong solvents, but often only after prolonged solvation or heating. The compact, sheet-like structure, unlike the random coil-like structures of branched or linear polymers, permits the development of extended physical interaction between the molecules. This mutual interaction is sufficiently strong to resist rapid dissolution by osmotic solvation forces.

Irradiation of montmorillonite–methyl acrylate complexes also can produce an insertion polymer having a sheet-like molecular structure, without the need for a cross-linking comonomer [25]. This results from the greater tendency of the acrylate esters to undergo radical-initiated chain branching reactions which, in the restricted interlamellar environment, results in the formation of a two-dimensional polymer; this is in contrast to the three-dimensional gel formed by irradiation of the bulk monomer.

Insertion polymers of acrylonitrile or methacrylonitrile can differ structurally from those formed in homogeneous media. The nitrile groups of the monomer are susceptible to radical attack, although conventional radical polymerization techniques result in predominant propagation via the vinylic groups alone. However, on heating the linear polymers, radical or anionic reactions can result in further condensation of the pendant nitrile groups to form highly colored ladder polymeric structures (Figure 6.5). Insertion polymers prepared at low temperatures have been found to contain a high proportion of ladder polymer sequences. This is ascribed to the preorientation of the monomer molecules in isotactic diads by ion–dipole interactions with the exchangeable cations. These diads have a planar arrangement of monomers, with their nitrile groups in close proximity, which favors the cyclization process [29].

6.3 PREBIOTIC SYNTHESES

Many of the reports in the scientific literature that refer to the organization of simple molecules on naturally occurring inorganic templates, such

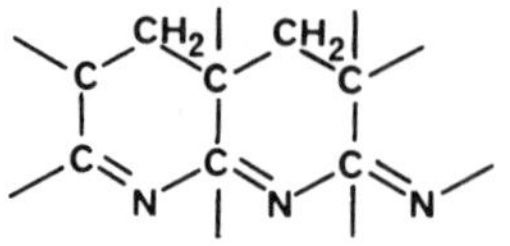

Figure 6.5 Ladder structure formed by condensation of the nitrile groups of a polyacrylonitrile.

as the kaolins and smectites present in the primordial sediments, are related to theories on the origin of life. In the following discussion, we are not outlining a case for origin via evolution, as distinct to origin by Divine Creation; rather we shall be discussing some chemical reactions leading to simple molecules, such as the aminoacids, sugars, lipids, and pyrimidine and purine heterocycles, or their condensation products, which are components, or potential precursors, of vital natural organic substances.

Many of the advocates of a terrestrial origin of life have proposed that the mineral surfaces can act as a means for concentration, by adsorption from the atmosphere or from the primordial seas, of simple inorganic molecules, such as ammonia, carbon dioxide, or hydrogen cyanide. Subsequent reactions, catalyzed by solar radiation, by electrical discharges, or by the mineral surface itself, could then convert the adsorbed species to the more complex precursors of biologically active molecules. Many of these reactions have been demonstrated, under simulated conditions, using the components of the various hypothetical prebiotic environments [30,31]. Some workers have also claimed that the mineral surfaces may have also enabled the chiral synthesis of optically active complex molecules such as polypeptides [32,33]. Considerable controversy still surrounds work in this field, and the various experimental techniques and assumptions used by the various schools are frequently subject to criticism by others. We are therefore reporting some of the published accounts, with the proviso that, as the supposed prebiotic conditions are still hypothetical, these reports cannot be definitive. Further work is required, in most cases, to substantiate the various claims.

The use of clay as a template to bring about the formation of living molecules goes back to Biblical times. The modern version of the use of clays originated in 1949 with J. D. Bernal's hypothesis that the minerals acted as concentrators and organizers of simple molecules prior to their conversion into the precursors of proteins and other biologically important molecules [34].

The starting ingredients of the natural syntheses are variously believed to include carbon compounds, such as methane, carbon dioxide, carbon monoxide, formaldehyde, and hydrogen cyanide. These could have been produced from inorganic materials by hydrolysis or oxidation and would have been present in volcanic gases or derived from extraterrestrial sources [35]. The nitrogen precursors would have included nitrogen itself, ammonia, and possibly hydrazine and hydroxylamine. The primordial atmosphere is believed to have been strongly reducing, with a very low oxygen content. The activation energy for the syntheses could have been

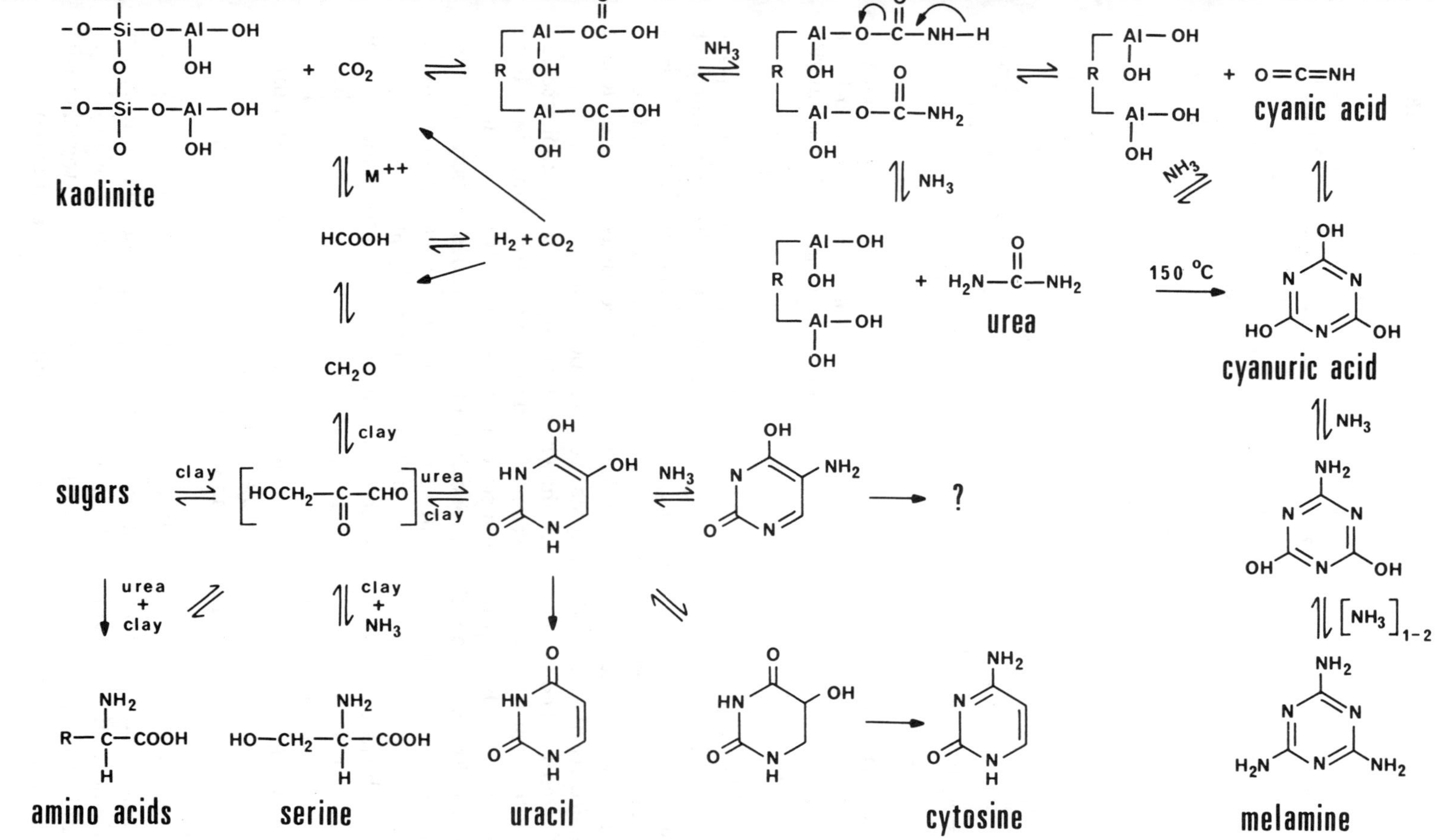

Figure 6.6 Possible mechanisms for the prebiotic formation of some precursors for living materials. (Reprinted, with permission, from G. R. Harvey, K. Mopper, and E. T. Degens, *Chemical Geology*, 9, p. 79, copyright © Elsevier Scientific Publishing Co., New York, 1972.)

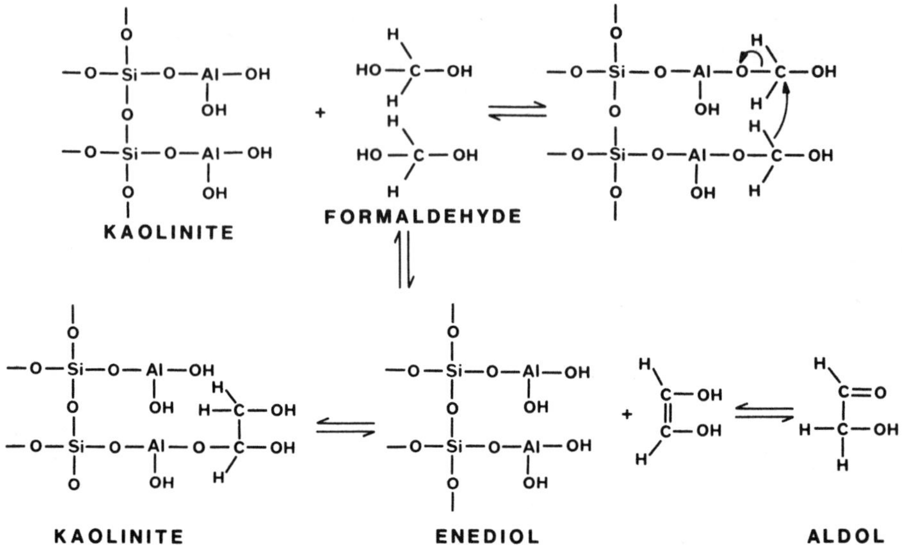

Figure 6.7   Formation of sugars from formaldehyde on a kaolinite surface. (Reprinted, with permission, from E. T. Degens, *Chemical Geology*, **13**, p. 1, copyright © Elsevier Scientific Publishing Co., New York, 1974.)

supplied by lightning discharges, or by solar ultraviolet light which, prior to the development of the ozone layer, would have included a large amount of highly actinic radiation having wavelengths of 100 to 300 nm. Modern workers have been able to simulate these hypothetical primordial conditions in reactors and have obtained a variety of condensation products; a few, arbitrarily selected examples are described below.

Degens and co-workers have shown that minerals, such as kaolinite, are capable of catalyzing the reaction of carbon dioxide and ammonia to form urea, and they have proposed a series of mineral-catalyzed reactions that lead to the formation of aminoacids and polypeptides (Figure 6.6) [36]. Related reactions were thought to produce cyanuric acid, the pyrimidine nucleoside precursors, cytosine, uracil, and probably also purine derivatives. The intermediates in these reactions are believed to include formaldehyde and formic acid. Degens has shown that kaolins or calcium silicates can promote the transformation of formaldehyde and paraformaldehyde to reducing sugars via aldol-like condensations (Figure 6.7) [37]. Further reactions between glycine, or other simple aminoacids, and the aldol condensates could result in the formation of the more complex natural amino acids.

Ponnamperuma and his colleagues have studied the effects of actinic radiation and electrical discharges on atmospheres containing nitrogen, methane, and water vapor. The products were found to include hydrogen cyanide, formaldehyde, adenine, and nitrogenous polymers, together with the α-aminoacids glycine, alanine, aminobutyric acid, and sarcosine, as components of a very dilute prebiotic soup. The amino acids may have been hydrolysis products of the heterocyclic nitrogenous polymers [38]. Clays were considered to be important as concentrators of the precursors from the prebiotic soup and as protective agents for the primitive materials against destruction by the actinic light [39]. Condensation of the adsorbed amino acids, to form more complex molecules, was believed to have been promoted by cyanamide and related species which are formed under these simulated prebiotic conditions. Clays can also promote the formation of C_2 to C_{35} hydrocarbons under these conditions [40]. Although clays may have assisted the formation of proteins and the earliest nucleic acids by the adsorption and ordering of the primitive materials, Ponnamperuma considered that the minerals did not change the reaction pathways to these species [39,41].

Fripiat has shown that amino acids and purines can be synthesized by the reactions of ammonia and carbon monoxide in the presence of outgassed, thermally activated zeolites at temperatures of 250°C or greater, which he considered to simulate the ambient conditions during this prebiotic period. The key intermediates in these reactions were believed to be formamide and its dehydration product, hydrogen cyanide. Fripiat described a scheme whereby subsequent hydrolysis of nitrogenous polymers could have led to various amino acids, such as glycine, alanine, and serine, identified in the products from the simulated prebiotic syntheses [42]. Hatanaka and Egami have found that about 40 species of amino acids are produced by the reaction of formaldehyde and hydroxylamine in saline media at pH 5.5 and 105°C; although their presence was not required for the syntheses, the addition of clays, or of the transition elements they contained, increased the yield of amino acids under these simulated primordial conditions [43].

The pathways to the various amino acids, sugars, and the nucleotide precursors are highly speculative and controversial, but there is general agreement that the concentration from the prebiotic soup and subsequent condensation of these species occurred at mineral–liquid interfaces, most probably those of kaolins or montmorillonites in marine suspensions or sediments [e.g., 3,30,31]. The genesis of the lamellar silicates and the polypeptides may well have been interrelated. Degens, for example, has proposed that the original polymerization of the amino acids may have occurred via the formation of coacervates between hydrated silica–alumina (protokaolin) sols and the organic molecules, which, after sedimentation, could undergo polymerization by epitaxis, or oriented growth, followed by natural chromatography, to yield both polypeptides and the extended lamellar silicate lattice of kaolin [44]. These processes would have been assisted by natural fluctuations in hydration, salinity, and temperature of the sediments, which could have provided the different conditions required for the adsorption, condensation, and desorption phases of the polypeptide synthesis [45,46].

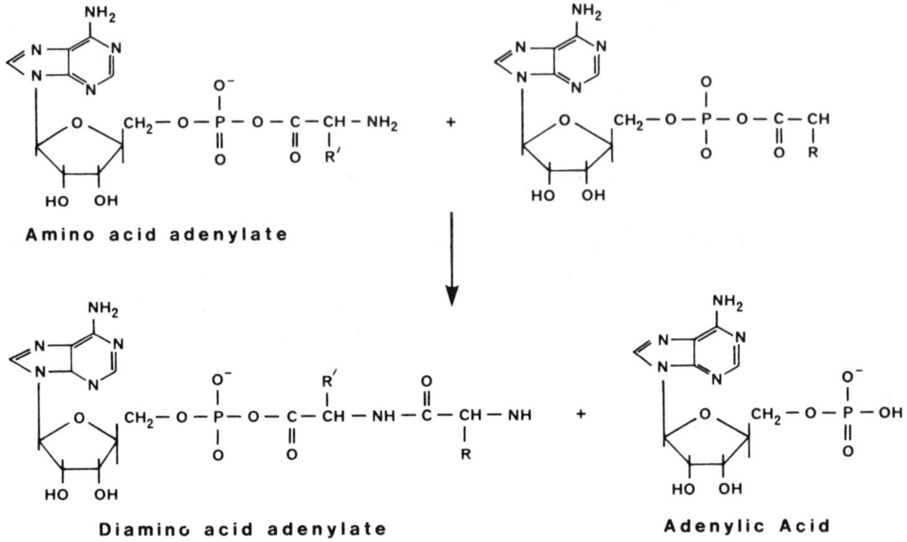

Figure 6.8 Protoenzymatic formation of polypeptides from precursors adsorbed on clay mineral surfaces. (Reprinted, with permission, from M. Paecht-Horowitz, *Origins of Life*, 7, p. 369, copyright © D. Reidel, Dordrecht-Holland, 1976.)

Although clays could have promoted the formation of polypeptides from the zwitterionic amino acids at high temperatures (as originally proposed by Fox in reference 31), or alternatively, in a fluctuating environment, they could also have been formed, under relatively mild conditions, by condensation of aminoacyl nucleotides such as the amino acid adenylates (Figure 6.8), a process that parallels the synthesis of polypeptides in living systems. The catalysis of these reactions by montmorillonites and kaolinites has been likened to that of a mineral "protoenzyme" [47]. Adenylates of serine, alanine, aspartine, and glycine copolymerize in homogeneous media to form random oligomers, with degrees of polymerization of 1 to 4. In the presence of montmorillonite suspensions, however, the various adenylates tend to form block copolymers. These have degrees of polymerization up to 40, the values forming a discrete spectrum based on multiples of 6 or 7. Adsorption on the clays allows interaction between the molecules of the amino acid adenylates and the formation of preferred sequences prior to condensation. This is believed to occur at the aluminosilicate layer edges, with movement of the polypeptide product into the interlamellar region. The transition metal-exchanged montmorillonites have been shown to preferentially adsorb particular nucleotides, and could thus favor their polymerization; some examples of specific coadsorption have also been reported [48].

Possibly the most controversial aspect of prebiotic synthesis is the origin of almost exclusive predominance of a particular optical isomer, for example, the L-form of the asymmetric amino acids, in the components of most living materials and byproducts. In the absence of some asymmetric agent, the various reactions discussed above would have produced racemic mixtures of optical isomers, that is, mixtures containing equal proportions of the dextro (D) and laevo (L) rotary components, which would have no net optical activity. Various classical hypotheses have included the effects of the asymmetry of the various radiation fields in the Universe, or to the preferential formation or destruction of a particular optical isomer on optically active crystals of quartz, or by solar ultraviolet light which had become circularly polarized through interaction with the Earth's magnetic field [30].

Degens, Matheja, and Jackson have proposed that optical activity may have originated from reactions of organic substances adsorbed on clay minerals. They claimed to have demonstrated that L-aspartic acid, for example, polymerized more rapidly than its D-isomer in the presence of kaolin [32,49]. They and others [50,52] have also reported that D-glucose and L-amino acids are bound to clay surfaces more tightly than their enantiomers. As their adsorption stabilizes these species against various natural forms of degradation, preferential adsorption of one particular isomer could indirectly lead to its eventual predominance. The relation between natural optical activity and clay mineral surfaces is still being disputed [e.g., 50], and at least one series of attempts to duplicate the asymmetric polymerization experiments has failed [51].

One important aspect of this work by Degens and others, which so far appears to have been overlooked, is the origin of the kaolins used in the various laboratory simulations. The kaolins, in all cases, had either been obtained from Cretaceous sedimentary deposits or from commercial sources that were most probably mined from similar deposits. These deposits are of relatively recent origin and are about 100 million years old. They were formed during the final days of the dinosaurs, rather than when the chemical evolution of protolife is thought to have occurred, about three billion years ago. In Section 3.1 we described the printing of molecular outlines of adsorbed organic materials onto precipitating silica sols. Degens has implicitly indicated that a similar process could occur during the epitaxis of the protokaolin coacervates. It is therefore quite possible that the asymmetric activity of the Cretaceous kaolins, if it in fact exists, might merely be the result of an imprinted memory of the already existing, highly developed, optically active natural materials that were adsorbed during the crystallization of the mineral lattice.

The field of study, in which clays serve to bring about these conversions of biological activity, is still one which is open to question and one for which a great deal of further careful detailed research is still required. Nevertheless, it seems clear from the examples discussed here and for other cases where clear-cut differences have been noted between reactions of adsorbed molecules and those in bulk that preorganization of the molecules can significantly alter the subsequent chemical conversions.

REFERENCES

1   G. F. Walker, *Clay Miner.*, **7**, 129 (1967).

2   G. Lagely, *Clay Clays Miner.*, **27**, 1 (1979).

3   M. Paecht-Horawitz, *Midland Macromolecular Monographs*, Vol. 3, Gordon and Breach, New York, 1977, p. 89.

4   N. C. Deno, R. Fishbein, and C. Pierson, *J. Amer. Chem. Soc.*, **92**, 1451 (1970).

5   S. M. Csicsery, Shape Selective Catalysis, in J. A. Rabo, Ed., *Zeolite Chemistry and Catalysis*, ACS Monograph No. 171, American Chemical Society, Washington, 1976.

6   J. van Dijk, J. J. van Daalen, and G. B. Paerels, *Rec. Trav. Chim. Pays-Bas*, **93**, 72 (1974).

7   P. A. Risbood and D. M. Ruthven, *J. Amer. Chem. Soc.*, **100**, 4919 (1978).

8   R. M. Dessau, *J. Amer. Chem. Soc.*, **101**, 1344 (1979).

9   A. L. J. Beckwith, C. L. Bodkin, and T. Duang, *Aust. J. Chem.*, **30**, 2177 (1977).

10   A. K. Galwey, *J. Catalysis*, **19**, 330 (1970).

11   A. K. Galwey, *J. Catalysis*, **12**, 352 (1968).

12   A. O. Allen and J. M. Caffrey (United States Atomic Energy Commission), U.S. Patent 2,955,997 (October 11, 1960).

13   T. J. Pinnavaia, R. Raythatha, J. G-S. Lee, L. J. Halloran, and J. F. Hoffmann, *J. Amer. Chem. Soc.*, **101**, 6891 (1979).

14   M. J. Rosen and C. Eden, *J. Phys. Chem.*, **74**, 2303 (1970).

15   H. Noller and W. Kladnig, *Catal. Rev. Sci. Eng.*, **13**, 149 (1976).

16   M. Misono, T. Takizawa, and Y. Yoneda, *J. Catalysis*, **52**, 397 (1978).

17   H. Knoezinger, H. Buehl, and K. Kochloefl, *J. Catalysis*, **24**, 57 (1972).

18   A. Blumstein, S. L. Malhotra, and A. C. Watterson, *J. Polym. Sci., Polym. Phys. Ed.*, **8**, 1599 (1970).

19   A. Blumstein, *Bull. Soc. Chim. Fr.*, 899 (1961).

20   O. L. Glavati, L. S. Polak, and V. V. Shchekin, *Neftekhimiya*, **3**, 905 (1963); C.S.I.R.O. Trans. No. 7307.

21   H. Z. Friedlander and C. R. Frink, *J. Polym. Sci., Pt. B*, **2**, 475 (1964).

22   Yu. S. Zaitsev, N. G. Kisel, V. D. Enalev, and A. I. Yurzhenko, *Colloid J. USSR*, **32**, 173 (1970).

23   A. Blumstein, *J. Polym. Sci.*, Pt. A, **3**, 2653 (1965).

24   A. Blumstein, K. K. Parikh, S. L. Malhotra, and R. Blumstein, *J. Polym. Sci., Polym. Phys. Ed.*, **9**, 1681 (1971).

25   A. Blumstein and F. W. Billmeyer, Jr., *J. Polym. Sci., Pt. A-2*, **4**, 465 (1966).

26   A. Blumstein, *J. Polym. Sci., Pt. A*, **3**, 2665 (1965).

27   S. L. Malhotra, K. K. Parikh, and A. Blumstein, *J. Colloid Interface Sci.*, **41**, 318 (1972).

28   A. Blumstein, R. Blumstein, and T. H. Vanderspurt, *J. Colloid Interface Sci.*, **31**, 236 (1969).

29   R. Blumstein, A. Blumstein, and K. K. Parikh, *J. Appl. Polym. Sci., Appl. Polym. Symp.*, **25**, 81 (1974).

30   R. M. Lemmon, Prebiotic Chemistry, in A. F. Scott, Ed., *Survey of Progress of Chemistry*, Vol. 6, Academic, New York, 1973.

31   A. Katchalsky, *Naturwiss.*, **60**, 215 (1973).

32   W. E. Elias, *J. Chem. Educ.*, **49**, 448 (1972).

33   E. T. Degens, J. Matheja, and T. A. Jackson, *Nature*, **227**, 492 (1970).

34   G. Cairns-Smith, *Chem. Brit.*, **15**, 576 (1979).

35 R. D. Brown, *Chem. Brit.*, **15**, 570 (1979).

36 G. R. Harvey, K. Mopper, and E. T. Degens, *Chem. Geol.*, **9**, 79 (1972).

37 E. T. Degens, *Chem. Geol.*, **13**, 1 (1974).

38 C. Ponnamperuma, *Chem. Brit.*, **15**, 560 (1979).

39 C. Ponnamperuma, *ACS Nat. Meet., 179th., Abstr.*, American Chemical Society, Washington, 1980, Paper COLL-38.

40 J. Levine, A. Shimoyama, C. Ponnamperuma, S. Crane, and P. Furness; synopsis supplied by C. Ponnamperuma.

41 A. Shimoyama, N. Blair, C. Ponnamperuma, *Origin Life, Proc. ISSOL Meet., 2nd,* 1977, p. 95; *Chem. Abs.*, **90**, 49865 (1979).

42 J. J. Fripiat, G. Poncelet, A. T. Van Assche, and J. Mayaudon, *Clays Clay Miner.*, **20**, 331 (1972).

43 H. Hatanaka and F. Egami, *Bull. Soc. Chem. Jap.*, **50**, 1147 (1977).

44 E. T. Degens, and J. Matheja, *Formation of Organic Polymers on Inorganic Templates* in A. P. Kimball and J. Oro, Eds., *Prebiotic and Biochemical Evolution*, North-Holland, Amsterdam, 1971.

45 N. Lahav, D. White, and S. Chang, *Science*, **201**, 67 (1978).

46 D. G. Odam, N. Lahav, and S. Chang, *J. Molec. Evol.*, **12**, 259, (1974).

47 M. Paecht-Horowitz, *Origin Life*, **7**, 369 (1976).

48 M. Tunzi, F. Church, J. Mazzurco, K. O'Brien, and J. G. Lawless, *ACS Nat. Meet., 179th.*, Abstr., American Chemical Society, Washington, 1980, Paper COLL-40.

49 T. A. Jackson, *Experientia*, **27**, 242 (1971); *Chem. Geol.*, **7**, 295 (1971).

50 T. A. Jackson, *Science*, **206**, 483 (1979); D. Wellner, *Science*, **206**, 483, (1979), p. 484; S. C. Bondy, *Science*, **206**, 483 (1979), p. 484.

51 S. C. Bondy and M. E. Harrington, *Science*, **203**, 1243 (1979).

52 J. J. Flores and W. A. Bonner, *J. Molec. Evol.*, **3**, 49 (1974).

AUTHOR INDEX

Numbers in parentheses are reference numbers and indicate that the author's work is referred to although his name may not be mentioned in the text.